Poultry Diseases: Diagnosis and Treatment

Poultry Diseases: Diagnosis and Treatment

Editor:

Preethi Kartan

Delve Publishing

www.delvepublishing.com

Poultry Diseases: Diagnosis and Treatment

Editor: Preethi Kartan

708 3rd Avenue, 6th Floor
New York, NY 10017
United States of America
www.delvepublishing.com

ISBN: 978-1-68095-764-8

Library of Congress Control Number: 2016951520

About the Editor

Preethi Kartan

Preethi is a postgraduate in Biotechnology from University of Leeds, UK. She is currently working as a Scientific Associate at one of the CRO's in India and her interest lie in life sciences related writing.

Contents

List of Contributors

Faruku Bande
Department of Veterinary Pathology and Microbiology, Faculty of Veterinary Medicine, Universiti Putra Malaysia (UPM), 43400 Serdang, Selangor, Malaysia
Department of Veterinary Services, Ministry of Animal Health and Fisheries Development, PMB 2109, Usman Faruk Secretariat, Sokoto 840221, Sokoto State, Nigeria

Siti Suri Arshad
Department of Veterinary Pathology and Microbiology, Faculty of Veterinary Medicine, Universiti Putra Malaysia (UPM), 43400 Serdang, Selangor, Malaysia

Abdul Rahman Omar
Department of Veterinary Pathology and Microbiology, Faculty of Veterinary Medicine, Universiti Putra Malaysia (UPM), 43400 Serdang, Selangor, Malaysia
Laboratory of Vaccine and Immunotherapeutics, Institute of Bioscience, Universiti Putra Malaysia (UPM), 43400 Serdang, Selangor, Malaysia

Mohd Hair Bejo
Department of Veterinary Pathology and Microbiology, Faculty of Veterinary Medicine, Universiti Putra Malaysia (UPM), 43400 Serdang, Selangor, Malaysia
Laboratory of Vaccine and Immunotherapeutics, Institute of Bioscience, Universiti Putra Malaysia (UPM), 43400 Serdang, Selangor, Malaysia

Muhammad Salisu Abubakar
Department of Veterinary Pathology and Microbiology, Faculty of Veterinary Medicine, Universiti Putra Malaysia (UPM), 43400 Serdang, Selangor, Malaysia

Yusuf Abba
Department of Veterinary Pathology and Microbiology, Faculty of Veterinary Medicine, Universiti Putra Malaysia (UPM), 43400 Serdang, Selangor, Malaysia

Erica Spackman
Southeast Poultry Research Laboratory, USDA-ARS, 934 College Station Rd. Athens, GA, 30605, USA

Jack Gelb Jr
Avian Biosciences Center, University of Delaware, Newark, DE, 19716, USA

Lauren A Preskenis
Avian Biosciences Center, University of Delaware, Newark, DE, 19716, USA

Brian S Ladman
Avian Biosciences Center, University of Delaware, Newark, DE, 19716, USA

Conrad R Pope
Avian Biosciences Center, University of Delaware, Newark, DE, 19716, USA

Mary J Pantin-Jackwood
Southeast Poultry Research Laboratory, USDA-ARS, 934 College Station Rd. Athens, GA, 30605, USA

Enid T Mckinley
Southeast Poultry Research Laboratory, USDA-ARS, 934 College Station Rd. Athens, GA, 30605, USA

El-Sayed M Abdelwhab
Friedrich-Loeffler-Institute, Greifswald-Insel Riems, Germany
Animal Health Research Institute, Dokki, Giza, Egypt
Institute of Poultry Diseases, Free University, Berlin, Germany

Ahmed M Erfan
Animal Health Research Institute, Dokki, Giza, Egypt

Christian Grund
Friedrich-Loeffler-Institute, Greifswald-Insel Riems, Germany

Mario Ziller
Friedrich-Loeffler-Institute, Greifswald-Insel Riems, Germany

Abdel-Satar Arafa
Animal Health Research Institute, Dokki, Giza, Egypt

Martin Beer
Friedrich-Loeffler-Institute, Greifswald-Insel Riems, Germany

Mona M Aly
Animal Health Research Institute, Dokki, Giza, Egypt

Hafez M Hafez
Institute of Poultry Diseases, Free University, Berlin, Germany

Timm C Harder
Friedrich-Loeffler-Institute, Greifswald-Insel Riems, Germany

Mirza Imran Shahzad
Department of Biochemistry, Pir Mehr Ali Shah Arid Agriculture University, Murree Rawalpindi-46300, Pakistan

Khalid Naeem
National Reference Laboratory for Poultry Diseases (NRLPD), Animal Sciences Institute, National Agricultural Research Center (NARC), Islamabad, Pakistan

Muhammad Mukhtar
Department of Biochemistry, Pir Mehr Ali Shah Arid Agriculture University, Murree Rawalpindi-46300, Pakistan

Azra Khanum
Department of Biochemistry, Pir Mehr Ali Shah Arid Agriculture University, Murree Rawalpindi-46300, Pakistan

Mar Costa-Hurtado
Exotic and Emerging Avian Viral Diseases Unit, Southeast Poultry Research Laboratory, United States National Poultry Research Center, Agricultural Research Service, U.S. Department of Agriculture, Athens, GA, USA

Claudio L. Afonso
Exotic and Emerging Avian Viral Diseases Unit, Southeast Poultry Research Laboratory, United States National Poultry Research Center, Agricultural Research Service, U.S. Department of Agriculture, Athens, GA, USA

Patti J. Miller
Exotic and Emerging Avian Viral Diseases Unit, Southeast Poultry Research Laboratory, United States National Poultry Research Center, Agricultural Research Service, U.S. Department of Agriculture, Athens, GA, USA

Eric Shepherd
Exotic and Emerging Avian Viral Diseases Unit, Southeast Poultry Research Laboratory, United States National Poultry Research Center, Agricultural Research Service, U.S. Department of Agriculture, Athens, GA, USA

Ra Mi Cha
Exotic and Emerging Avian Viral Diseases Unit, Southeast Poultry Research Laboratory, United States National Poultry Research Center, Agricultural Research Service, U.S. Department of Agriculture, Athens, GA, USA

Diane Smith
Exotic and Emerging Avian Viral Diseases Unit, Southeast Poultry Research Laboratory, United States National Poultry Research Center, Agricultural Research Service, U.S. Department of Agriculture, Athens, GA, USA

Erica Spackman
Exotic and Emerging Avian Viral Diseases Unit, Southeast Poultry Research Laboratory, United States National Poultry Research Center, Agricultural Research Service, U.S. Department of Agriculture, Athens, GA, USA

Darrell R. Kapczynski
Exotic and Emerging Avian Viral Diseases Unit, Southeast Poultry Research Laboratory, United States National Poultry Research Center, Agricultural Research Service, U.S. Department of Agriculture, Athens, GA, USA

David L. Suarez
Exotic and Emerging Avian Viral Diseases Unit, Southeast Poultry Research Laboratory, United States National Poultry Research Center, Agricultural Research Service, U.S. Department of Agriculture, Athens, GA, USA

David E. Swayne
Exotic and Emerging Avian Viral Diseases Unit, Southeast Poultry Research Laboratory, United States National Poultry Research Center, Agricultural Research Service, U.S. Department of Agriculture, Athens, GA, USA

Mary J. Pantin-Jackwood
Exotic and Emerging Avian Viral Diseases Unit, Southeast Poultry Research Laboratory, United States National Poultry Research Center, Agricultural Research Service, U.S. Department of Agriculture, Athens, GA, USA

Katherine E Atkins
Centre for Infectious Diseases, University Of Edinburgh, West Mains Road, EH9 3JT, UK
Department of Epidemiology and Public Health, Yale School of Medicine, New Haven, CT 06510

Andrew F Read
Center for Infectious Disease Dynamics, Departments of Biology and Entomology, 208 Mueller Laboratory, The Pennsylvania State University, University Park, PA 16802, USA

Nicholas J Savill
Centre for Infectious Diseases, University Of Edinburgh, West Mains Road, EH9 3JT, UK

Katrin G Renz
Centre for Animal Health and Welfare, School of Environmental and Rural Science, The University of New England, Armidale NSW 2351, Australia

Stephen W Walkden-Brown
Centre for Animal Health and Welfare, School of Environmental and Rural Science, The University of New England, Armidale NSW 2351, Australia

Mark EJ Woolhouse
Centre for Infectious Diseases, University Of Edinburgh, West Mains Road, EH9 3JT, UK

Apratim Mitra
Department of Animal & Avian Sciences, University of Maryland, College Park, MD, USA

Juan Luo
Department of Animal & Avian Sciences, University of Maryland, College Park, MD, USA

Huanming Zhang
USDA, ARS, Avian Disease and Oncology Laboratory, East Lansing, MI, USA

Kairong Cui
Laboratory of Molecular Immunology, National Heart, Lung and Blood Institute, National Institutes of Health, Bethesda, MD, USA

Keji Zhao
Laboratory of Molecular Immunology, National Heart, Lung and Blood Institute, National Institutes of Health, Bethesda, MD, USA

Jiuzhou Song
Department of Animal & Avian Sciences, University of Maryland, College Park, MD, USA

Grzegorz Woz´niakowski
Department of Poultry Viral Diseases, National Veterinary Research Institute, Partyzanto´w 57 Avenue, 24-100 Puławy, Poland

Elzbieta Samorek-Salamonowicz
Department of Poultry Viral Diseases, National Veterinary Research Institute, Partyzanto´w 57 Avenue, 24-100 Puławy, Poland

Zerihun Hailemariam
Faculty of Veterinary Medicine, Haramaya University, Haramaya, Ethiopia
Faculty of Veterinary Medicine, Universiti Putra Malaysia, 43400 UPM Serdang, Selangor, Darul Ehsan, Malaysia

Abdul Rahman Omar
Institute of BioScience, Universiti Putra Malaysia, 43400 UPM Serdang, Selangor Darul Ehsan, Malaysia
Faculty of Veterinary Medicine, Universiti Putra Malaysia, 43400 UPM Serdang, Selangor, Darul Ehsan, Malaysia

Mohd Hair-Bejo
Faculty of Veterinary Medicine, Universiti Putra Malaysia, 43400 UPM Serdang, Selangor, Darul Ehsan, Malaysia

Tan Ching Giap
Faculty of Veterinary Medicine, Universiti Putra Malaysia, 43400 UPM Serdang, Selangor, Darul Ehsan, Malaysia

Yang Li
College of Veterinary Medicine, Shandong Agricultural University, Tai'an 271018, China

Yixin Wang
College of Veterinary Medicine, Shandong Agricultural University, Tai'an 271018, China

Lichun Fang
College of Veterinary Medicine, Shandong Agricultural University, Tai'an 271018, China

Jiayuan Fu
College of Veterinary Medicine, Shandong Agricultural University, Tai'an 271018, China

Shuai Cui
College of Veterinary Medicine, Shandong Agricultural University, Tai'an 271018, China

Yingjie Zhao
College of Veterinary Medicine, Shandong Agricultural University, Tai'an 271018, China

Zhizhong Cui
College of Veterinary Medicine, Shandong Agricultural University, Tai'an 271018, China

Shuang Chang
College of Veterinary Medicine, Shandong Agricultural University, Tai'an 271018, China

Peng Zhao
College of Veterinary Medicine, Shandong Agricultural University, Tai'an 271018, China

Jing Zhao
Key Laboratory of Animal Epidemiology and Zoonosis, Ministry of Agriculture, College of Veterinary Medicine, China Agricultural University, Beijing, People's Republic of China

Qi Zhong
Key Laboratory of Animal Epidemiology and Zoonosis, Ministry of Agriculture, College of Veterinary Medicine, China Agricultural University, Beijing, People's Republic of China

Ye Zhao
Key Laboratory of Animal Epidemiology and Zoonosis, Ministry of Agriculture, College of Veterinary Medicine, China Agricultural University, Beijing, People's Republic of China

Yan-xin Hu
Key Laboratory of Animal Epidemiology and Zoonosis, Ministry of Agriculture, College of Veterinary Medicine, China Agricultural University, Beijing, People's Republic of China

Guo-zhong Zhang
Key Laboratory of Animal Epidemiology and Zoonosis, Ministry of Agriculture, College of Veterinary Medicine, China Agricultural University, Beijing, People's Republic of China

Helena Grgić
Department of Pathobiology, Ontario Veterinary College, University of Guelph, Guelph, Ontario, Canada

Zvonimir Poljak
Department of Population Medicine, Ontario Veterinary College, University of Guelph, Guelph, Ontario, Canada

Shayan Sharif
Department of Pathobiology, Ontario Veterinary College, University of Guelph, Guelph, Ontario, Canada

Éva **Nagy**
Department of Pathobiology, Ontario Veterinary College, University of Guelph, Guelph, Ontario, Canada

Olimpia Kursa
Department of Poultry Diseases, National Veterinary Research Institute, Partyzantów 57 Avenue, 24-100 Puławy, Poland

Grzegorz Woz´niakowski
Department of Poultry Diseases, National Veterinary Research Institute, Partyzantów 57 Avenue, 24-100 Puławy, Poland

Grzegorz Tomczyk
Department of Poultry Diseases, National Veterinary Research Institute, Partyzantów 57 Avenue, 24-100 Puławy, Poland

Anna Sawicka, Zenon Mint
Department of Poultry Diseases, National Veterinary Research Institute, Partyzantów 57 Avenue, 24-100 Puławy, Poland

Li Huang
Guangxi Key Laboratory of Animal Vaccines and New Technology, Guangxi Veterinary Research Institute, Nanning 530001, PR China
College of Life Science and Technology, Guangxi University, Nanning 530004, PR China

Zhixun Xie
Guangxi Key Laboratory of Animal Vaccines and New Technology, Guangxi Veterinary Research Institute, Nanning 530001, PR China

Liji Xie
Guangxi Key Laboratory of Animal Vaccines and New Technology, Guangxi Veterinary Research Institute, Nanning 530001, PR China

Xianwen Deng
Guangxi Key Laboratory of Animal Vaccines and New Technology, Guangxi Veterinary Research Institute, Nanning 530001, PR China

Zhiqin Xie
Guangxi Key Laboratory of Animal Vaccines and New Technology, Guangxi Veterinary Research Institute, Nanning 530001, PR China

Sisi Luo
Guangxi Key Laboratory of Animal Vaccines and New Technology, Guangxi Veterinary Research Institute, Nanning 530001, PR China

Jiaoling Huang
Guangxi Key Laboratory of Animal Vaccines and New Technology, Guangxi Veterinary Research Institute, Nanning 530001, PR China

Tingting Zeng
Guangxi Key Laboratory of Animal Vaccines and New Technology, Guangxi Veterinary Research Institute, Nanning 530001, PR China

Jiaxun Feng
College of Life Science and Technology, Guangxi University, Nanning 530004, PR China

Hao Chen
Poultry Diseases Lab, College of Animal Science and Technology, Shandong Agricultural University, Tai'an, Shandong, China

Quanbin Ou
Poultry Diseases Lab, College of Animal Science and Technology, Shandong Agricultural University, Tai'an, Shandong, China

Yi Tang
Poultry Diseases Lab, College of Animal Science and Technology, Shandong Agricultural University, Tai'an, Shandong, China

Xuhui Gao
Poultry Diseases Lab, College of Animal Science and Technology, Shandong Agricultural University, Tai'an, Shandong, China

Lili Wu
Poultry Diseases Lab, College of Animal Science and Technology, Shandong Agricultural University, Tai'an, Shandong, China

Cong Xue
Poultry Diseases Lab, College of Animal Science and Technology, Shandong Agricultural University, Tai'an, Shandong, China

Chunmei Yu
Poultry Diseases Lab, College of Animal Science and Technology, Shandong Agricultural University, Tai'an, Shandong, China

Jingteng Cui
Poultry Diseases Lab, College of Animal Science and Technology, Shandong Agricultural University, Tai'an, Shandong, China

Youxiang Diao
Poultry Diseases Lab, College of Animal Science and Technology, Shandong Agricultural University, Tai'an, Shandong, China

Marcin Smialek
Department of Poultry Diseases, Faculty of Veterinary Medicine, University of Warmia and Mazury, Oczapowskiego 13/14, 10-719 Olsztyn, Poland

Daria Pestka
Department of Poultry Diseases, Faculty of Veterinary Medicine, University of Warmia and Mazury, Oczapowskiego 13/14, 10-719 Olsztyn, Poland

Bartlomiej Tykalowski
Department of Poultry Diseases, Faculty of Veterinary Medicine, University of Warmia and Mazury, Oczapowskiego 13/14, 10-719 Olsztyn, Poland

Tomasz Stenzel
Department of Poultry Diseases, Faculty of Veterinary Medicine, University of Warmia and Mazury, Oczapowskiego 13/14, 10-719 Olsztyn, Poland

Andrzej Koncicki
Department of Poultry Diseases, Faculty of Veterinary Medicine, University of Warmia and Mazury, Oczapowskiego 13/14, 10-719 Olsztyn, Poland

Sayuri Nakamura
Laboratory of Comparative Pathology, Graduate School of Veterinary Medicine, Hokkaido University, Sapporo, Japan

Kenji Ochiai
Laboratory of Comparative Pathology, Graduate School of Veterinary Medicine, Hokkaido University, Sapporo, Japan

Akihiro Ochi
Miho Training Center, Japan Racing Association, Utsunomiya, Japan

Hiroki Yabushita
Medical School, Okayama University, Okayama, Japan

Asumi Abe
Kenpoku Meat Hygiene Inspection Institute, Tochigi Prefecture, Ohtawara, Japan

Sayaka Kishi
Laboratory of Comparative Pathology, Graduate School of Veterinary Medicine, Hokkaido University, Sapporo, Japan

Yuji Sunden
Laboratory of Comparative Pathology, Graduate School of Veterinary Medicine, Hokkaido University, Sapporo, Japan

Takashi Umemura
Laboratory of Comparative Pathology, Graduate School of Veterinary Medicine, Hokkaido University, Sapporo, Japan

Xiuzhen Wang
College of Veterinary Medicine, Shandong Agricultural University, Taian, Shandong, PR China

Bo Wang
College of Veterinary Medicine, Shandong Agricultural University, Taian, Shandong, PR China

Peipei Zhang
College of Veterinary Medicine, Shandong Agricultural University, Taian, Shandong, PR China

Hegang Cheng
College of Veterinary Medicine, Shandong Agricultural University, Taian, Shandong, PR China

Shuhong Sun
College of Veterinary Medicine, Shandong Agricultural University, Taian, Shandong, PR China

Preface

Poultry are a diverse group of species of birds that are raised primarily for meat and eggs but sometimes for feathers, skin and oil also such as ratites. These species comprise of chickens, turkeys, ducks, geese, pheasants, quail, squabs (young pigeons), Guinea fowl, partridges and ratites (ostrich, rhea and emu). Knowledge about the type of birds, their anatomy and how they are managed helps one to understand the type and kind of diseases different birds are susceptible. In certain types of bird species that are raised for egg production or meat, such as commercial poultry, infectious diseases can spread rapidly among birds housed in a confined space. Poultry can also be raised in small numbers as backyard flocks for eggs and meat, as hobby and pet birds. Their numbers are increasing rapidly. They are often exposed to natural elements and are often not vaccinated, may lack proper nutrition and lax biosecurity that can lead to frequent viral, bacterial, parasitic and nutritional diseases. Backyard poultry can also be a source of diseases to the commercial poultry. In addtion to the different management practices that are used for raising poultry, genetics and nutrition play a significant role in the initiation and outcome of a disease. There is also increased demand for poultry raised as antibiotic free and organic which can lead to unintended consquences. These outlines on prevention and differential diagnoses of diseases in poultry.

Editor
Preethi Kartan

INTRODUCTION

Poultry Diseases is to acquaint veterinarians and poultry health professionals with current information on the diagnosis and prevention of poultry disease in commercial broiler and egg production flocks in emerging and established industries. Productivity and profitability are enhanced by application of sound principles of biosecurity, vaccination, and management. Improving efficiency increases the availability of eggs and poultry meat to supply the protein needs of populations in countries with expanding demand.

During the past two decades, primary breeders of broiler, egg and laying strains have eliminated vertically-transmitted diseases from their elite and great-grandparent generations. Unfortunately, infection of grandparent and parent flocks occurs in many developing countries resulting in dissemination of diseases including mycoplasmosis, salmonellosis and reoviral infection. Improved biosecurity and an awareness of the need for appropriate vaccination programs, reduces the potential losses caused by both catastrophic and erosive infections on commercial-scale farms, village cooperatives and in integrated operations.

Angara disease, virulent infectious bursal disease, highly pathogenic influenza, reoviral stunting syndrome and swollen head syndrome are examples of emerging diseases affecting flocks in Asia, Africa, and Latin America. In addition, chronic, low-intensity infections such as coryza, pasteurellosis, and salmonellosis continue to erode profit margins.

Prevention of disease depends on a comprehensive program incorporating a sequence of planning, implementing and control in a repetitive cycle (Figure 1). Strategies to prevent infection are based on the purchase of breeding stock free of vertically-transmitted disease. Vaccination of parent flocks and progeny and appropriate levels of biosecurity represent the components of disease prevention subject to direct managemental control. The relative importance and contribution of these strategies can be calculated using simulation studies. These should incorporate projections of risk of infection and compare the production parameters and costs for diseased and healthy flocks. The components of biosecurity comprise a hierarchy with each of 3 levels influencing the cost and effectiveness of the entire program (Figure 2):

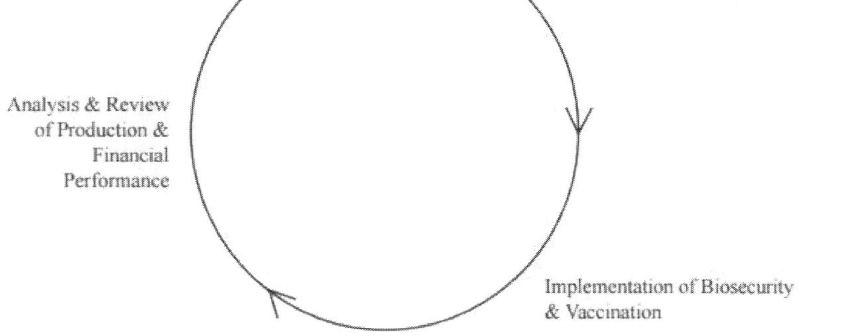

Figure 1: Cycle of Management.

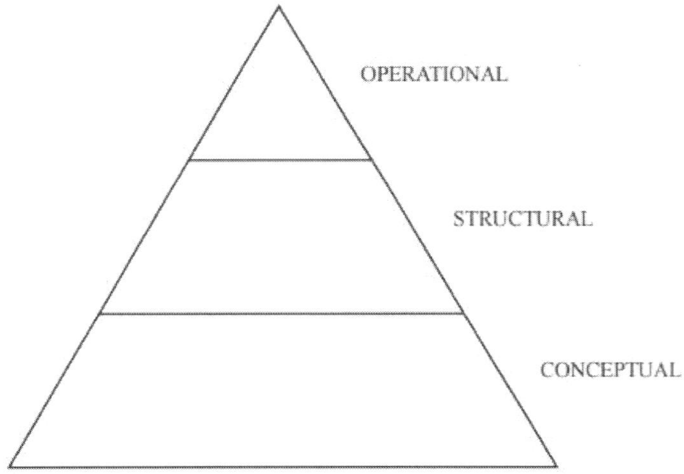

Figure 2: Hierarchy of Biosecurity.

- *Conceptual Biosecurity*: The primary level represents the basis of all programs to prevent disease. Conceptual biosecurity includes selecting the location of a complex or operation in a specific area to separate different types of poultry, reduce biodensity, and avoid contact with free-living birds. Siting of farms in relation to public roads and service facilities such as hatcheries, feed mills, and processing plants has a profound impact on the effectiveness of a program to maintain optimal standards of production. Decisions concerning conceptual biosecurity influence all subsequent activities relating to prevention and control of disease. Generally, defects in conceptual biosecurity cannot be changed in response to the emergence of new diseases which may result in severe losses or even failure of an enterprise.
- *Structural Biosecurity*: The second level of biosecurity includes considerations such as the layout of farms, erection of fences, construction of drainage, allweather roads, equipment for decontamination, bulk feed installations, change rooms, exclusion of rodents and wild birds, and the interior finishes in houses. Structural biosecurity can be enhanced in the intermediate term with appropriate capital investment. Remedial action may often be too late to respond to the emergence of a new disease or an epornitic of a catastrophic infection such as highly pathogenic avian influenza.

Operational Biosecurity: The third level comprises routine managemental procedures intended to prevent introduction and spread of infection within a complex or enterprise. These activities can be modified at short notice to respond to disease emergencies. Constant review of procedures, participation by all levels of management and labor and appropriate monitoring of the health status and immunity of flocks contributes to effective operational biosecurity.

Aspergillosis (Fungal Pneumonia)

Cause

The principal fungus causing Aspergillosis in poultry is Aspergillus fumigatus.

Transmission

Transmission is by inhalation of fungus spores from contaminated litter (e.g. wood shavings, straw) or contaminated feed. Hatcheries may also contribute to infection of chicks.

Species affected

Young chickens are very susceptible. Older chickens are more resistant to infection. Turkey poults, pheasants, quails, ducklings, and goslings may also become infected.

Clinical signs

Infected chickens are depressed and thirsty. Gasping and rapid breathing can be observed. Mortality is variable, from 5 to 50 %. Gross lesions involve the lungs and airsacs primarily. Yellow-white pin head sized lesions can be found. Sometimes all body cavities are filled with small yellow-green granular fungus growth.

Diagnosis

The presence of Aspergillus fumigatus can be identified microscopically or sometimes even with the naked eye in the air passages of the lungs, in the airsacs or in lesions of the abdominal cavity. Aspergillosis can be confirmed by isolation and identification of the fungus from lesions.

Treatment and control

There no specific treatment for infected birds. The best is to remove and destroy affected birds. Strict hygiene in breeder (hatching eggs) and hatchery management is necessary. Choice and quality of litter is also important to prevent that spore bearing wood shavings or straw are used. Hatchery control with anti-fungal disinfectant may be critical to cleaning and disinfection procedures to control fungus infection.

Avian Influenza (AI)

Cause

Avian Influenza is caused by an Orthomyxovirus; there are several serotypes.

Currently we know there are 16 H- types and 9 N-types and they can show up in all kinds of combinations. For poultry the most important ones are H5, H7 and H9. Pathogenicity varies with the strains HPAI and LPAI (high or low pathogenic AI).

Transmission

AI virus is excreted from nares, mouth, conjunctiva and cloaca. Airborne virus particles from the respiratory tract, droppings, and people carrying virus on their clothing and equipment are the main routes of transmission. Migratory water fowl and other wild birds infected with AI virus may be a source of infection.

Species affected

Avian Influenza viruses have been shown to naturally infect a wide variety of wild and domestic birds. In poultry production main problems are in chickens, turkeys and ducks.

Clinical signs

Clinical signs will vary, depending on the pathogenicity (HPAI and LPAI) of AI virus involved and other factors as host species, sex, concurrent infections, acquired immunity and environmental factors. LPAI shows generally mild symptoms: respiratory coughing sneezing, wet eyes, nasal discharge depression, lethargy limited reduction of feed intake and limited drop in egg production; low mortality rate. HPAI shows fast onset with increased mortality even before clinical signs are seen, depression, drop in feed and water intake, severe drop in egg production and mortality can vary between 50-90%.

Diagnosis

Clinical signs are indicative for AI; final confirmation by laboratory testing:

- Direct detection of AI proteins or Nucleic acids(RNA) using PCR.
- Virus isolation from infected organs, tracheal or cloacal swabs.
- Serology from blood samples after infection and for routine monitoring showing specific AI antibodies.

Treatment

There is no treatment for Avian Influenza. Antibiotics will help to control secondary bacterial infections.

Prevention and control

In many countries AI is a notifiable disease with specific local regulations on its control. In AI free areas the disease (LPAI and HPAI) is controlled by monitoring and stamping out. In case of LPAI infected areas countries can decide to allow vaccination only for LPAI. In case of endemic HPAI and/or LPAI vaccination might be allowed. Vaccination is generally done with inactivated AI vaccines based on the strain H-type causing the outbreaks.

Avian Metapneumovirus (aMPV) "Rhinotracheitis" (RT) Turkey Rhinotracheitis (TRT) "Swollen Head Syndrome (SHS)"

Cause

Avian Metapneumovirus, belongs to the subfamily Pneumovirus from the family Paramyxoviridae. Within the aMPV 4 subtypes (A-B-C-D) are identified.

Transmission

The virus spreads horizontally via direct contact from bird to bird and via contaminated personnel, water and equipment.

Species affected

Turkeys and chickens.

Clinical signs and lesions

- In both turkeys and chickens.
- In young birds the respiratory clinical signs; snicking, rales, sneezing, nasal discharge , foamy conjunctivitis, swollen infraorbital and periorbital sinus "swollen head" are most prominent; when birds become older head shaking and coughing can be added. In adult laying hens there may a drop in egg production up to 70% with an increased incidence of poor shell quality.
- Morbidity is general high (up to 100%); mortality varies from 0.5% up to 50% depending on secondary infections (such as IBV, E.coli) and farm climate conditions.
- At post mortem the lesions vary. In an uncomplicated infection: rhinitis, tracheitis and sinusitis will dominate , in case of Swollen Head Syndrome, oedema at the head often in combination with purulent or caseous subcutaneous exudate. In complicated cases with E.coli polyserositis including fibrinous exudate in abdominal cavity can be found.

Diagnosis

Clinical signs can give a good indication for aMPV but need to be confirmed by laboratory testing. Identification of virus by isolation or PCR from nasal secretions and/or tracheal or sinus scraping. After infection several serological methods (VN, Elisa, IFT) can be used to detect antibodies. Be aware that vaccines also induce antibodies.

Treatment

There is no treatment for aMPV infections. Treatment with antibiotics can be given to control secondary bacterial infections.

Prevention

Vaccination is the most effective prevention method. Over the years live and inactivated vaccines have been developed and are very efficacious in controlling infections. Short living birds will only be vaccinated with a live vaccine(s). Long living birds are advised to be primed with live- and boosted with inactivated vaccines. In this way the birds will have good long lasting local and systemic protection.

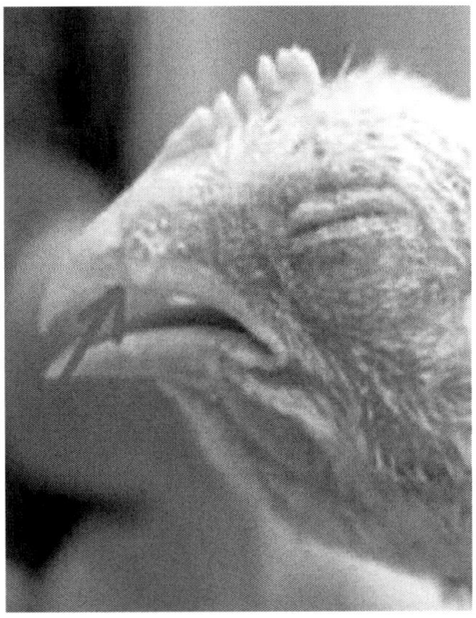

Infectious Coryza

Cause

This is a bacterial disease caused by Avibacterium paragallinarum, (in the past known as Haemophilus paragallinarum). There are 3 common serotypes, representing different immunotypes: A, B, C.

Transmission

The disease spreads from bird to bird and flock to flock by contact and airborne infected dust particles and via the drinking water. Spread by equipment and personnel have also been reported. The incubation period varies from 1 to 3 days.

Species affected

The chicken is the natural host for Avibacterium paragallinarum. All ages are susceptible, but the disease is usually less severe in juvenile birds.

Clinical signs

The main clinical signs are due to an acute inflammation around the eyes and upper respiratory tract. Signs include a serous to mucoid discharge in the nasal passage and sinuses, facial edema and conjunctivitis. Feed and water consumption will be decreased resulting in loss of weight gain and loss of egg production (10-40%) in laying birds. In affected breeders, the hatchability and day-old chick quality might decline. Mortality will vary with the virulence of the infection, but is generally low. Complicating infections with eg IB, ILT, MG, MS and ND can make it worse.

Diagnosis

A field infection produces similar symptoms to chronic respiratory disease, therefore a diagnosis based only on clinical signs is difficult to establish. The most certain diagnosis may be obtained by the isolation of the organism from the sinus or airsac exudate from affected birds. This procedure must be carried out in the laboratory. There is no practical serological test.

Treatment

Treatment with various antibiotics (erythromycin and tetracycline are commonly used) will alleviate the severity and course of the disease. Relapse often occurs after treatment is discontinued and recovered birds will remain carriers. Because of noted drug resistance of Avibacterium paragallinarum, an antimicrobial sensitivity test is recommended.

Prevention

Vaccination is the preferred control method and is standard in most endemic Coryza areas. There is no cross protection between the serotypes A, B, C. There is cross protection within the serotypes A and C, but recent outbreaks due to B serotype strains showed there is partial cross protection within the B- serotype. Therefore, besides vaccines made of A, B, C serotype strains, a broader multivalent vaccine based on serotypes A, B, B-variant and C was developed. Vaccination with multi-serotype inactivated vaccines during the rearing period will reduce clinical signs/control Infectious Coryza.

Infectious Laryngotracheitis (ILT)

Cause

ILT is caused by a Herpesvirus, only one serotype is known.

Transmission

The natural entry of ILT is via the upper respiratory tract and ocular route. Field spread occurs via direct contact from bird to bird and/or transmission by contaminated people or equipment (visitors, shoes, clothing, egg boxes, transport crates). Incubation period varies from 4-12 days.

Species affected

Chickens are the primary natural host but other species (pheasants) can also be affected.

Clinical signs

An acute respiratory disease with nasal discharge and moist rales followed by gasping, marked respiratory distress and expectoration of blood-stained mucus in laying birds. Egg production can drop 10-50% but will return to normal after 3-4 weeks. Mortality can vary from 5-70%. Spread through a chicken house is slower compared to IB and ND.

Post mortem lesions

Lesion are found throughout the respiratory tract but most pronounced in the larynx and trachea.

Depending on the severity of the infection you can find tracheitis with haemorrhagic and/or diphteric changes.

Diagnosis

Clinical picture with birds showing respiratory distress and expectoration of bloody mucus are indicative for ILT. Laboratory confirmation with: histopathology showing intranuclear inclusion bodies in tracheal epithelial cells, virus isolation from tracheal swabs on embryonated chicken eggs, virus detection with PCR or IFT on tracheal samples. Detecting antibodies from blood samples after infection.

Treatment

There is no treatment for ILT; emergency vaccination in the early stage of an infected flock may reduce the spread and limit the outbreak.

Prevention and control

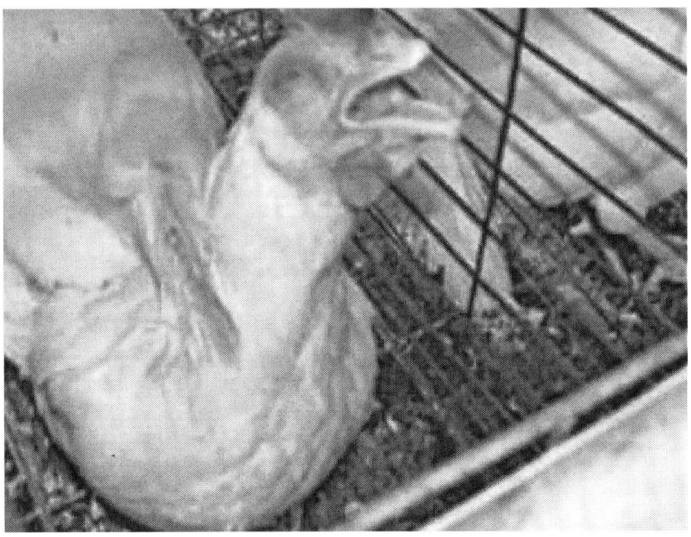

In many countries vaccination is the preferred control method. Though in some countries it is not allowed to vaccinate or only under restriction. The existing live conventional CEO ILT vaccines are effective in controlling clinical problems but have the risk of spreading and reversion to virulence after multiple passage through chickens. Recent outbreaks show often the relation to the live vaccine strains used in an area. Therefore, the new generation of recombinant vector vaccines are more suitable in the control and prevention of ILT. Recombinant vaccines based on HVT-vector carrying inserts of important immunogenic ILT proteins show good efficacy and do not spread and cannot revert to virulence because there is not a full ILT virus involved.

CHAPTER 1 PATHOGENESIS AND DIAGNOSTIC APPROACHES OF AVIAN INFECTIOUS BRONCHITIS

Faruku Bande, Siti Suri Arshad, Abdul Rahman Omar, Mohd Hair Bejo, Muhammad Salisu Abubakar, and Yusuf Abba

INTRODUCTION

Infectious bronchitis (IB) causes significant economic losses to the poultry industry worldwide [1, 2]. The disease was first identified in North Dakota, USA, when Schalk and Hawn reported a new respiratory disease in young chickens [3]. Since then, IBV has been recognized widely, especially in countries with large commercial poultry populations. Apart from respiratory infections, IB affects the kidney and reproductive tract, causing renal dysfunction and decreased egg production, respectively. Although the disease first was believed to occur primarily in young chickens, however, chickens of all age are also susceptible [1].

AETIOLOGY AND MOLECULAR BIOLOGY

Infectious bronchitis is caused by infectious bronchitis virus (IBV), a single stranded positive sense, enveloped RNA virus of 27–32 kb length [4]. The virus has been classified under the Gammacoronavirus genus in the family Coronaviridae, order Nidovirales. Like other members of coronavirus family, the IBV genome is composed of structural and nonstructural proteins. Structural proteins include the spike [S] glycoprotein, envelope [E], matrix [M], and nucleocapsid [N]. These proteins together play different roles in viral attachment, replication, and inducing clinical disease. Of major structural proteins, the M protein is the most abundant transmembrane protein, which play vital role in coronavirus assembly through interaction with viral ribonucleocapsid and spike glycoprotein [5, 6]. IBV E protein is, however, scant and contains highly hydrophobic transmembrane N-terminal and cytoplasmic C-terminal domains. Studies have shown that the E protein is localized to the Golgi complex in IBV infected cells and is integrally associated with viral envelope formation, assembly, budding, ion channel activity, and apoptosis [7, 8]. Similar to other coronaviruses, the phosphorylated 409 amino acid of IBV-N protein is highly conserved between amino acid residues 238 and 293 [9]. IBV-N protein binds with the genomic RNA to form a helical ribonucleoprotein complex (RNP), thus aiding transcription, replication, translation, and packaging of the viral genome during replication [10]. The S1 portion of the spike glycoprotein plays important role in the attachment and entry of the virus into the cell via sialic acid receptors and has been considered as the determinant for viral diversity and immune protection [11]. This protein has been targeted for genotypic characterization as well as recombinant IBV serotypes vaccines [6,12, 13].

PATHOGENESIS

Infectious bronchitis virus infects primarily the respiratory system. However, some variants and several field isolates affect the reproductive, renal, and digestive systems of chickens. Disease pathogenesis differs according to the system involved, as well as the strain of the virus [1].

Host Susceptibility

Although domestic fowl (Gallus gallus) and pheasant (Phasianus spp.) are considered to be natural hosts for IBV [14], other IBV-like coronaviruses have been identified in nondomestic avian species including pheasant, peafowl, turkey, teal, geese, pigeon, penguins quail, duck, and Amazon parrot [15–18]. Antigenic similarities between turkey coronavirus (TCoV) and avian infectious bronchitis virus (AIBV) have also been demonstrated [19]. Antibodies to IBV have been demonstrated in humans with close contact to poultry, but the virus has not been reported to cause human clinical disease [20].

Age and Breed Predisposition

Chickens of all ages and breed types are susceptible to IBV infection, but the extent and severity of the disease is pronounced in young chicks, compared to adults. Similarly, resistance to infection was suggested to increase with increasing age [21]. Experimental evidence suggests that line C white leghorn chickens are more resistant to M41 IBV challenge, compared to line 151, although both lines had similar virus shedding rate [22, 23], perhaps influenced by genetic polymorphism in the chicken major histocompatibility complex (MHC), as observed between $B^{*}15$, $B^{*}13$, or $B^{*}21$ chicken haplotypes [24].

Citation: Faruku Bande, Siti Suri Arshad, Abdul Rahman Omar, Mohd Hair Bejo, Muhammad Salisu Abubakar, and Yusuf Abba, "Pathogenesis and Diagnostic Approaches of Avian Infectious Bronchitis," Advances in Virology, vol. 2016, Article ID 4621659, 11 pages, 2016. doi:10.1155/2016/4621659.

Receptor and Entry

The IBV receptor-binding domain (RBD) in the S1-spike plays a major role in attachment of the virus to host cells [25, 26]. Thus, variation in the S1 glycoprotein partly determines tissue tropism and virulence [27, 28]. IBV affects trachea, kidney, and reproductive tract through interaction of S1 glycoproteins RBD (AAs 19–69 in M41) with α-2,3-sialic acid receptors on the surface of the cells [29, 30]. In addition to the sialic acid receptor, attenuated Baudette-IBV strain has been shown to interact with a putative heparan sulfate- (HS-) binding site that might contribute to its wide host range [31]. Following viral attachment, conformational changes occurring in the S1 glycoprotein mediate the membrane fusion activity of the S2 carboxylic acid terminal of the spike glycoprotein [1]. Subsequently, IBV enters the cell and releases its nucleocapsid into the cell's cytoplasm, thus triggering replication, virus budding, and release [32].

Infection and Transmission

The virus is transmitted via the respiratory secretions, as well as faecal droplets from infected poultry. Contaminated objects and utensils may aid transmission and spread of the virus from one flock to another. Evidence of virus was shown in trachea, kidney, and Bursa of Fabricius 24 hrs following aerosol transmission [33]. The nature of IBV persistence remains to be elucidated; however, detection of the virus in the caecal tonsils (up to 14 weeks) and from faeces (20 weeks) after infection might suggest a role of faecal shedding in viral transmission and persistence [34].

Incubation Period

Generally the short incubation period for IBV varies with infective dose and route of infection. For example, while infection via the tracheal route may take a course as short as 18 hours, ocular inoculation leads to an incubation period of 36 hours [33].

Clinical Course and Manifestations

In the host, initial infection occurs at epithelia of Harderian gland, trachea, lungs, and air sacs. The virus then moves to the kidney and urogenital tract, to establish systemic infection [33, 35]. In this regard, the severity and clinical features of IB depend on the organ or system involved. Infection of the respiratory system may result in clinical signs such as gasping, sneezing, tracheal rales, listlessness, and nasal discharges. Affected birds appeared listless and dull with ruffled feathers (Figure 1). Other signs may include weight loss and huddling of birds together under a common heat source [33].

Figure 1: Dullness exhibited in chickens infected following experimental infection with IBV (courtesy: Siti Suri Arshad).

Other clinical outcomes associated with IB infection include frothy conjunctivitis, profuse lacrimation, oedema, and cellulitis of periorbital tissues. Infected birds may also appear lethargic, with evidence of dyspnoea and reluctance to move [36]. Nephropathogenic IBV strains are most described in broiler-type chickens. Clinical signs include depression, wet droppings, and excessive water intake. Infection of reproductive tract is associated with lesions of the oviduct, leading to decreased egg production and quality. Eggs may appear misshapen, rough-shelled, or soft with watery egg yolk (Figure 2). Unless effective measures are instituted, decline in egg production does not return to normal laying, thus contributing to high economic loss [1, 37].

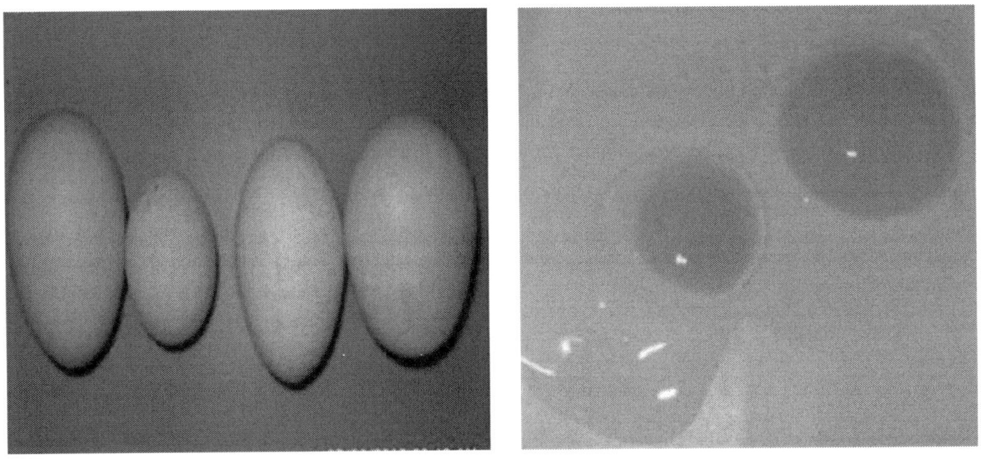

(a) (b)

Figure 2: Irregularity in the shape and sizes of eggs from natural IBV infected breeder chickens (a). Watery albumen from IBV infected chicken ((b) left) compared to normal egg ((b) right).

Gross and Histopathology

Pathological changes observed grossly at necropsy include congestion and oedema of tracheal mucosa and extrapulmonary bronchi (Figure 3) [38, 39].

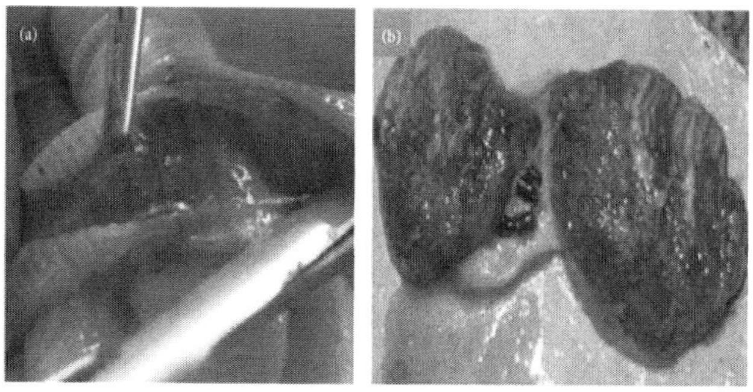

Figure 3: Gross lesions observed on respiratory organs of chicken naturally infected with IBV. Presence of mucoid secretion, congestion, and hyperaemia in the trachea (a); mild focal areas of lung consolidation (b).

Histopathological changes include loss of cilia, oedema, rounding and sloughing of epithelial cells, and infiltration by lymphocytes (Figure 4). Presence of Russell bodies in Harderian cells has been observed following infection with H120 IBV serotype [40].

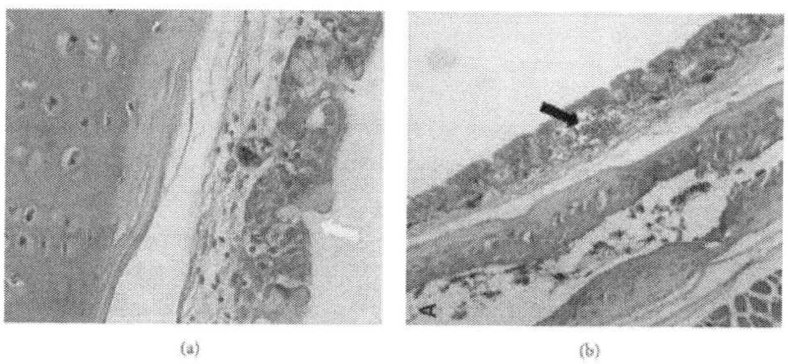

Figure 4: Histopathological changes in the trachea of naturally IBV infected chicken. Note: the marked infiltration of lymphocytes within the epithelia (black arrow (b)) and evidence of mucosal secretions of goblet cells (yellow arrow (a)).Nephropathogenic IBV strains cause nephritis characterized by swelling and congestion of the kidney (Figure5), sometimes with pallor of ureters that contain urate deposits. Coinfection with bacterial pathogens such as E. coli may lead to a more complex outcome, usually associated with high morbidity and mortality. Similarly, infection with nephropathogenic IBV strains may result in pale, swollen, and mottled kidneys [39, 41]. Histological findings include interstitial nephritis, tubular degeneration, and infiltration by heterophils. In some cases, necrotic and dilated tubules are filled with urates and casts [33]. Experimental studies have shown that IBV-T-strain causes necrosis of the proximal convoluted tubule and distension of distal convoluted tubule. In addition, necrotic foci, heterophils, and lymphocytes are observed in the interstitial spaces. Oedema of Bowman's capsule and granulocytic infiltration has been reported in the collecting ducts and spheroids [42, 43].

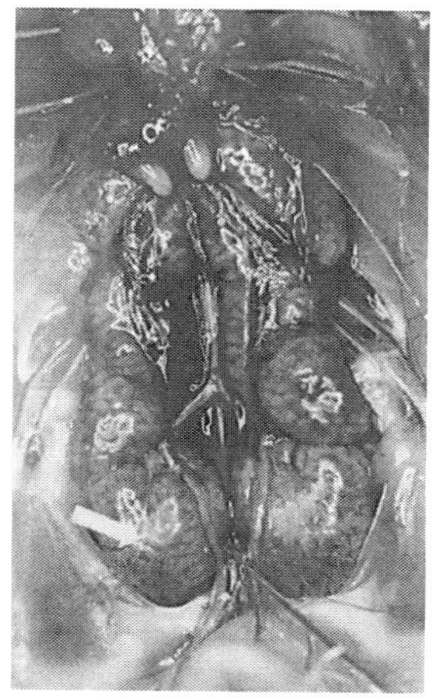

Figure 5: Gross lesions in kidney of chicken following experimental infection with a nephropathogenic infectious bronchitis virus. Note: swelling and congestion of the kidney (arrow) (courtesy: Siti Suri Arshad).

When the reproductive system is affected, there may be nonpatent and hypoglandular oviduct, especially in severely affected chickens [43, 44]. Large accumulation of yolk fluid may be seen in the abdominal cavity (Figure 6), often associated with bacterial infection in laying hens [45, 46]. Cystic oviduct has also been observed in young layers following infection with certain IBV strains (Figure 7).

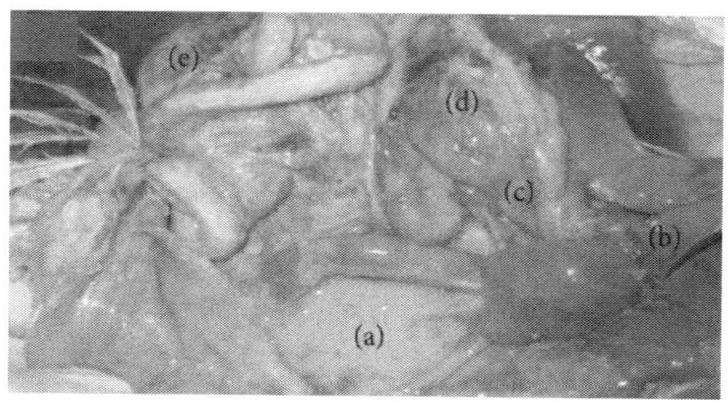

Figure 6: Chicken showing natural IBV infection. Accumulation of egg yolk in abdominal cavity (a); slightly enlarged, pale, friable liver (b) and multiple petechial haemorrhages on the serosal surfaces of proventriculus (c), gizzard (d), and small intestine (e).

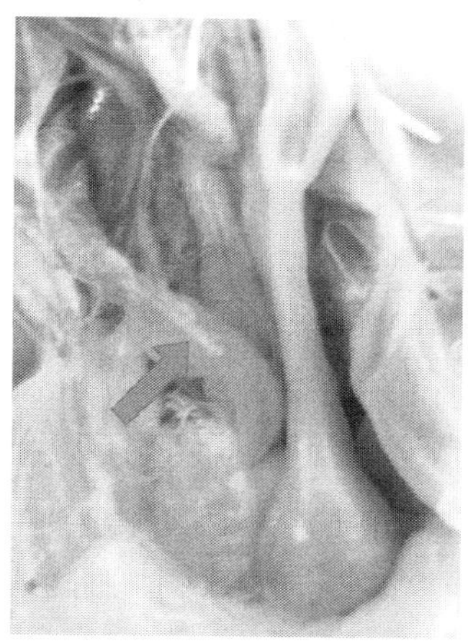

Figure 7: Cystic oviduct in 11-week-old chicken experimentally infected with a CR88 infectious bronchitis virus strain. Note the distention of the entire oviduct and fluid accumulation (arrow).

Morbidity and Mortality

Morbidity due to IBV infection can reach up to 100%. Mortality rate may range from 25 to 30% in young chicks but may increase to 80% as a result of factors that are host-associated (age, immune status), virus-associated (strain, pathogenicity, virulence, and tissue tropism), or environmental (cold and heat stresses, dust, and presence of ammonia). Secondary bacterial infections (E. coli) or coinfection with immunosuppressive viruses such as Marek's disease virus, infectious bursal disease virus (IBDV) [33, 47, 48], may worsen the outcomes of IBV infection. Generally, nephropathogenic IBV strain causes high mortality, compared with strains infecting only the respiratory or reproductive systems [49].

DIAGNOSIS

Conventional and more advanced methods have been used for the diagnosis of IBV infection. The choice of one test over another is guided by type of sample, availability of test materials and facilities, test reporting time, purpose of the test, and whether the test is carried out in the field or at the laboratory. Selected testing procedures are discussed below.

Serology

In the past, serological assays such as virus neutralization (VN) and haemagglutination inhibition (HI) were used widely for detecting and serotyping IBV strains. These tests also have been used to measure flock protection following vaccination [50, 51]. Serotype-specific antibodies usually are detected using HI, even though the HI test is less reliable [51]. On the other hand, ELISA assays are more sensitive and easily applied for field use and in monitoring antibody response following vaccination or exposure. However, emergence of different IBV serotypes that do not cross-react with commonly available antisera generally made serological tests less applicable and nonconclusive in classifying new or emerging IBV isolates [52, 53].

Virus Isolation and Identification

Virus isolation has been the gold standard for the diagnosis of IBV [54, 55]. Taking samples during early onset of the disease and ensuring the right sampling techniques are important keys for successful isolation of IBV. To support successful virus isolation from swabs, recommended to place swab sample in buffered solutions of media or PBS before transporting them to the laboratory. If tissue samples are to be collected, recommended tissues are trachea, kidney, proventriculus, tonsil, and oviduct. Tissue samples must be collected aseptically from scarified chickens or immediately upon death, placed in sterile, tightly sealed plastic specimen bags, and transported to the laboratory on ice for further processing [56]. The stringent technique requirements and factors, such as the time required for several passages of virus in egg or cell culture, limit the use of virus isolation as a diagnostic method of choice for IBV infection. Notwithstanding, different laboratories use various isolation methods, as described below.

Embryonated Chicken Egg

Most IBV strains grow well when inoculated into the allantoic cavity of a 9–11-day-old chicken embryo. Clinical samples from tracheal swab, broth, or tissue homogenate (10% w/v) are inoculated into the allantoic cavity of specific pathogen-free eggs and incubated at 34–37°C, after inoculation. Eggs are candled daily to monitor embryo viability; death within 24 hrs is considered nonspecific. After 48–72 hrs, allantoic fluid (AF) is harvested from representative eggs that were chilled overnight and tested for the presence of IBV using serological tests or RT-PCR assay. Sometimes the allantoic fluid needs to be subjected to several passages to allow the virus to adapt and replicate to high titre, thus increasing the period that is needed to obtain results. The latter may vary among viral strains [54]. After 5–7 days, inoculated eggs are opened and observed for characteristic IB lesions such as curling and dwarfism of the infected embryo (Figure 8). It is important to note that such findings are suggestive, but not pathognomonic [57].

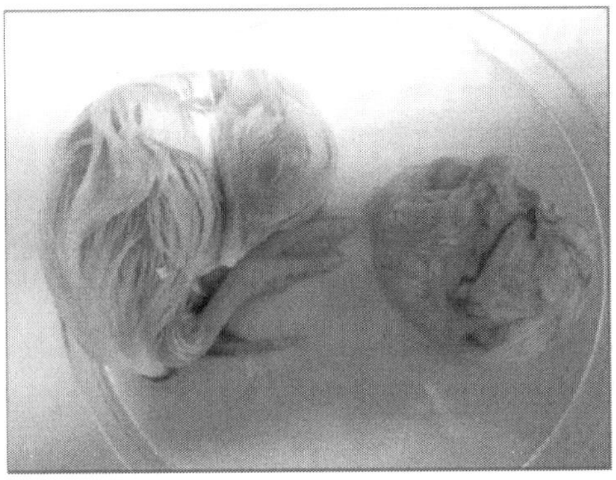

Figure 8: Embryo development at 17 days old following inoculations with IBV-CR88 strain. Note evidence of dwarfism and curling of the toes in IBV infected embryo (right) compared to a noninfected control embryo (left).

Cell Cultures

Isolation of IBV has been attempted in various primary and secondary cells, such as chicken embryo kidney fibroblast and Vero cells, respectively [58, 59]. Infected cultures are characterized by rounding, development of syncytia, and subsequent detachment from the surface of the plate [59]. A major limitation of cell culture methods for IBV isolation is that not all strains of IBV are easily adapted in cell culture. Even for some cell culture adaptable IBV (M41, Iowa 97, and NZ) strains, growth of the virus often requires primary isolation in embryonated eggs and several passages, prior to adaptation. In some cases, attempts to grow IBV in various cell lines either failed or resulted in very low viral titre [58].

Organ Cultures

Tracheal organ culture (TOC) can be used to propagate both embryo-adapted and non-embryo-adapted IBV strains. TOC is prepared from tracheal rings of 20-day-old chicken embryo. The tracheal rings are maintained in a roller bottle and infected with IBV-suspected samples. The culture is observed microscopically for evidence of ciliostasis under light microscope. Complete impairment of ciliary activity usually is considered as a positive culture [60]. Successful growth of IBV has been demonstrated in organ cultures derived from kidney, intestine, proventriculus, and oviduct. However, susceptibility of these organs to IBV can be influenced by the strain of the virus and the amount of virus presence in the sample (infective dose). While a study suggested the universality of using kidney, bursa, and proventriculus in growing IBV, a poor result was obtained when IBV was propagated in cultures derived from different intestinal segments [61]. An advantage of this method includes easy titration and serotyping of IBV, since no virus adaptation is required [62]. Possible constraints include lack of affinity of some IBV strain for some organ cells and difficulty in differentiating ciliostasis arising from other viruses, such as Newcastle disease virus and avian adenovirus [33].

Electron Microscopy

Electron microscopy provides a direct means of detecting and identifying IBV in biological samples based on morphological characteristics of coronavirus. Positive cultures are confirmed based on the presence of coronavirus-like pleomorphic structures with spike projections, following negative staining with phosphotungstic acid (Figure 9). Importantly, the shape and diameter (120 nm) of the virus are taken into consideration when making diagnostic judgements. Apart from the negative staining method, transmission electron microscopy (TEM) is also a useful tool which enables the visualization of virus-like particles in infected cells [59, 63]. However, this method is often applied to understand viral attachment and entry into the cell but is not a specific diagnostic test [35].

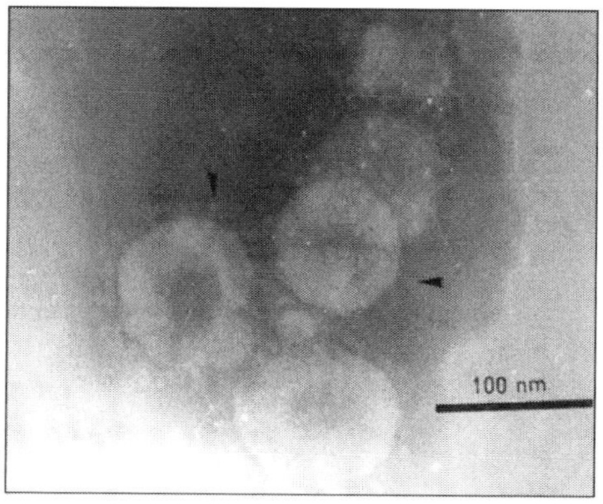

Figure 9: Negative staining electron microscope showing spherical shape of virus with typical spike projections (arrow) surrounding the virion of avian infectious bronchitis virus (courtesy: Siti Suri Arshad).

Immunohistochemistry

Immunoperoxidase and immunofluorescence are two important histochemistry methods for detection and confirmation of IBV antigen from infected tissue and/or cells. These methods work based on antigen-antibody reactions [64, 65]. Immunoperoxidase methods such as the avidin-biotin complex (ABC) have been used successfully to localize IBV antigen in tissue samples [66]. Likewise, indirect immunofluorescent assay is the most frequently used fluorescent technique [66, 67].

Molecular Diagnostic Assays

In view of their increased sensitivity and reduced reporting time, molecular methods, such as Reverse Transcriptase Polymerase Chain Reaction (RT-PCR), real-time PCR, Restriction Fragment Length Polymorphism (RFLP), and genome sequencing, have nearly replaced conventional serology and virus cultivation methods of IBV diagnosis [68, 69].

RT-PCR Methods

This approach uses viral RNA, amplified either directly (one-step RT-PCR) or following cDNA synthesis (two-step RT-PCR). An RT-PCR assay was designed and introduced in 1991 for detecting the IBV-S2 gene [70]. Subsequently, general and serotype-specific RT-PCR assays were designed to target different regions and/or fragments (Figure 10) in the IBV viral genome [71–73]. The UTR and N-gene-based RT-PCR are used for universal detection, because of the conserved nature of the target region in many IBV serotypes [68, 71]. A pan-coronavirus primer, targeting a conserved region of different coronavirus isolates, could also be used in one-step RT-PCR amplification of IBV strains [55]. However, amplification and sequencing of the S1 gene provide a reliable means for genotypic classification of new IBV strains [74]. A serotype-specific PCR assay has been designed to enable differentiation of Massachusetts, Connecticut, Arkansas, and Delaware field isolates [73].

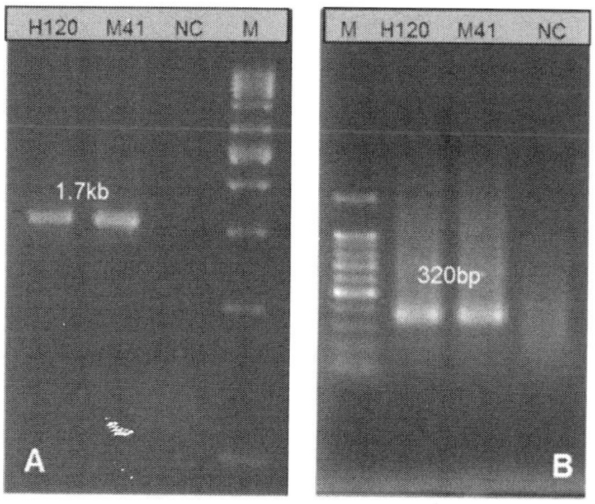

(a) (b)

Figure 10: Electropherogram showing 1.7 kb RT-PCR amplified S1 genes from vaccine (H120) and virulent (M41) IBV strains (a) compared to a 320 bp RT-PCR amplified N-gene (b) of H120 and M41 IBV serotypes. Lane M = 1 kb molecular ladder (a) and 100 bp ladder (b); lane NC: negative control (no template control).

Restriction Fragment Length Polymorphism (RFLP)

This is an IBV genotyping method carried out to differentiate different known strains of IBV and to identify new variants following RT-PCR amplification [75]. Full-length sequence of IBV S1 glycoprotein could be targeted for amplification and enzymes analysis [72, 76]. RFLP allows differentiation of various known IBV strains, based on their unique electrophoresis banding patterns defined by restriction enzyme digestion [72,77]. The assay was found to be comparable with traditional virus neutralization assay, although some strains such as the Gray and JMK strains were reportedly difficult to differentiate using arrays of restriction enzymes, thus limiting the universal application of this method [72].

Real-Time RT-PCR and Other Forms of PCR Assays

For increased test sensitivity and specificity, real-time RT-PCR assays [78, 79] have been introduced for detecting IBV. Apart from detection, it is possible to quantify IBV viral load from tissue and/or clinical samples by real-time RT-PCR assays based on viral copy number or fold changes [80, 81]. Likewise, differentiation of Massachusetts from non-Massachusetts is possible by real-time RT-PCR assay targeting S1 glycoprotein gene [79, 82]. Recently, a high resolution melt curve analysis (HRM) was also developed to allow differentiation of field from vaccine IBV strains as well as for rapid and sensitive detection of recombinant variants [83, 84]. Meir et al. [85] reported that real-time RT-PCR was comparable to virus isolation and one or two times more sensitive in detecting M41 IBV than ordinary N-gene and S1 gene specific RT-PCR assays. On the other hand, real-time RT-PCR was tenfold more sensitive compared to virus isolation and 30- or 40-fold compared to N-gene or S1 gene-based RT-PCR, respectively. The authors, however, reported variations in sensitivity when either N-gene or S1 genes were targeted as well as when different samples are used for viral amplification. Other forms of PCR methods used in detecting IBV include nested PCR [68]; multiplex PCR [86]; and reverse transcription loop-mediated isothermal amplification (RT-LAMP) [87]. While these methods are more sensitive than standard RT-PCR, they are more expensive as well and might be beyond the financial capacity of many producers.

Sequence and Phylogenetic Analyses

For genotyping, S1 gene usually is amplified using RT-PCR, sequenced, and subjected to bioinformatics analyses [88, 89]. Following S1 gene sequencing, isolates are characterized through bioinformatics analyses based on their phylogenetic relatedness with reference sequences available in sequence databases such as the NCBI, EMBL, and DDBJ (Figure 11). Lack of method standardization among laboratories, particularly with respect to the S1 gene segment length that is used in phylogenetic analysis, limits genotyping to some extent. Currently, molecular methods such as next generation sequencing (NGS) have been introduced to sequence whole genomes within limited periods of time, though this approach has been used only in the laboratory.

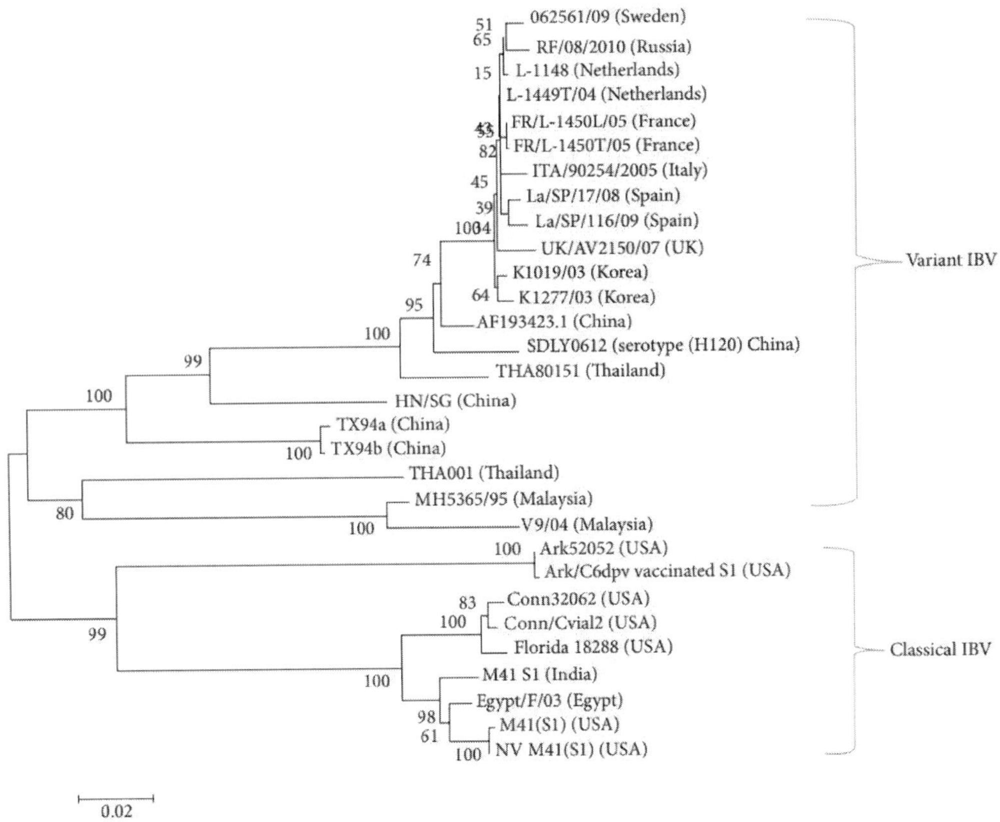

Figure 11: Neighbour joining phylogenetic analysis based on nucleotide acid sequence of S1-spike gene of classical and variant IBV strains identified in different countries. The tree was drawn with MEGA5 software using 1000 bootstrap replicates.

DIFFERENTIAL DIAGNOSIS

Several respiratory diseases, such as Newcastle disease (ND), infectious laryngotracheitis, infectious coryza, avian metapneumovirus (aMPV), and avian influenza (AI), may produce clinical signs similar to avian infectious bronchitis. However, certain clinical features, including neurological signs and diarrhoea in ND, high mortality in AI, and pronounced head swelling in coryza, are uncommon in IBV infection and thus may be used in ruling out or arriving at narrowed tentative differential list [33, 90].

CONCLUSION

Ever since the first identification of IBV in 1930s, the poultry industry has suffered a growing number of emerging IBV serotypes. Importantly, the newly evolved strains have been favoured by selection pressure, mutation, and/or recombinations, thus allowing them to avoid detection, evade host immune response, and cause diverse pathological outcomes. Lack of effective diagnostic methods and vaccines that could easily tackle the menace caused by multiple IBV serotypes is partly blamed for the serious economic losses as results of infectious bronchitis disease. Conventional detection assays such as virus neutralization and virus isolation have been used extensively, but, due to lack of sensitivity and specificity of serological assays and laborious nature of virus isolation methods, these assays have gradually been replaced by the new sensitive and specific assays such as RT-PCR, RFLP, and qRT-PCR that enable rapid genotyping and identification of new IBV strains. However, there is a need for standardization across laboratories with respect to the type and length of target gene to be considered for genotyping so as to ensure common understanding of genotype distributions in order to guide vaccine selection for prevention and control.

CONFLICT OF INTERESTS

The authors declared no conflict of interests regarding the present paper.

AUTHORS' CONTRIBUTION

Faruku Bande conceived the idea, collected and studied published papers, drafted the review paper, and made all uncited photos from cases handled in his Ph.D. work available. Siti Suri Arshad, Abdul Rahman Omar, Mohd Hair Bejo, Yusuf Abba, and Muhammad Salisu Abubakar all participated in conceptualization of the idea, study design, review, and editing of paper. Siti Suri Arshad provided photos used in Figures 1, 5, and 9. All authors have read and agreed with submission of final paper to the journal.

ACKNOWLEDGMENTS

The authors would like to thank the Universiti Putra Malaysia and Ministry of Science, Technology and Innovation (MOSTI), Project no. 02-01-04-SF1070 for funding supports. Dennis F. Lawler also provided paper-editing assistance.

REFERENCES

1. D. Cavanagh, "Coronavirus avian infectious bronchitis virus," Veterinary Research, vol. 38, no. 2, pp. 281–297, 2007.
2. S. S. Arshad, "Infectious bronchitis," in Diseases of Poultry in South East Asia, M. Zamri-Saad, Ed., pp. 199–206, Universiti Putra Malaysia Press, Serdang, Malaysia, 2006.
3. Schalk and M. Hawn, "An apparently new respiratory disease of baby chicks," Journal of the American Veterinary Medical Association, vol. 78, pp. 413–422, 1931.
4. M. M. Lai and D. Cavanagh, "The molecular biology of coronaviruses," Advances in Virus Research, vol. 48, pp. 1–100, 1997.
5. C. A. M. de Haan, H. Vennema, and P. J. M. Rottier, "Assembly of the coronavirus envelope: homotypic interactions between the M proteins," Journal of Virology, vol. 74, no. 11, pp. 4967–4978, 2000.
6. F. Bande, S. S. Arshad, M. Hair Bejo, H. Moeini, and A. R. Omar, "Progress and challenges toward the development of vaccines against avian infectious bronchitis," Journal of Immunology Research, vol. 2015, Article ID 424860, 12 pages, 2015, 10.1155/2015/424860.
7. E. Corse and C. E. Machamer, "The cytoplasmic tails of infectious bronchitis virus E and M proteins mediate their interaction," Virology, vol. 312, no. 1, pp. 25–34, 2003.
8. L. Wilson, P. Gage, and G. Ewart, "Hexamethylene amiloride blocks E protein ion channels and inhibits coronavirus replication," Virology, vol. 353, no. 2, pp. 294–306, 2006.
9. . Williams, W. Li, L. W. Sneed, and E. W. Collisson, "Comparative analyses of the nucleocapsid genes of several strains of infectious bronchitis virus and other coronaviruses," Virus Research, vol. 25, no. 3, pp. 213–222, 1992.
10. J. Jayaram, S. Youn, and E. W. Collisson, "The virion N protein of infectious bronchitis virus is more phosphorylated than the N protein from infected cell lysates," Virology, vol. 339, no. 1, pp. 127–135, 2005.
11. M. W. Jackwood, D. Hall, and A. Handel, "Molecular evolution and emergence of avian gammacoronaviruses," Infection, Genetics and Evolution, vol. 12, no. 6, pp. 1305–1311, 2012.
12. Z. J. Wei, P. Wei, M. L. Mo, M. Li, T. C. Wei, and K. R. Li, "Genetic variation of S1 gene hypervariable region I of infectious bronchitis viruses isolated in different periods in Guangxi," Chinese Journal of Virology, vol. 24, no. 2, pp. 126–132, 2008.
13. X.-M. Shi, Y. Zhao, H.-B. Gao et al., "Evaluation of recombinant fowlpox virus expressing infectious bronchitis virus S1 gene and chicken interferon-7 gene for immune protection against heterologous strains," Vaccine, vol. 29, no. 8, pp. 1576–1582, 2011.
14. D. Cavanagh, K. Mawditt, D. D. B. Welchman, P. Britton, and R. E. Gough, "Coronaviruses from pheasants (Phasianus colchicus) are genetically closely related to coronaviruses of domestic fowl (infectious bronchitis virus) and turkeys," Avian Pathology, vol. 31, no. 1, pp. 81–93, 2002.
15. S. Dea and P. Tijssen, "Detection of turkey enteric coronavirus by enzyme-linked immunosorbent assay and differentiation from other coronaviruses," American Journal of Veterinary Research, vol. 50, no. 2, pp. 226–231, 1989.
16. C. M. Jonassen, T. Kofstad, I.-L. Larsen et al., "Molecular identification and characterization of novel coronaviruses infecting graylag geese (Anser anser), feral pigeons (Columbia livia) and mallards (Anas platyrhynchos)," Journal of General Virology, vol. 86,

no. 6, pp. 1597–1607, 2005.

17. D. Cavanagh, "Coronaviridae: a review of coronaviruses and toroviruses," in Coronaviruses with Special Emphasis on First Insights Concerning SARS, pp. 1–54, Birkhäuser, Basel, Switzerland, 2005.

18. E. Circella, A. Camarda, V. Martella, G. Bruni, A. Lavazza, and C. Buonavoglia, "Coronavirus associated with an enteric syndrome on a quail farm," Avian Pathology, vol. 36, no. 3, pp. 251–258, 2007.

19. J. S. Guy, "Turkey coronavirus is more closely related to avian infectious bronchitis virus than to mammalian coronaviruses: a review," Avian Pathology, vol. 29, no. 3, pp. 207–212, 2000.

20. L. T. Miller and V. J. Yates, "Neutralization of infectious bronchitis virus human sera," American Journal of Epidemiology, vol. 88, no. 3, pp. 406–409, 1968.

21. R. A. Crinion and M. S. Hofstad, "Pathogenicity of four serotypes of avian infectious bronchitis virus for the oviduct of young chickens of various ages," Avian Diseases, vol. 16, no. 2, pp. 351–363, 1972.

22. K. Otsuki, M. Huggins, and J. K. Cook, "Comparison of the susceptibility to avian infectious bronchitis virus infection of two inbred lines of white leghorn chickens," Avian Pathology, vol. 19, no. 3, pp. 467–475, 1990.

23. N. Bumstead, "Genetic resistance to avian viruses," OIE Revue Scientifique et Technique, vol. 17, no. 1, pp. 249–255, 1998.

24. L. D. Bacon, D. B. Hunter, H. M. Zhang, K. Brand, and R. Etches, "Retrospective evidence that the MHC (B haplotype) of chickens influences genetic resistance to attenuated infectious bronchitis vaccine strains in chickens," Avian Pathology, vol. 33, no. 6, pp. 605–609, 2004.

25. G. J. Babcock, D. J. Esshaki, W. D. Thomas Jr., and D. M. Ambrosino, "Amino acids 270 to 510 of the severe acute respiratory syndrome coronavirus spike protein are required for interaction with receptor,"Journal of Virology, vol. 78, no. 9, pp. 4552–4560, 2004.

26. N. Promkuntod, R. E. W. van Eijndhoven, G. de Vrieze, A. Gröne, and M. H. Verheije, "Mapping of the receptor-binding domain and amino acids critical for attachment in the spike protein of avian coronavirus infectious bronchitis virus," Virology, vol. 448, pp. 26–32, 2014.

27. R. Casais, B. Dove, D. Cavanagh, and P. Britton, "Recombinant avian infectious bronchitis virus expressing a heterologous spike gene demonstrates that the spike protein is a determinant of cell tropism," Journal of Virology, vol. 77, no. 16, pp. 9084–9089, 2003.

28. D. E. Wentworth and K. V. Holmes, Coronavirus Binding and Entry. Coronaviruses: Molecular and Cellular Biology, Caister Academic Press, Norfolk, UK, 2007.

29. C. Winter, C. Schwegmann-Weßels, D. Cavanagh, U. Neumann, and G. Herrler, "Sialic acid is a receptor determinant for infection of cells by avian infectious bronchitis virus," Journal of General Virology, vol. 87, no. 5, pp. 1209–1216, 2006.

30. K. Shahwan, M. Hesse, A.-K. Mork, G. Herrler, and C. Winter, "Sialic acid binding properties of soluble coronavirus spike (S1) proteins: differences between infectious bronchitis virus and transmissible gastroenteritis virus," Viruses, vol. 5, no. 8, pp. 1924–1933, 2013.

31. G. Madu, V. C. Chu, H. Lee, A. D. Regan, B. E. Bauman, and G. R. Whittaker, "Heparan sulfate is a selective attachment factor for the avian coronavirus infectious bronchitis virus Beaudette," Avian Diseases, vol. 51, no. 1, pp. 45–51, 2007.

32. V. C. Chu, L. J. McElroy, V. Chu, B. E. Bauman, and G. R. Whittaker, "The avian coronavirus infectious bronchitis virus undergoes direct low-pH-dependent fusion activation during entry into host cells,"Journal of Virology, vol. 80, no. 7, pp. 3180–3188, 2006.

33. D. Cavanagh and J. Gelb, "Infectious bronchitis," in Diseases of Poultry, pp. 117–135, Wiley-Blackwell, 12th edition, 2008.

34. D. J. Alexander and R. E. Gough, "Isolation of avian infectious bronchitis virus from experimentally infected chickens," Research in Veterinary Science, vol. 23, no. 3, pp. 344–347, 1977.

35. S. Arshad, K. Al-Salihi, and M. Noordin, "Ultrastructural pathology of trachea in chicken experimentally infected with infectious bronchitis Virus-MH-5365/95," Annals of Microscopy, vol. 3, pp. 43–47, 2002.

36. C. Terregino, A. Toffan, M. Serena Beato et al., "Pathogenicity of a QX strain of infectious bronchitis virus in specific pathogen free and commercial broiler chickens, and evaluation of protection induced by a vaccination programme based on the Ma5 and 4/91 serotypes," Avian Pathology, vol. 37, no. 5, pp. 487–493, 2008.

37. R. W. Winterfield, H. L. Thacker, and S. F. Badylak, "Effects of subtype variations in the Holland strain of infectious bronchitis virus when applied as a vaccine," Poultry Science, vol. 63, no. 2, pp. 246–250, 1984.

38. D. A. Purcell and J. B. McFerran, "The histopathology of infectious bronchitis in the domestic fowl,"Research in Veterinary Science, vol. 13, no. 2, pp. 116–122, 1972.

39. Z. Boroomand, K. Asasi, and A. Mohammadi, "Pathogenesis and tissue distribution of avian infectious bronchitis virus isolate IRFIBV32 (793/B serotype) in experimentally infected broiler chickens," The Scientific World Journal, vol. 2012, Article ID 402537, 6 pages, 2012.

40. H. Toro, V. Godoy, J. Larenas, E. Reyes, and E. F. Kaleta, "Avian infectious bronchitis: viral persistence in the harderian gland and histological changes after eyedrop vaccination," Avian Diseases, vol. 40, no. 1, pp. 114–120, 1996.

41. F. Cong, X. Liu, Z. Han, Y. Shao, X. Kong, and S. Liu, "Transcriptome analysis of chicken kidney tissues following coronavirus avian infectious bronchitis virus infection," BMC Genomics, vol. 14, no. 1, article 743, 2013.

42. K. Chousalkar, J. R. Roberts, and R. Reece, "Histopathology of two serotypes of infectious

bronchitis virus in laying hens vaccinated in the rearing phase," Poultry Science, vol. 86, no. 1, pp. 59–62, 2007.K

43. . Chousalkar and J. R. Roberts, "Ultrastructural study of infectious bronchitis virus infection in infundibulum and magnum of commercial laying hens," Veterinary Microbiology, vol. 122, no. 3-4, pp. 223–236, 2007.
44. . Chousalkar, J. R. Roberts, and R. Reece, "Comparative histopathology of two serotypes of infectious bronchitis virus (T and N1/88) in laying hens and cockerels," Poultry Science, vol. 86, no. 1, pp. 50–58, 2007.
45. R. W. Winterfield and S. B. Hitchner, "Etiology of an infectious nephritis-nephrosis syndrome of chickens," American Journal of Veterinary Research, vol. 23, pp. 1273–1279, 1962.
46. J. J. de Wit, J. Nieuwenhuisen-van Wilgen, A. Hoogkamer, H. vande Sande, G. J. Zuidam, and T. H. F. Fabri, "Induction of cystic oviducts and protection against early challenge with infectious bronchitis virus serotype D388 (genotype QX) by maternally derived antibodies and by early vaccination," Avian Pathology, vol. 40, no. 5, pp. 463–471, 2011.
47. P. Ariaans, M. G. R. Matthijs, D. van Haarlem et al., "The role of phagocytic cells in enhanced susceptibility of broilers to colibacillosis after infectious bronchitis virus infection," Veterinary Immunology and Immunopathology, vol. 123, no. 3-4, pp. 240–250, 2008.
48. R. A. Gallardo, V. L. van Santen, and H. Toro, "Effects of chicken anaemia virus and infectious bursal disease virus-induced immunodeficiency on infectious bronchitis virus replication and genotypic drift,"Avian Pathology, vol. 41, no. 5, pp. 451–458, 2012.
49. J. Ignjatovic, D. F. Ashton, R. Reece, P. Scott, and P. Hooper, "Pathogenicity of Australian strains of avian infectious bronchitis virus," Journal of Comparative Pathology, vol. 126, no. 2-3, pp. 115–123, 2002.
50. D. King and D. Cavanagh, "Infectious bronchitis," Diseases of Poultry, vol. 9, pp. 471–484, 1991.
51. OIE, Avian Infectious Bronchitis, chapter 2. 3. 2., 2008.
52. D. Cavanagh, P. J. Davis, and J. K. Cook, "Infectious bronchitis virus: evidence for recombination within the Massachusetts serotype," Avian Pathology, vol. 21, no. 3, pp. 401–408, 1992.
53. Kant, G. Koch, D. J. Van Roozelaar, J. G. Kusters, F. A. J. Poelwijk, and B. A. M. Van der Zeijst, "Location of antigenic sites defined by neutralizing monoclonal antibodies on the S1 avian infectious bronchitis virus glycopolypeptide," Journal of General Virology, vol. 73, no. 3, pp. 591–596, 1992.
54. F. Beaudette and C. Hudson, "Cultivation of the virus of infectious bronchitis," Journal of the American Veterinary Medical Association, vol. 90, pp. 51–60, 1937.
55. C. B. Stephensen, D. B. Casebolt, and N. N. Gangopadhyay, "Phylogenetic analysis of a highly conserved region of the polymerase gene from 11 coronaviruses and development of a consensus polymerase chain reaction assay," Virus Research, vol. 60, no. 2, pp. 181–189, 1999.
56. J. Gelb Jr., W. A. Nix, and S. D. Gellman, "Infectious bronchitis virus antibodies in tears and their relationship to immunity," Avian Diseases, vol. 42, no. 2, pp. 364–374, 1998.
57. L. N. Loomis, C. H. Cunningham, M. L. Gray, and F. Thorp Jr., "Pathology of the chicken embryo infected with infectious bronchitis virus," American Journal of Veterinary Research, vol. 11, no. 40, pp. 245–251, 1950.
58. K. Otsuki, H. Yamamoto, and M. Tsubokura, "Studies on avian infectious bronchitis virus (IBV)—I. Resistance of IBV to chemical and physical treatments," Archives of Virology, vol. 60, no. 1, pp. 25–32, 1979.
59. S. S. Arshad, A study on two malaysian isolates of infectious bronchitis virus [Ph.D. thesis], Universiti Pertanian Malaysia, 1993.
60. B. V. Jones and R. M. Hennion, "The preparation of chicken tracheal organ cultures for virus isolation, propagation, and titration," Methods in Molecular Biology, vol. 454, pp. 103–107, 2008.·
61. P. S. Bhattacharjee and R. C. Jones, "Susceptibility of organ cultures from chicken tissues for strains of infectious bronchitis virus isolated from the intestine," Avian Pathology, vol. 26, no. 3, pp. 553–563, 1997.
62. Armesto, S. Evans, D. Cavanagh, A.-B. Abu-Median, S. Keep, and P. Britton, "A recombinant Avian infectious bronchitis virus expressing a heterologous spike gene belonging to the 4/91 serotype," PLoS ONE, vol. 6, no. 8, Article ID e24352, 2011.
63. S. Patterson and R. W. Bingham, "Electron microscope observations on the entry of avian infectious bronchitis virus into susceptible cells," Archives of Virology, vol. 52, no. 3, pp. 191–200, 1976.
64. Bezuidenhout, S. P. Mondal, and E. L. Buckles, "Histopathological and immunohistochemical study of air sac lesions induced by two strains of infectious bronchitis virus," Journal of Comparative Pathology, vol. 145, no. 4, pp. 319–326, 2011.
65. S. S. Arshad and K. A. Al-Salihi, "Immunohistochemical detection of infectious bronchitis virus antigen in chicken respiratory and kidney tissues," in Proceedings of the 12th Federation of Asian Veterinary Associations Congress/14th Veterinary Association Malaysia Scientific Congress, p. 51, August 2002.
66. S. Abdel-Moneim, P. Zlotowski, J. Veits, G. M. Keil, and J. P. Teifke, "Immunohistochemistry for detection of avian infectious bronchitis virus strain M41 in the proventriculus and nervous system of experimentally infected chicken embryos," Virology Journal, vol. 6, article 15, 2009.
67. K. Yagyu and S. Ohta, "Detection of infectious bronchitis virus antigen from experimentally infected chickens by indirect immunofluorescent assay with monoclonal antibody," Avian Diseases, vol. 34, no. 2, pp. 246–252, 1990.
68. Adzhar, K. Shaw, P. Britton, and D. Cavanagh, "Universal oligonucleotides for the detection of infectious bronchitis

virus by the polymerase chain reaction," Avian Pathology, vol. 25, no. 4, pp. 817–836, 1996.

69. Adzhar, R. E. Gough, D. Haydon, K. Shaw, P. Britton, and D. Cavanagh, "Molecular analysis of the 793/B serotype of infectious bronchitis virus in Great Britain," Avian Pathology, vol. 26, no. 3, pp. 625–640, 1997.

70. Z. Lin, A. Kato, Y. Kudou, K. Umeda, and S. Ueda, "Typing of recent infectious bronchitis virus isolates causing nephritis in chicken," Archives of Virology, vol. 120, no. 1-2, pp. 145–149, 1991.

71. K. A. Zwaagstra, B. A. M. van der Zeijst, and J. G. Kusters, "Rapid detection and identification of avian infectious bronchitis virus," Journal of Clinical Microbiology, vol. 30, no. 1, pp. 79–84, 1992.

72. H. M. Kwon, M. W. Jackwood, and J. Gelb Jr., "Differentiation of infectious bronchitis virus serotypes using polymerase chain reaction and restriction fragment length polymorphism analysis," Avian Diseases, vol. 37, no. 1, pp. 194–202, 1993.

73. L. Keeler Jr., K. L. Reed, W. A. Nix, and J. Gelb Jr., "Serotype identification of avian infectious bronchitis virus by RT-PCR of the peplomer (S-1) gene," Avian Diseases, vol. 42, no. 2, pp. 275–284, 1998.

74. J. G. Zhu, H. D. Qian, Y. L. Zhang, X. G. Hua, and Z. L. Wu, "Analysis of similarity of the S1 gene in infectious bronchitis virus (IBV) isolates in Shanghai, China," Archivos de Medicina Veterinaria, vol. 39, no. 3, pp. 223–228, 2007.

75. Z. Lin, A. Kato, Y. Kudou, and S. Ueda, "A new typing method for the avian infectious bronchitis virus using polymerase chain reaction and restriction enzyme fragment length polymorphism," Archives of Virology, vol. 116, no. 1–4, pp. 19–31, 1991.

76. K. Mardani, A. H. Noormohammadi, J. Ignatovic, and G. F. Browning, "Typing infectious bronchitis virus strains using reverse transcription-polymerase chain reaction and restriction fragment length polymorphism analysis to compare the 3′ 7.5 kb of their genomes," Avian Pathology, vol. 35, no. 1, pp. 63–69, 2006.

77. M. D. F. S. Montassier, L. Brentano, H. J. Montassier, and L. J. Richtzenhain, "Genetic grouping of avian infectious bronchitis virus isolated in Brazil based on RT-PCR/RFLP analysis of the S1 gene," Pesquisa Veterinaria Brasileira, vol. 28, no. 3, pp. 190–194, 2008.

78. K. K. Chousalkar, B. F. Cheetham, and J. R. Roberts, "LNA probe-based real-time RT-PCR for the detection of infectious bronchitis virus from the oviduct of unvaccinated and vaccinated laying hens,"Journal of Virological Methods, vol. 155, no. 1, pp. 67–71, 2009.

79. M. Acevedo, C. L. Perera, A. Vega et al., "A duplex SYBR Green I-based real-time RT-PCR assay for the simultaneous detection and differentiation of Massachusetts and non-Massachusetts serotypes of infectious bronchitis virus," Molecular and Cellular Probes, vol. 27, no. 5-6, pp. 184–192, 2013.

80. M. W. Jackwood, D. A. Hilt, and S. A. Callison, "Detection of infectious bronchitis virus by real-time reverse transcriptase-polymerase chain reaction and identification of a quasispecies in the Beaudette strain," Avian Diseases, vol. 47, no. 3, pp. 718–724, 2003.

81. S. A. Callison, D. A. Hilt, T. O. Boynton et al., "Development and evaluation of a real-time Taqman RT-PCR assay for the detection of infectious bronchitis virus from infected chickens," Journal of Virological Methods, vol. 138, no. 1-2, pp. 60–65, 2006.

82. R. M. Jones, R. J. Ellis, W. J. Cox et al., "Development and validation of RT-PCR tests for the detection and S1 genotyping of infectious bronchitis virus and other closely related gammacoronaviruses within clinical samples," Transboundary and Emerging Diseases, vol. 58, no. 5, pp. 411–420, 2011.

83. K. Hewson, A. H. Noormohammadi, J. M. Devlin, K. Mardani, and J. Ignjatovic, "Rapid detection and non-subjective characterisation of infectious bronchitis virus isolates using high-resolution melt curve analysis and a mathematical model," Archives of Virology, vol. 154, no. 4, pp. 649–660, 2009.

84. K. A. Hewson, G. F. Browning, J. M. Devlin, J. Ignjatovic, and A. H. Noormohammadi, "Application of high-resolution melt curve analysis for classification of infectious bronchitis viruses in field specimens,"Australian Veterinary Journal, vol. 88, no. 10, pp. 408–413, 2010.

85. R. Meir, O. Maharat, Y. Farnushi, and L. Simanov, "Development of a real-time TaqMan® RT-PCR assay for the detection of infectious bronchitis virus in chickens, and comparison of RT-PCR and virus isolation," Journal of Virological Methods, vol. 163, no. 2, pp. 190–194, 2010.

86. H.-W. Chen and C.-H. Wang, "A multiplex reverse transcriptase-PCR assay for the genotyping of avian infectious bronchitis viruses," Avian Diseases, vol. 54, no. 1, pp. 104–108, 2010.

87. H.-T. Chen, J. Zhang, Y.-P. Ma et al., "Reverse transcription loop-mediated isothermal amplification for the rapid detection of infectious bronchitis virus in infected chicken tissues," Molecular and Cellular Probes, vol. 24, no. 2, pp. 104–106, 2010.

88. Z. M. Zulperi, A. R. Omar, and S. S. Arshad, "Sequence and phylogenetic analysis of S1, S2, M, and N genes of infectious bronchitis virus isolates from Malaysia," Virus Genes, vol. 38, no. 3, pp. 383–391, 2009.

89. S. H. Abro, L. H. M. Renström, K. Ullman et al., "Emergence of novel strains of avian infectious bronchitis virus in Sweden," Veterinary Microbiology, vol. 155, no. 2–4, pp. 237–246, 2012.

90. R. Droual and P. Woolcock, "Swollen head syndrome associated with E. coli and infectious bronchitis virus in the Central Valley of California," Avian Pathology, vol. 23, no. 4, pp. 733–742, 1994.

CHAPTER 2 THE PATHOGENESIS OF LOW PATHOGENICITY H7 AVIAN INFLUENZA VIRUSES IN CHICKENS, DUCKS AND TURKEYS

Erica Spackman, Jack Gelb Jr, Lauren A Preskenis, Brian S Ladman, Conrad R Pope, Mary J Pantin-Jackwood, Enid T Mckinley

BACKGROUND

Avian influenza (AI) virus causes one of the most economically important diseases of poultry worldwide. AI is classified by the world organization for animal health (OIE) into two forms, low pathogenicity (LP) and high pathogenicity (HP), based on virulence in chickens [1]. H7 is one of the two most economically important AI virus subtypes because historically all HP AI viruses have been either the H7 or H5 subtype and it is among the most common subtypes in commercial poultry in the world [1,2]. In numerous cases the HP form mutated from a LPAI H7 (or H5) virus that was circulating in chickens or turkeys [3-6]. However, not all H7 LPAI viruses become HP. In the U.S., H7 AI viruses are sporadically recovered from wild birds (WB) and commercial poultry. Many of the outbreaks in commercial poultry [7-9] can be traced to the live bird market (LBM) system of New York and New Jersey where a single LP H7 genetic lineage persisted from 1994 to 2006 [10,11]. Few studies have directly compared the pathogenesis of AI virus in the three primary poultry species: chickens, ducks and turkeys. The aim of this work was to characterize the pathogenesis of selected North American H7 LPAI virus isolates from WB, commercial poultry, and the LBMs in the three primary domestic poultry species; chickens, ducks and turkeys.

Citation: Erica Spackman, Jack Gelb Jr, Lauren A Preskenis, Brian S Ladman, Conrad R Pope, Mary J Pantin-Jackwood, Enid T Mckinley, The pathogenesis of low pathogenicity H7 avian influenza viruses in chickens, ducks and turkeys, doi:10.1186/1743-422X-7-331.

RESULTS

Clinical Disease

Clinical disease signs varied in severity among the virushost combinations. Mean maximum clinical disease scores (the mean of the maximum clinical scores for each bird) ranged from 0 to 0.7 in chickens (Figure 1). Disease was only observed in chickens with 6 isolates (Figure 1) and not all chickens in these groups were affected. Disease signs in chickens were primarily conjunctivitis and lacrimation, which generally occurred from 2-4 days post infection. Only 3 isolates caused observable clinical disease in ducks (CK/NY/30749, ML/ OH/421 and RT/ DE/1538 (abbreviations defined in table 1)). Mean maximum clinical disease scores for ducks ranged from 0 to 0.5. The primary clinical sign presented by ducks was nasal discharge at 2 days PI and conjunctivitis. All 12 isolates caused observable clinical disease in turkeys with mean maximum scores ranging from 0.7 to 2.6, which were significantly higher than chickens and ducks with 8 of the 12 viruses. Clinical disease in turkeys included mild to severe conjunctivitis, nasal discharge, swollen sinuses as well as lethargy. Turkeys were the only species where mortality was observed, which ranged from 10-60% with 9 isolates (Table 2). The turkeys that died had severe sinusitis (a bacteriological examination was not conducted). Only one isolate, CK/NY/30749, caused clinical disease in all three species, although shed titers and clinical disease was most severe in turkeys. At no time were clinical disease signs observed in any of the sham inoculated birds. Microscopic lesions were consistent with what has been previously reported for AI virus in chickens, ducks and turkeys. Briefly, lesions observed in tissues from turkeys included viral tracheitis, cilial loss, heterophilic cecitis, and heterophilic rhinitis in the nasal cavity. Patchy cilial loss in the trachea, heterophilic tracheitis, and serositis of the kidney was observed in tissues from ducks. Rare to minimal rhinitis and the presence of heterophils, excessive mucus and slight lymphocytic accumulation were observed in the nasal cavity as well as rare focal heterophilic bronchitis was seen in tissues from chickens.

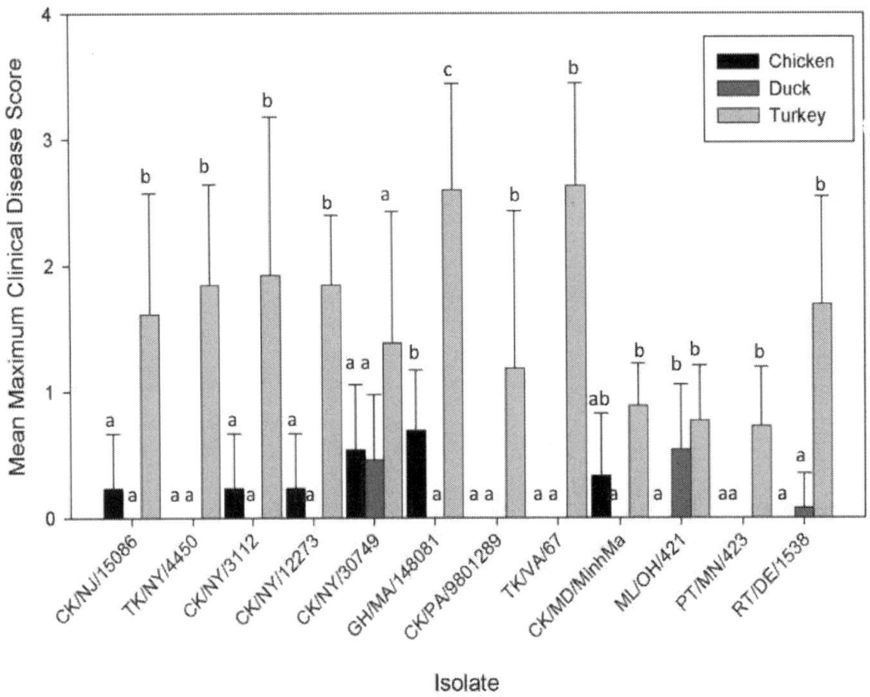

Figure 1: Mean maximum clinical disease score for chickens, ducks and turkeys infected with H7 LPAI virus. Clinical signs were scored as: 0 = no signs, 1 = mild to moderate respiratory signs (mild depression in ducks), 2 = moderate to severe (i.e. depressed, not eating, neurological signs), 3 = Dead. Letters denote statistical groups among species for each virus. Error bars indicate standard deviation of clinical disease scores.

Table 1: Low pathogenicity avian influenza virus isolates evaluated for pathogenesis in chickens, ducks and turkeys

Isolate	Subtype	Source	HA cleavage site	NA stalk deletion	NS subtype	Abbreviation
A/chicken/NJ/15086-3/1994	H7N3	LBM[A]	PENPKT/R	None	A	CK/NJ/15086
A/turkey/NY/4450-4/1994	H7N2	LBM	PENPKT/R	50-73[B]	B	TK/NY/4450
A/chicken/NY/3112-1/1995	H7N2	LBM	PENPKP/R	58-73	B	CK/NY/3112
A/chicken/NY/12273-11/1999	H7N3	LBM	PENPKT/R	None	A	CK/NY/12273
A/chicken/NY/30749-3/2000	H7N2	LBM	PEKPKP/R	None	B	CK/NY/30749
A/guinea hen/MA/148081-11/2002	H7N2	LBM	PEKPKK/R	58-73	B	GH/MA/148081
A/chicken/PA/9801289/1998	H7N2	Commercial-poultry (LBM lineage)	PENPKP/R	58-73	B	CK/PA/9801289
A/turkey/VA/SEP-67/2002	H7N2	Commercial-poultry (LBM lineage)	PEKPKP/R	58-73	B	TK/VA/67
A/chicken/MD/MinhMa/2004	H7N2	Commercial-poultry (LBM lineage)	PEKPKP/R	58-73	B	CK/MD/ MinhMa
A/mallard/OH/421/1987	H7N8	WB	PESPKT/R	None	A	ML/OH/421
A/pintail/MN/423/1999	H7N3	WB	PENPKT/R	None	A	PT/MN/423
A/ruddy turnstone/DE/1538/2000	H7N9	WB	PENPKT/R	None	A	RT/DE/1538

A. LBM = live bird market; WB = wild bird B. Range of residues deleted.

Virus Shed

At days 2 PI, oro-pharyngeal (OP) and cloacal (CL) swabs from all species were positive for AI viruses by quantitative real-time RT-PCR (qrRT-PCR) with the exception of CL swabs from ducks and turkeys inoculated with ML/OH/421 (Figure 2). However, by day 4 PI CL swabs from ML/OH/421 infected ducks and turkeys were positive. The highest OP titers were observed 2-4 days PI, although in ducks, viral shed tended to peak at 2 days PI, whereas in chickens and turkeys, shed tended to peak at 4 days PI. The highest mean OP titer for turkeys corresponds to a titer of $10^{6.3}$ 50% egg infectious doses (EID_{50}) with CK/NY/30749 at 2 days PI and GH/ MA/148081 at 4 days PI (Figure 2). The highest mean OP titer from chickens was $10^{5.8}EID_{50}$ at 4 days PI with CK/PA/9801289 (Figure 2) and the highest mean duck OP titer was $10^{4.9}EID_{50}$ at 2 days PI with PT/MN/423. Titers from CL swabs were consistently lower than OP titers and were generally higher from ducks and turkeys than from chickens with peak titers observed between 4-7 days PI. The highest CL shed titers were $10^{3.8}EID_{50}$ with RT/DE/1538 for turkeys at 7 days PI, $10^{2.9}EID_{50}$ with TK/VA/67 at 7 days PI in chickens, and $10^{5.1}EID_{50}$ with PT/MN/423 at 4 days PI from ducks. Oral and CL shed persisted 'through 10 and 14 days PI with most isolate and species combinations (Figure 2).

Table 2: Mortality and mean death time in turkeys inoculated with North American H7 LPAI virus Isolates

Isolate	Mortality	Mean death time (days)
CK/NJ/15086	40 (4/10)[A]	7.0
TK/NY/4450	10 (1/10)	12.5
CK/NY/3112	60 (6/10)	9.1
CK/NY/12273	10 (1/10)	13.0
CK/NY/30749	20 (2/10)	9.0
GH/MA/148081	50 (4/8)	7.75
CK/PA/9801289	0 (0/8)	NA[B]
TK/VA/67	33.3 (3/9)	10.0
CK/MD/MinhMa	12.5 (1/8)	5.0
ML/OH/421	0 (0/10)	NA
PT/MN/423	0 (0/8)	NA
RT/DE/1538	10 (1/10)	5.0

A. Percent (Number dead/total).
B. NA = not applicable.

When comparing species for shed of all viruses collectively, turkeys shed significantly ($p < 0.05$) more virus orally than either chickens or ducks at 2, 7 and 14 days PI. At 4 days PI the amount of virus shed orally by both chickens and turkeys was significantly higher than ducks. Ten days PI OP viral shed by turkeys was significantly higher than chickens. The only statistically significant differences among species in CL shed for all viruses collectively was at 14 days PI when turkeys shed higher titers than either chickens or ducks. All sham inoculated birds were negative for AI virus OP and CL shedding throughout the experiment.

Immunohistochemistry

In order to evaluate sites of virus replication, immunohistochemical (IHC) staining for AI virus antigen with an antibody to the influenza A NP protein was conducted with tissues collected 3 days PI selected from birds shedding more than 104 EID50 at 2 days PI (Figure 3). The bronchial epithelium of turkeys inoculated with CK/MD/ MinhMa, CK/NJ/15086 or TK/VA/67, and the airsac epithelium of a turkey exposed to TK/NY/4450 were positive for AI virus staining. Avian influenza virus antigen was observed in macrophage in spleens from ducks exposed to GH/MA/148081 or TK/VA/67, and both the bronchus and spleen from ducks infected with PT/MN/ 423 were positive. Staining of macrophage in the spleens of AIV infected ducks has been reported before [12]. However it is not clear whether the virus is replicating or whether the antigen is present from non-specific uptake of the virus. The intestines and cecal tonsils of chickens infected with CK/NJ/15086 or CK/NY/3112 were also positive for AI virus antigen by immunohistochemical staining

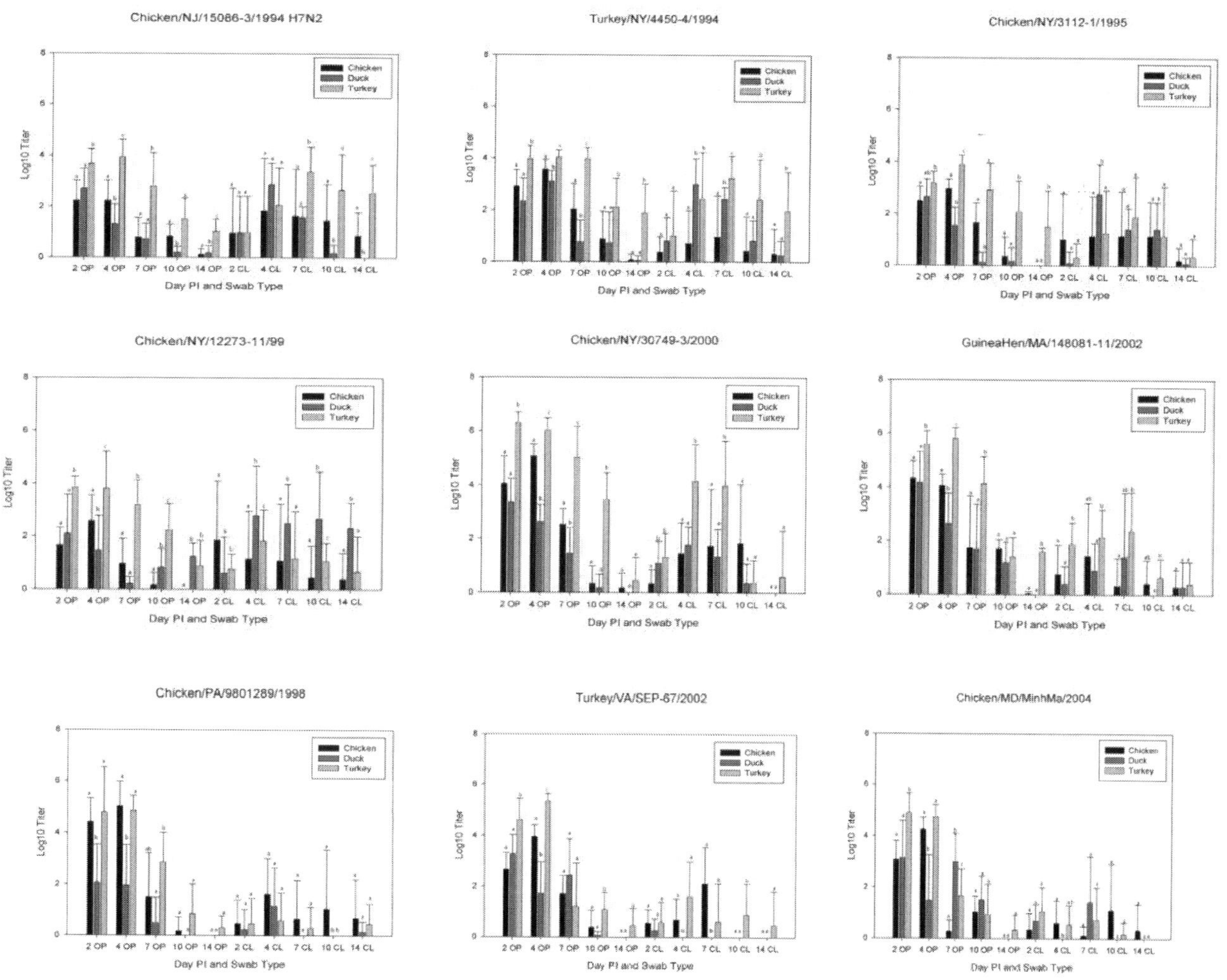

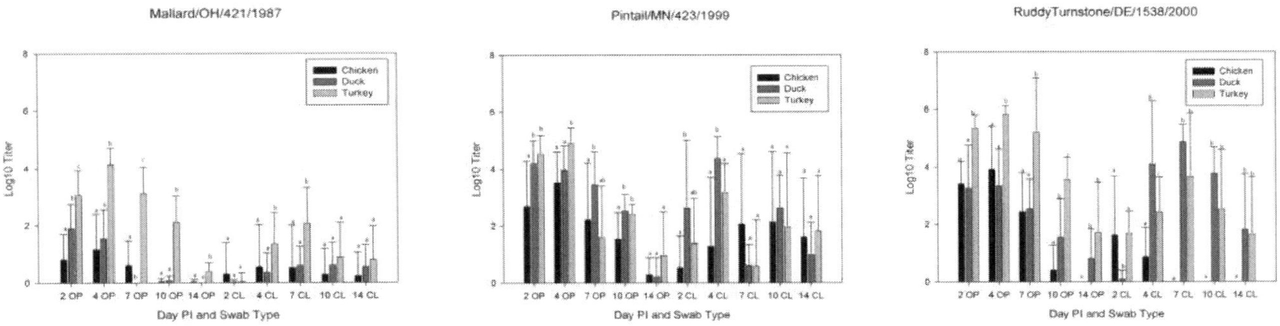

Figure 2: Mean OP and CL virus shed titers from white leghorn chickens, Pekin ducks and broad-breasted white chickens by day PI as determined by quantitative real-time RT-PCR for the influenza M gene. Letters indicate statistical groupings among the 3 species for each day PI and swab type. Error bars indicate standard deviation of titers. Abbreviations CL = cloacal, OP = oropharyngeal.

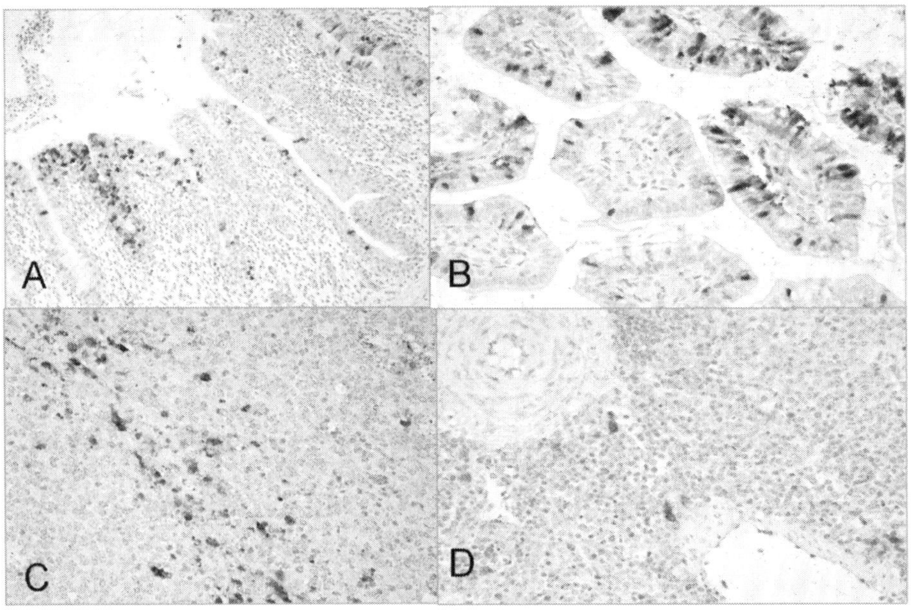

Figure 3: Immunohistochemical staining for avian influenza virus antigen in tissues of chickens, turkeys and ducks infected with H7 AI viruses, 2 days PI. A. Viral antigen (red staining) in bronchial epithelial cells from a turkey infected with TK/VA/67. B. and C. Viral antigen (red staining) in the intestinal epithelium and cecal tonsils of a chicken infected with CK/NY/3112. D. Viral antigen (red staining) in macrophages in the spleen of ducks infected with TK/VA/67.

Serology

At the termination of the experiments, 18-21 days PI, blood was collected from surviving birds and AI virus antibody was evaluated by commercial ELISA for all 3 species. All surviving turkeys except one exposed to CK/ MD/MinhMa had detectable AI virus antibody at termination (Table 3). All of the chickens seroconverted to all isolates except PT/MN/423 (89%), CK/PA/9801289 (90%), ML/OH/421 (70%) and RT/DE/1538 (90%). The proportion of ducks which serocoverted varied by isolate, and 100% of exposed ducks seroconverted to only 5 isolates, seroconversion to the remaining isolates were between 20% (CK/NY/30749) and 90% (TK/NY/4450) (Table 3).

Sequencing and Phylogenetic Analysis

Full genome sequence was generated for all 12 isolates to evaluate their genetic diversity and origins. Genetic diversity of the HA gene of domestic H7 isolates was one criteria used to select the isolates for this study, therefore the sequences of the HA genes of many of the isolates used here have been previously reported [10,11]. Based on these reports and the known epidemiology of the viruses, the H7 genes had been classified into 2 groups: 1). LBM and LBM-poultry, and 2). WB, which contained one isolate from the LBM system CK/NY/ 12273 (Figure 4). The isolates in this study have a range of nucleotide (nt) identity of 91.1 to 98.9% among their HA genes. Three proteolytic cleavage sites (PCS) were observed among the 12 isolates, all of which are consistent with LPAIV (Table 1). Five isolates (CK/NY/30749, GH/MA/148081, CK/MD/MinhMa, TK/VA/67, CK/PA/ 9801289) have a deletion in HA1 from amino acid 230 to 238, which has been previously described as a feature of the LBM lineage [10,11].

Table 3: Avian influenza virus antibody in sera from chickens, ducks and turkeys by commercial ELISA at termination of the experiment

Species	Isolate	ELISA	
		Percent	Pos/total
Chicken	CK/NJ/15086	100	10/10
	TK/NY/4450	100	10/10
	CK/NY/3112	100	10/10
	CK/NY/12273	100	10/10
	CK/NY/30749	100	9/9
	GH/MA/148081	100	10/10
	CK/PA/9801289	90	9/10
	TK/VA/67	100	8/8
	CK/MD/MinhMa	100	10/10
	ML/OH/421	70	7/10
	PT/MN/423	89	8/9
	RT/DE/1538	90	9/10
Duck	CK/NJ/15086	100	10/10
	TK/NY/4450	90	9/10
	CK/NY/3112	70	7/10
	CK/NY/12273	100	10/10
	CK/NY/30749	20	2/10
	GH/MA/148081	50	5/10
	CK/PA/9801289	44	4/9
	TK/VA/67	63	5/8
	CK/MD/MinhMa	33	2/6
	ML/OH/421	100	10/10
	PT/MN/423	100	9/9
	RT/DE/1538	100	10/10
Turkey	CK/NJ/15086	100	10/10
	TK/NY/4450	100	8/8
	CK/NY/3112	100	4/4
	CK/NY/12273	100	9/9
	CK/NY/30749	100	8/8
	GH/MA/148081	100	4/4
	CK/PA/9801289	100	8/8
	TK/VA/67	100	6/6
	CK/MD/MinhMa	86	6/7
	ML/OH/421	100	10/10
	PT/MN/423	100	6/6
	RT/DE/1538	100	9/9

Seven of the 12 isolates were the N2 NA subtype (Table 1) with range of nucleotide (nt) identity of 99.0 to 91.3% (additional file 1). A stalk deletion was observed in five isolates (CK/NY/3112, CK/PA/9801289, GH/MA/ 148081, TK/VA/67, and CK/MD/MinMa) from amino acids 58 to 73, and one isolate, TK/NY/4450, had a stalk deletion from 50-73 (Table 1). Three N3 isolates shared 95.0 to 97.7% nt identity. All N3 NA genes phylogenetically assort with other North American N3 genes (additional file 1). The viruses with NA types 8 and 9, both were found to be similar to other North American wild bird origin NA genes of the same subtype (additional file 1). All internal protein genes were similar to those of other North American AI virus isolates, but showed some phylogenetic diversity among the isolates (additional file 2). Among the M genes of the 12 isolates there was 92.5 to 99.4% nt identity. Seven isolates had subtype B NS gene (93.5 to 99.3% nt identity) and five were subtype A (96.2 to 98.3% nt identity) (table 1). The NP genes ranged in nt identity from 91.7 to 99.2%. The PA, PB1 and PB2 genes had 86.3 to 99.0%, 91.5 to 99.1% and 88.6 to 99.2% nt identity, respectively.

DISCUSSION

Twelve H7 LPAI viruses were evaluated for their pathogenesis in chickens, ducks and turkeys. The general patterns of virus shed observed here; OP shed peak at 2-4 days PI and overall higher tittered OP than CL shed which peaked later, at 4-7 days PI, is typical of LPAI virus infections in chickens and turkeys following respiratory inoculation [13]. Similarly, clinical signs, microscopic lesions, and patterns of IHC staining were consistent with what has been reported previously [13]. The comparison among the species did show consistent differences among the species in disease severity and virus shed. Although there were no statistically significant trends in disease severity by virus isolate (although most of the genes in 9 of the 12 isolates were relatively closely related to each other) there was a clear and significant trend for clinical disease to be more severe in turkeys than in either chickens or ducks. In turkeys the mean maximum clinical scores, which included mortality, were significantly higher than those of either chickens or ducks with 8 of 12 isolates. Mortality in turkeys possibly had a secondary bacterial component, which is typical in the field with respiratory viruses. Importantly, mortality was not observed in the controls, varied among the AIV isolates, and the birds which died shed the highest virus titers, therefore it appears that AIV was a critical factor for turkey mortality. Additionally, the turkeys generally shed significantly more virus from both the OP and CL routes at numerous time points. This suggests that turkeys may be more susceptible to disease from LPAI virus than chickens or ducks. This is consistent with a report by Tumpey et al. where turkeys were reported to be more susceptible than chickens to LP H7N2 AIV from the 2002 Shenandoah Valley outbreak [14]. In contrast, Ladman et al. [15] reported that chickens were more susceptible to disease, when inoculated with LP H7N2 AI viruses isolated from chickens when exposed by the conjunctival sac route. Although all of the individual birds did not seroconvert, serology indicated that each of the species did become infected at the dose administered and not all birds that seroconverted developed disease. Ducks had the lowest seroconversion rates, interestingly three of the five isolates with 100% seroconversion rates were the WB isolates. In contrast turkeys had the highest rates of seroconversion, which correlated with the trend for disease in turkeys to be more severe than in either chickens or ducks. This suggests that virus dose may have been a factor in disease development.

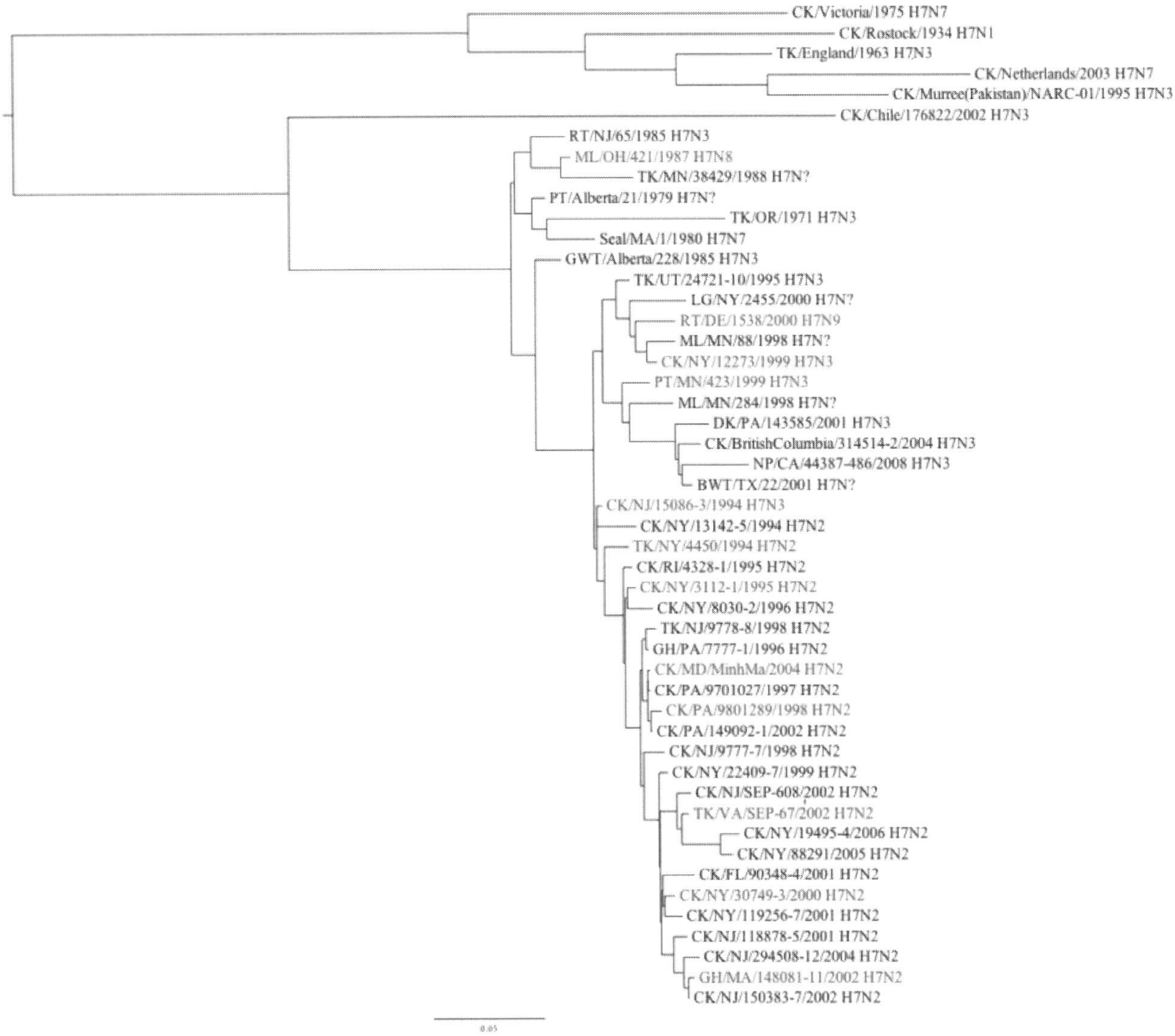

Figure 4: Phylogenetic tree of the H7 HA gene of influenza virus. Isolates included in this study appear in red type

Also, consistent with the differences in disease severity among the species, susceptibility studies have reported the 50% bird infectious dose (BID_{50}) for A/Turkey/VA/ 158512/2002 H7N2, which was isolated from the same outbreak as TK/VA/67, to be $10^{2.8}EID_{50}$ in chickens, $10^{0.8}EID_{50}$ in turkeys and $10^{3.5}EID_{50}$ in Pekin ducks [16], indicating that it is best adapted to turkeys. In contrast the BID_{50} for ML/OH/421 has been determined to be $10^{6.6}EID_{50}$ in chickens, and$10^{1.0}EID_{50}$ in mallard ducks (data not published). Interestingly, ML/OH/421 had some of the lowest shed titers of all the viruses in all three species and was shed at the highest titers by turkeys. Turkeys also had the highest clinical scores of all three species with ML/OH/421. Importantly, it is unknown how the chicken egg passage that was used to propagate these viruses may affect adaptation to chickens and turkeys. These results are similar to recent comparable studies with 20 H5 subtype LPAIV viruses [17], where turkeys were more susceptible to infection and disease, however mortality was not observed with the H5 isolates. Another comparable study with 16 H4, H6 and H9 subtype viruses [18] did not show differences in disease between chickens and turkeys, but based on serology, turkeys could be infected with more viruses that chickens. In contrast, taking into account data from Halvorson et al. [19] that showed that not all duck origin AI viruses will readily transmit to turkeys, it is unclear whether turkeys are more susceptible to infection with influenza or are just more susceptible to disease when they do become infected. The details of species susceptibility to LPAIV infection and disease needs to be explored with reverse genetics which can more definitively identify markers of host restriction and virulence. Additionally, minimum infectious dose studies would be necessary to establish host adaptation of the isolates. In the case of the wild bird isolates one may conclude that the isolates are more duck or shorebird adapted which seems to be supported by the virus shed levels by different species in this study. However, since numerous avian species are housed in close proximity in the live-bird markets, neither the exact host passage history nor the host adaptation can be inferred from the species of origin. One practical implication of this is that infection would likely be detected more easily in turkeys, whereas sub-clinical infection may spread unnoticed until the birds are tested prior to slaughter (currently 100% of turkey and chicken flocks are tested for AI virus prior to slaughter in the US). Since both LPAI viruses and HPAI viruses are reported to primarily cause sub-clinical disease in wild mallard ducks, the induction of mild disease by LPAI virus in Pekin ducks which are most closely related to mallard ducks but which are bred for rapid growth, is important for the commercial duck industry. It has been reported that some Asian lineage H5N1 HPAI virus isolates can cause disease in ducks, but the severity of disease depends on duck age and species [20,21]. Twoweek old Pekin ducks, like what was used here, were among the most susceptible to disease with the Asian H5N1 HPAI virus [20]. To complement the clinical data and to provide a more complete characterization of the isolates, full genome sequence was produced for all 12 isolates and a basic analysis was conducted. In depth analysis of the HA and NA genes of the H7N2 isolates has already been reported [10,11]. There was no clear correlation between gene constellation, or a particular gene and any biological or pathogenic characteristic evaluated here. Further work with reverse genetics would need to be conducted to identify markers for species adaptation and virulence.

CONCLUSIONS

Twelve LPAI viruses of the H7 subtype were evaluated in chickens, turkeys, and ducks for clinical disease and virus shed. All 12 viruses could infect all three species at the dose of 10^6 EID_{50} by the simulated respiratory route based on seroconversion. Turkeys consistently presented with the most severe disease and highest OP and CL shed titers, which indicate that, broad breasted white turkeys may be more susceptible to disease from some LPAI virus than white leghorn chickens or Pekin ducks. Methods Viruses Twelve North American origin LPAI viruses were selected to represent different H7 HA genetic groups, different species of origin and different dates of origin within the available H7 subtype domestic LPAI viruses (Table 1). Viruses were propagated and titrated in 9 to 11 day-old embryonated chicken eggs by standard procedures [22].

Pathogenesis Studies

Pathogenesis studies were conducted with specific pathogen free (SPF) white leghorn chickens (Gallus gallus domesticus), broad breasted white turkeys (Meleagris galopova) and Pekin ducks (Anas platyrhynchos domesticus). Chickens were obtained at 2 weeks of age from a commercial supplier of SPF animals (Charles-River SPAFAS, Franklin, CT) and were housed in isolators until they were exposed to the virus at 4 weeks of age. There are no sources of turkeys and ducks which are maintained as SPF or free from viral or bacterial respiratory diseases therefore turkeys and ducks were obtained from commercial hatcheries at hatch and were housed in isolators until they were exposed to virus at 2 weeks of age. All birds were obtained from flocks with no antibody or prior exposure to AI virus. The birds were housed in glove-port isolators (Allentown Caging, Allentown, NJ) with ad libitum access to feed and water before and after exposure to the viruses. Ducks and turkeys were exposed to the virus at 2 weeks instead of 4 weeks of age to accommodate the larger size and faster growth rates of these species in the animal facilities. Since the immune systems of all three species at these ages are considered to be relatively immature, this difference is not expected to impact species associated differences in susceptibility to LPAIV infection and disease. Animals were cared for in accordance with established humane procedures and biosecurity guidelines.

Thirteen to 15 of each species were inoculated with 10^6 EID_{50} per bird in 0.1 ml by the intrachoanal route. The birds were monitored daily for clinical disease signs which were scored as follows: 0 = no signs, 1 = mild to moderate respiratory signs (mild depression in ducks), 2 = moderate to severe (i.e. depressed, not eating, neurological signs), 3 = Dead. Oro-pharyngeal and CL swabs were each collected at days 2, 4, 7, 10 and 14 post inoculation (PI) to evaluate virus shed by quantitative real-time RT-PCR (qr-RT-PCR). Three days PI, 3-5 birds from each group were euthanized and necropsied to evaluate gross lesions. Tissues (heart, lung, pancreas/duodemun, kidney, liver, ileum, jejunum, ceca, bursa, thymus, spleen, breast muscle, thigh muscle, brain, nasal cavity, adrenal glands, cecal tonsils, trachea, and reproductive organs) were collected for microscopic evaluation. Serum was collected from ducks at 18 days PI and from chickens and turkeys at 21 days PI to confirm infection status.

qrRT-PCR

RNA was extracted with the MagMAX-96 Viral Isolation Kit (Ambion Inc. Austin, TX) with the KingFisher (Thermo-Fisher Scientific, Waltham, MA) magnetic particle processor in accordance with the manufacturer's instructions. Quantitative real-time RT-PCR was conducted with a primer-probe set that targeted the matrix gene as described previously [23] using the AB 7500 FAST (Applied Biosystems, Foster City, CA) instrument and the AgPathID (Ambion) one-step RT-PCR kit in accordance with kit instructions. Standard curves for virus quantification were established with RNA extracted from dilutions of the same titrated stock of the virus being evaluated.

Immunohistochemistry

Because of the sensitivity limitations of immunohisotochemical (IHC) staining for AI virus antigen, tissues were only processed for IHC from birds with OP or CL shed titers greater than 104 EID50 2 days PI. Tissue sections were cut (4 μm thick) from paraffin-embedded tissue samples and mounted on charged glass slides (Superfrost/Plus; Fisher Scientific). Deparaffinization, antigen retrieval and blocking procedures have been previously described [24]. A 1:2,000 dilution of a mouse-derived monoclonal antibody (P13C11) specific for a type A influenza virus nucleoprotein (developed at Southeast Poultry Research Laboratory, USDA) was applied and allowed to incubate for 2 hours at 37°C. The primary antibody was then detected by the application of biotinylated goat anti-mouse IgG secondary antibody using a biotin-streptavidin detection system (Supersensitive Multilink Immunodetection System, Biogenex). Fast Red TR (Biogenex) served as the substrate chromagen, and hematoxylin was used as a counterstain.

Serology

Antibody collected at 18 (ducks) or 21 (chickens and turkeys) days PI from surviving birds was used to confirm infection status. Sera were tested by commercial ELISA (FlockCheck, IDEXX Inc., Westbrook ME). Sera were tested at the manufacturer's recommended dilution of 1:500 and also at 1:100 and 1:50. The dilution of 1:100 was selected for final analysis because at this dilution there were no false-positives among the sera from negative control birds and there appeared to be better sensitivity.

Sequencing and Phylogenetic Analysis

Full genome sequencing of all isolates was performed as previously described [11]. Genbank accession numbers for new sequence generated for this study are and previously reported sequences are provided in Additional File 3. Phylogenetic analysis was performed using either ClustalV or ClustalW (Lasergene 7.1, DNASTAR, Madison, WI). Trees were constructed with BEAST v. 1.4.8 [25] using HKY substitution, empirical base frequency, Gamma heterogeneity, codon 2 partitions, relaxed lognormal clock, Yule Process tree prior with default operators with UPGMA starting tree and MCMC length of 10^6.

Statistical Analysis

Mean maximum clinical scores were compared by poultry species for each isolate and virus shed was evaluated by species for each day PI and swab type (OP or CL). All comparisons were conducted with the Student's Ttest and the Mann-Whitney rank-sum test if the normality test failed (Sigmaplot 11.0, Systat Inc. San Jose, CA). All statistics were evaluated with a significance threshold of $p \leq 0.05$

ACKNOWLEDGEMENTS

The authors gratefully acknowledge Scott Lee, Melissa Scott, Barbara Stancil, Joyce Bennett, Joanne Kramer, Kyle LeStrange and Michelle Wood for technical assistance with this work. This work was supported by USDA-ARS CRIS Project #6612-32000-048 and USDA-ARS cooperative agreement #58- 6612-6-245.

AUTHORS' CONTRIBUTIONS

ES was involved in virus selection and experimental design and conducted sequence and phylogenetic analysis. JG was involved in virus selection and experimental design. BL and LP conducted animal experiments, collected specimens, ran qrRT-PCR, and serological assays. CP evaluated tissue sections for microscopic lesions. MJP conducted the immuno-histochemical staining and evaluation. ETM contributed to data analysis and conducted minimum infectious dose studies. All authors contributed to data analysis and manuscript preparation.

COMPETING INTERESTS

The authors declare that they have no competing interests.

REFERENCES

1. Senne DA: Avian influenza in North and South America, the Caribbean, and Australia, 2006-2008. Avian Dis 2010, 54:179-186.
2. Brown IH: Summary of avian influenza activity in Europe, Asia, and Africa, 2006-2009. Avian Dis 2010, 54:187-193.
3. Elbers AR, Fabri TH, de Vries TS, de Wit JJ, Pijpers A, Koch G: The highly pathogenic avian influenza A (H7N7) virus epidemic in The Netherlands in 2003–lessons learned from the first five outbreaks. Avian Dis 2004, 48:691-705.
4. Naeem K, Siddique N, Ayaz M, Jalalee MA: Avian influenza in Pakistan: outbreaks of low- and high-pathogenicity avian influenza in Pakistan during 2003-2006. Avian Dis 2007, 51:189-193.
5. Pasick J, Handel K, Robinson J, Copps J, Ridd D, Hills K, Kehler H, CottamBirt C, Neufeld J, Berhane Y, Czub S: Intersegmental recombination between the haemagglutinin and matrix genes was responsible for the emergence of a highly pathogenic H7N3 avian influenza virus in British Columbia. J Gen Virol 2005, 86:727-731.
6. Suarez DL, Senne DA, Banks J, Brown IH, Essen SC, Lee C, Manvell RJ, Mathieu-Benson C, Moreno V, Pedersen JC, Panigrahy B, Rojas H, Spackman E, Alexander DJ: Recombination resulting in virulence shift in avian influenza outbreak in, Chile. Emerging Infectious Diseases 2004, 10:693-699.
7. Ziegler AF, Davdison S, Acland H, Eckroade RJ: Characterization of H7N2 (Non-pathogenic) Avian Influenza Virus Infections in Commercial Layer, in Pennsylvania, 1997-98. Avian Dis 1999, 43:142-149.
8. Akey BL: Low-pathogenicity H7N2 avian influenza outbreak in Virgnia during 2002. Avian Dis 2003, 47:1099-1103.
9. Senne DA: Avian influenza in North and South America, 2002-2005. Avian Dis 2007, 51:167-173.
10. Spackman E, Senne DA, Davison S, Suarez DL: Sequence analysis of recent H7 avian influenza viruses associated with three different outbreaks in commercial poultry in the United States. J Virol 2003, 77:13399-13402.
11. Suarez DL, Garcia M, Latimer J, Senne D, Perdue M: Phylogenetic analysis of H7 avian influenza viruses isolated from the live bird markets of the Northeast United States. J Virol 1999, 73:3567-3573.
12. Pillai SP, Pantin-Jackwood M, Suarez DL, Saif YM, Lee CW: Pathobiological characterization of low-pathogenicity H5 avian influenza viruses of diverse origins in chickens, ducks and turkeys. Arch Virol 2010, 155:1439-1451.
13. Swayne DE, Halvorson DA: Influenza. In Diseases of Poultry. 12 edition. Edited by: Saif YM, Fadly A, Glisson J, McDougald L, Nolan L, Swayne DE. Ames, IA: Blackwell; 2008:1324.
14. Tumpey TM, Kapczynski DR, Swayne DE: Comparative susceptibility of chickens and turkeys to avian influenza A H7N2 virus infection and protective efficacy of a commercial avian influenza H7N2 virus vaccine. Avian Dis 2004, 48:167-176.
15. Ladman BS, Rosenberger SC, Rosenberger JK, Pope CR, Gelb J Jr: Virulence of low pathogenicity H7N2 avian influenza viruses from the Delmarva peninsula for broiler and leghorn chickens and turkeys. Avian Dis 2008, 52:623-631.

16. Swayne DE, Slemons RD: Using mean infectious dose of high- and lowpathogenicity avian influenza viruses originating from wild duck and poultry as one measure of infectivity and adaptation to poultry. Avian Dis 2008, 52:455-460.

17. Pillai SP, Pantin-Jackwood M, Yassine HM, Saif YM, Lee CW: The high susceptibility of turkeys to influenza viruses of different origins implies their importance as potential intermediate hosts. Avian Dis 2010, 54:522-526.

18. Morales AC Jr, Hilt DA, Williams SM, Pantin-Jackwood MJ, Suarez DL, Spackman E, Stallknecht DE, Jackwood MW: Biologic characterization of H4, H6, and H9 type low pathogenicity avian influenza viruses from wild birds in chickens and turkeys. Avian Dis 2009, 53:552-562.

19. Halvorson DA, Kelleher CJ, Senne DA: Epizootiology of avian influenza: effect of season on incidence in sentinel ducks and domestic turkeys in Minnesota. Appl Environ Microbiol 1985, 49:914-919.

20. Pantin-Jackwood MJ, Suarez DL, Spackman E, Swayne DE: Age at infection affects the pathogenicity of Asian highly pathogenic avian influenza H5N1 viruses in ducks. Virus Res 2007, 130:151-161.

21. Brown JD, Stallknecht DE, Beck JR, Suarez DL, Swayne DE: Susceptibility of North American ducks and gulls to H5N1 highly pathogenic avian influenza viruses. Emerg Infect Dis 2006, 12:1663-1670.

22. Woolcock PR: Avian influenza virus isolation and propagation in chicken eggs. Methods Mol Biol 2008, 436:35-46.

23. Spackman E, Senne DA, Myers TJ, Bulaga LL, Garber LP, Perdue ML, Lohman K, Daum LT, Suarez DL: Development of a real-time reverse transcriptase PCR assay for type A influenza virus and the avian H5 and H7 hemagglutinin subtypes. J Clin Microbiol 2002, 40:3256-3260.

24. Pantin-Jackwood MJ, Spackman E, Day JM: Pathology and virus tissue distribution of Turkey origin reoviruses in experimentally infected Turkey poults. Vet Pathol 2007, 44:185-195.

25. Drummond AJ, Rambaut A: BEAST: Bayesian evolutionary analysis by sampling trees. BMC Evol Biol 2007, 7:214.

CHAPTER 3

SIMULTANEOUS DETECTION AND DIFFERENTIATION BY MULTIPLEX REAL TIME RT-PCR OF HIGHLY PATHOGENIC AVIAN INFLUENZA SUBTYPE H5N1 CLASSIC (CLADE 2.2.1 PROPER) AND ESCAPE MUTANT (CLADE 2.2.1 VARIANT) LINEAGES

El-Sayed M Abdelwhab, Ahmed M Erfan, Christian Grund, Mario Ziller, Abdel-Satar Arafa, Martin Beer, Mona M Aly, Hafez M Hafez, Timm C Harder

BACKGROUND

The incursion of highly pathogenic avian influenza virus (HPAIV) of subtype H5N1 of phlyogenetic clade 2.2, subclade 2.2.1 [1], into Egypt in 2005/2006 caused severe economic losses in the commercial (previous total annual production of 850 million birds) and backyard sectors (250 million birds) of poultry production in this country [2]. The virus also possesses considerable zoonotic potential. Human cases of HPAIV H5N1 infection, characterized by a high fatality rate, started to occur due to virus exposure of humans at the poultryhuman interface which is highly fissured in Egypt [3]. In order to restore poultry production capacities and to mitigate risks of an emergence of new virus variants with increased pandemic potential in the human population, efforts to control HPAIV H5N1 were given a high priority [4].

Despite intense control measures including blanket vaccination, surveillance and depopulation of infected poultry holdings, HPAI H5N1 has gained endemic status in Egyptian poultry populations [5] and continuous, year-round circulation of HPAI H5N1 virus has been reported [6-8]. This is at least in part due to the highly divergent evolution of H5N1 viruses in Egypt which seems to be accelerated and shaped by vaccine-induced selection pressure leading to the emergence of genetically and antigenically distinct viruses [9]. Strains of the parent subclade 2.2.1 proper (2.2.1p) are reported to circulate nationwide mainly in unvaccinated birds, particularly waterfowl from backyard holdings [7]. The vast majority of human infections (34 fatalities out of 109 infected cases until 22nd June 2010 [10]) is attributable to viruses of this group. Since 2007, viruses of a variant sublineage emerged from clade 2.2.1. These viruses, which will here be referred to as lineage 2.2.1v, originated from and circulate predominantly in vaccinated commercial chickens [8]. These antigenically drifted strains were shown to escape immunity induced by standard H5 vaccination and are prevalent mainly in Lower Egypt, particularly in the Nile Delta [7,8,11,12]. Knowledge of the epidemiology, especially the transmission pathways, of those two lineages between commercial farms, backyard birds, feral birds and humans is incomplete but urgently required to improve control measures. So far, assignment of viruses to either lineage requires virus isolation and antigenic characterization by hemagglutination inhibition or sequencing and molecular analysis. No rapid typing tools are currently available.

A number of RT-qPCR assays for diagnosis and characterization with respect to subtype and pathogenicity of HPAIV H5N1 have been published. These assays target the matrix gene [13,14], the nucleoprotein gene [15,16], the neuraminidase and the hemagglutinin [17-23]. Egypt's surveillance program embarked on the H5-specific RT-qPCR assay [23], which is recommended by the World Organization

Citation: El-Sayed M Abdelwhab, Ahmed M Erfan, Christian Grund, Mario Ziller, Abdel-Satar Arafa, Martin Beer, Mona M Aly, Hafez M Hafez3 , Timm C Harder, Simultaneous detection and differentiation by multiplex real time RT-PCR of highly pathogenic avian influenza subtype H5N1 classic (clade 2.2.1 proper) and escape mutant (clade 2.2.1 variant) lineages, doi:10.1186/1743-422X-7-260.

of Animal Health (OIE). The assay was initially highly successful in detecting H5N1 infections in Egypt [5,7]. Since 2007, however, an increasing number of strains in Egypt escaping detection by this assay was reported [24]. Therefore, the aim of this study was to develop a sensitive multiplex RT-qPCR able to detect all HPAIV H5N1 variants of clade 2.2 currently circulating in Egypt and, simultaneously, to distinguish between conventional 2.2.1p strains and the 2.2.1v lineage of vaccine-driven escape variants.

RESULTS

A multiplex RT-qPCR assay was designed to detect HPAIV H5N1 viruses and simultaneously differentiate the main two major lineages circulating in Egypt; classic 2.2.1p viruses and the newly evolving clade 2.2.1v of antigenically drifted variant viruses. A single reaction assay using multiplexed primers and probes (three colours) was developed. Analytical characterization of the assay was carried out using 50 Egyptian HPAIV H5N1 strains and other avian pathogens. Diagnostic performance was examined with 63 swab samples from experimentally infected chickens and 50 avian influenza virus (AIV)-positive swab samples obtained from different host species in the field in Egypt.

Real-Time RT-PCR Optimization

Two strategies were combined to sensitively detect and distinguish viruses of clade 2.2.1p from 2.2.1v escape strains (Figure 1). Lineage-specific primers had disparate nucleotides at their most 3'-positions whereas lineage specific-probes were distinguished by two nucleotide positions towards the centre of the probe. Locked nucleotide chemistry was used to further increase the specificity of certain probes. Optimization runs were carried out using log10 dilution series of RNA extracted from A/chicken/Egypt/0879-NLQP/2008 (clade 2.2.1v), A/chicken/Egypt/NLQP-0918Q/2009 (clade 2.2.1p) and A/Whooper swan/Germany/R65/2006 (European clade 2.2) (data not shown). The concentration of primers and probes was optimized to increase the efficiency and sensitivity of amplification to final values shown in table 1. The lowest Ct values and highest ΔR_n values for the multiplex RT-qPCR were observed using the following thermoprofile and the SuperScript III RT/Platinum Taq Mix chemistry: 30 min at 50°C and 2 min at 94°C, then 42 cycles of 94°C, 56°C and 68°C for 30 seconds each. No significant differences in sensitivity were evident when running the three RT-qPCRs separately indicating that the multiplex approach produced no relevant detrimental effects on amplification efficacy (data not shown).

Analytical Characteristics

The current protocol was found to have a detection limit of approximately 2 - 5 RNA copies/reaction which amounts to 400 - 1000 copies per ml sample matrix when using cRNA as a copy-based standard. The dynamic ranges of target detection are summarized in figure 2. A total of 50 HPAIV H5N1 isolates from Egypt obtained between 2006 and 2010 for which nucleotide sequences of the HA gene were available, were examined by the multiplex assay which assigned 33 of them to clade 2.2.1p (HEX-positive) whereas 15 isolates reacted like 2.2.1v strains (FAM-positive) (Additional file 1, Table S1). An exactly similar clustering was achieved for these strains in a phylogenetic analysis based on full length HA sequences (7, 8 and Additional file 1, Table S1). In addition, two isolates reacted positive for both 2.2.1p and 2.2.1v lineages. All 50 isolates were also detected by the ROX probe of the multiplex RT-qPCR (Additional file 1, Table S1). For 40 isolates a comparison with the generic H5 RT-qPCR validated by Slomka et al. [23] was performed showing markedly lower Ct values, corresponding to a higher sensitivity, for the ROX probe assay. In addition, several, but not all, non-Egyptian H5 viruses could also be detected by the multiplex RT-qPCR with equal or slightly lower sensitivity compared to the generic H5 protocol described by Slomka et al. [23]. Negative results (no measurable Ct obtained [> > 40]) were generated with all non-H5 AI viruses as well as with other avian viral or bacterial pathogens (Additional file 2, Table S2).

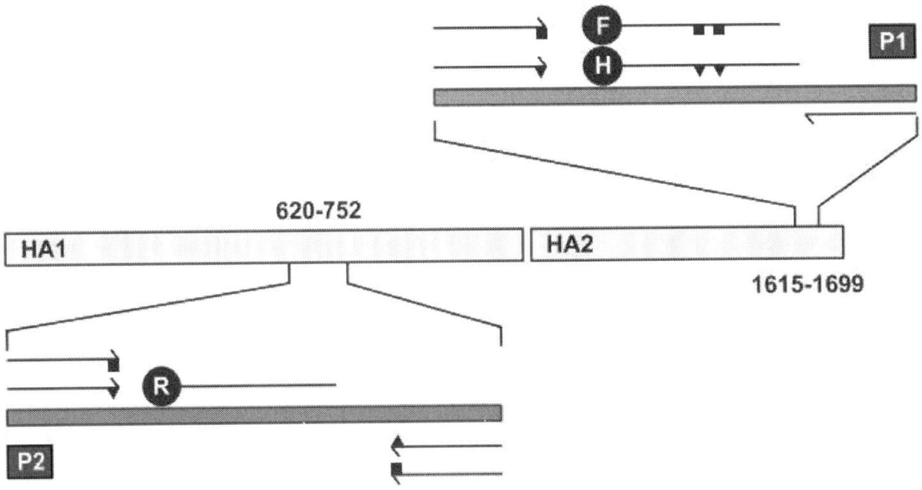

Figure 1: Localization of primers and probes. Localization of primers and probes as mentioned in table 1 along the HA gene is depicted. Two target regions (P1, P2) were chosen. Primers and probes were selected with specificity for the HA of Egyptian HPAIV H5N1 viruses either of lineage 2.2.1p (proper) or 2.2.1v (variant). Triangles (2.2.1p) and squares (2.2.1v) distinguish nucleotide positions specific for either lineage. Labelling of hydrolysis probes was with FAM (F, 2.2.1v), HEX (H, 2.2.1p) or ROX (2.2.1p+v).

Diagnostic Performance

Swab samples (n = 63) originating from SPF chickens which were experimentally infected with A/chicken/Egypt/ 0879-NLQP/2008 (clade 2.2.1v) or A/chicken/Egypt/ NLQP-0918Q/2009 (clade 2.2.1p) were examined by the multiplex RT-qPCR and compared to Ct values obtained with an H5-specific assay described by Slomka et al. [23]. Results are shown in Additional file 3, Table S3, and Figure 3. Swabs collected from individual birds were selected on basis of Ct values obtained by a generic M gene-targeted RT-qPCR (14, not shown) so as to represent a wide range of samples with low to high concentrations of AIV RNA. Special emphasis was put on samples with Ct-values around 35. The multiplex assay assigned the correct lineage within clade 2.2.1 for all samples. One sample (#15, Additional file 3, Table S3) yielded a weak false-positive signal (Ct 39.03) in the 2.2.1v specific assay although the sample came from a chicken that was infected by a 2.2.1p proper virus. However, upon repeated RNA extraction and analysis of the sample, this false-positive reaction was not reproducible which indicates a possible spill-over contamination possibly during the previous extraction procedure. No significant differences in Ct values were evident between the 2.2.1p specific assay (HEX) of the multiplex RT-qPCR and the Slomka H5 RT-qPCRs (Figure 3a). However, both the 2.2.1v specific assay (FAM) and the new Egyptian H5.specific assay (ROX) of the multiplex mixture were significantly more sensitive ($p = 0{,}027$ and $p \leq 0{,}001$) than the generic H5 RT-qPCR (Figures 3b-c). Oropharyngeal or cloacal swabs were also sampled in commercial farms (n = 11) and backyard poultry holdings (n = 39) in Egypt. A total of 50 samples pretested by M-specific RT-qPCR [14] to be positive for AIV was examined using the H5 multiplex RT-qPCR assay and the generic H5 protocol [23]. The multiplex RT-qPCR found all 50 samples positive for subtype H5 (ROX). Among them, 32 were assigned to clade 2.2.1p and 15 to lineage 2.2.1v (Additional file 4, Table S4, and figure 4) while three samples were positive for both. Only 45 samples tested positive in the generic H5 RTqPCR using the original protocol by Slomka et al. [23], and on average the Ct values produced with the newly developed multiplex assay were significantly lower ($p < < 0{,}001$) than those obtained in both the M and the H5 generic RT-qPCRs (Figures 4a-c).

Table 1: Oligonucleotide primers and probes designed for this study

No.	Primers/probes	Sequence 5'- 3'	Conc.[1] (nM)	Position[2]	Amplificate size (bp)
1	P1FW_2.2.1p	GAR TCA ATA GGA AYT TAC CAA ATA CT**G**	400	1615-1641	
2	P1FW_2.2.1v	GAA TCA ATA GGA ACT TAC CAA ATA CTA T**C**	800	1615-1643	
3	P1RV_2.2.1	AGA CCA GCC ACC ATT GAT TGC	400	1699-1679	85
4	PRO1.1_2.2.1v	FAM-ACA G*T*G GC***A*** AG***T*** TCC CT-BHQ-1	64	1654-1670	
5	PRO1.2_2.2.1p	HEX-ACA GTG GCG AGC TCC CTA GC-BHQ-1	64	1652-1675	
6	P2FW_2.2.1p	GGA TTC ACC ATC C**RA** ATG ATG**C**	1600	620-641	
7	P2FW_2.2.1v	GGG ATT CAC CAT CCA AAT GAT G**A**	1600	619-641	
8	P2RV_2.2.1p	CCG TTT ACC TTA GAT CTA GTA GCT AT**T**	1600	752-726	106
9	P2RV_2.2.1v	CCG TTT ACC TTA GAT CTA GT**R** GCT AT**C**	1600	752-726	
10	PRO2_2.2.1	ROX -TAC CTA TAT TTC CGT TGG GAC ATC AAC ACT AAA-BHQ-2	64	675-707	

[1] Nanomolar concentration of a 25 µl reaction.

[2] Position relative to the initiating codon of A/chicken/Egypt/06541-NLQP/2006 (H5N1), GenBank accession no. EU372946.1.

Bold face nucleotides indicate mismatch positions between clade 2.2.1p proper and lineage 2.2.1v variant viruses. Italics indicate use of locked nucleic acid nucleotides.

The specificity of primers and probes for either or both of the 2.2.1 lineages is indicated with their designation.

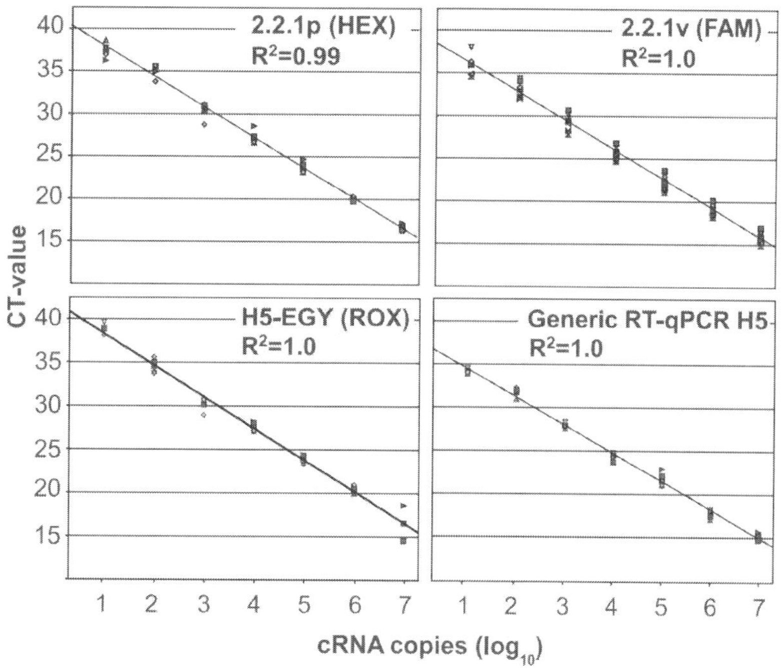

Figure 2: Detection limits and amplification efficiency. Detection limits and amplification efficiency of the multiplex RT-qPCR for detection and differentiation of clade 2.2.1p proper and 2.2.1v variant HPAIV H5N1 strains from Egypt using cRNA as a copy-based target. The standard curves were established by up to five independent runs. Average Ct values and variations are plotted against the cRNA copy numbers.

Discussion

We report here the development of a multiplex RTqPCR assay for detection and differentiation of Egyptian H5N1 HPAI viruses of clade 2.2.1 proper (p) and the emerging 2.2.1 lineage of viruses which escape standard vaccine-induced immunity (designated here 2.2.1v). The assay was shown to have a detection limit of 2-5 cRNA copies per reaction. Based on a phylogenetic analysis of 50 isolates tested [7,8], the assay is fully specific with regard to assigning the Egyptian H5N1 isolates to either phylogenetic cluster. No unspecific reactivity with either

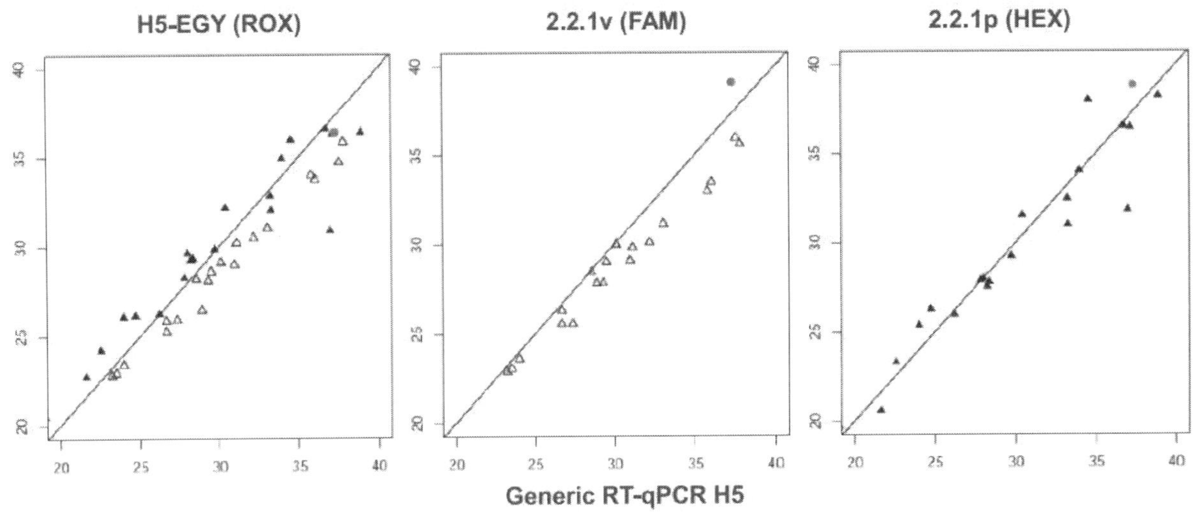

Figure 3: Diagnostic performance characteristics (experimental infections). Diagnostic performance characteristics of a newly developed H5-specific multiplex RT-qPCR for detection and differentiation of clade 2.2.1p proper and 2.2.1v variant HPAIV H5N1 strains from Egypt compared to standard H5 RT-qPCR (Slomka et al. [20]). The analysis comprises swab samples obtained from SPF chickens after experimental infection with A/chicken/Egypt/0879-NLQP/2008 (clade 2.2.1v, open triangles, n = 34 samples) or A/chicken/Egypt/NLQP-0918Q/2009 (clade 2.2.1p, black triangles, n = 29 samples). See Additional file 3, Table S3, for individual values. Grey dots represent samples in which both clades were detected. X-axis: Ct-values obtained by the generic H5 RT-qPCR [20]. Y-axis: Ct-values generated by the specific component of the multiplex RT-qPCR as indicated above each graph.

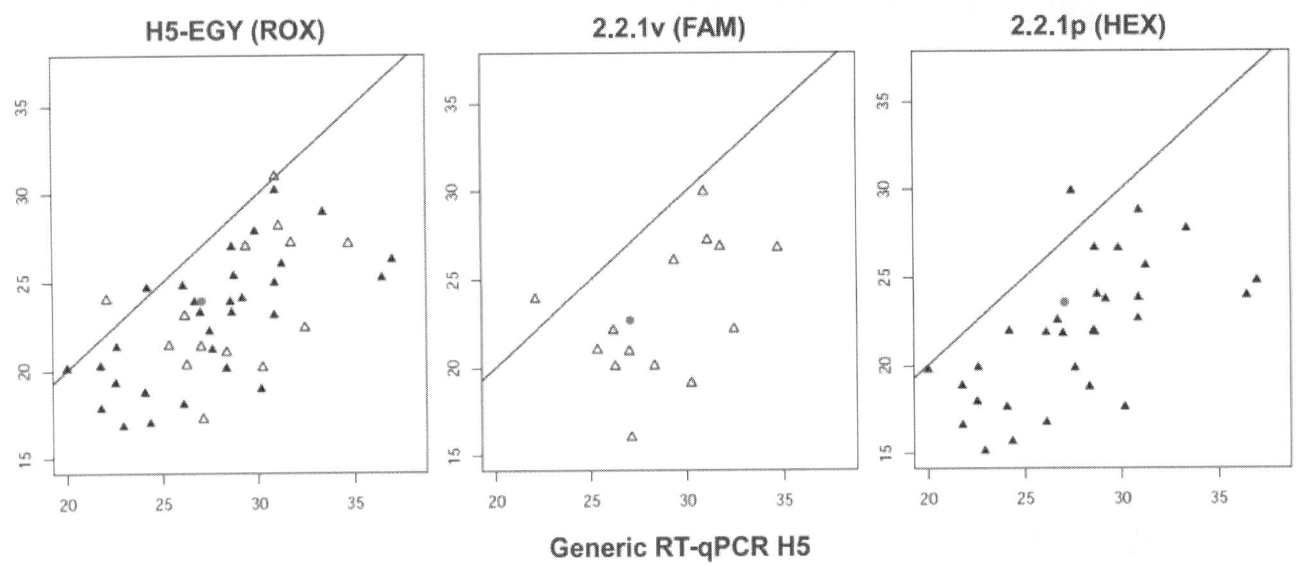

Figure 4: Diagnostic performance characteristics (field samples). Diagnostic performance characteristics of a newly developed H5-specific multiplex RT-qPCR for detection and differentiation of clade 2.2.1p proper (black triangles) and 2.2.1v variant HPAIV H5N1 strains (open triangles) from Egypt compared to standard H5 RT-qPCR [20]. The analysis comprises 50 M-PCR positive field samples obtained within the frame of the Egyptian poultry surveillance program from commercial poultry farms (n = 11) and backyard holdings (n = 39). See Additional file 4, Table S4, for individual values. Grey dots represent samples in which both clades were detected. X-axis: Ct-values obtained by the generic H5 RT-qPCR [20]. Y-axis: Ct-values generated by the specific component of the multiplex RT-qPCR as indicated above each graph.

Non-H5 AIV or other avian pathogens was evident. However, the assay can not be used for generic detection of subtype H5 viruses as a large percentage of nonEgyptian H5 subtype strains could not be detected. In terms of diagnostic performance regarding the HPAIV H5N1 strains currently circulating in Egypt, the multiplex assay was at least equal to standard RT-qPCRs targeting the M gene of AIV and superior to a generic H5 RT-qPCR [23] when examining swabs which originated from experimental infections or from field samples of poultry holdings in Egypt. The generic H5- specific RT-qPCR assay described by Slomka et al. [23] apparently missed five Egyptian field samples, possibly due to mismatches in binding regions of primers and/or probes; one of the missed samples was assigned to lineage 2.2.1v while four of them belonged to 2.2.1p. In three field samples the multiplex assay detected both clades with almost similar Ct values (Additional file 4, Table S4, #9, #17, #18). The field samples analysed were derived from pooled swabs of five birds of each holding; as such it can not be excluded that infections with both clades occurred simultaneously at these holdings. Work is in progress to clarify these cases by sequencing clones of HA gene fragments. A similar situation was also encountered with two isolates (Additional file 1, Table S1, #3, #18); alignment of primer and probe sequences with the published sequences of these isolates, however, did not yield any hint for an unspecific reactivity. A contamination of these isolates can only be excluded by sequencing clones of HA gene fragments. Reducing the amount of circulating HPAIV H5N1 virus by concerted actions of rapid and specific testing, culling and vaccination of poultry is the key to mitigate the risk of human infections and fatalities in Egypt. Controlling the endemic HPAIV H5N1 situation in Egypt is particularly painstaking because of [1] the concentration of the majority of commercial and backyard poultry business in a very small part of the whole country (Nile valley and, particularly, Nile delta), [2] the integration/contacts of backyard birds within small commercial poultry farms (farms with 5.000-20.000 birds represent circa 75% of the poultry production), [3] the marketing system (random uncontrolled movement of birds to/from live bird markets), and [4] day labourers at commercial farms usually raise backyard birds in their houses. In addition, continuing viral evolution which is even further accelerated and skewed by vaccination pressure remains a daily challenge for diagnostic measures which are at the root of all efforts to control the situation. Characterization of currently circulating strains and, if required, adaptation of amplification-based diagnostic tools, such as introduced here, is essential to improve the situation.

CONCLUSIONS

The necessity to update commercial and generic H5-specific RT-qPCRs for the Egyptian situation has been stressed recently [24]. The current assay provides this update. The assay is tailored to suite the special Egyptian situation. Therefore, the multiplex assay is not recommended for use elsewhere, particularly in areas where non-clade 2.2 HPAIV H5N1 are prevalent. Also, should new lineages of HPAIV H5N1 be introduced into Egypt, such as the 2.3.2 subclade viruses which already escaped from Central and South-eastern Asia to South-eastern Europe earlier in 2010, the current assay will need updating again. In addition to detection of clade 2.2.1p H5 HPAIV the multiplex assay also allows the positive identification of the 2.2.1v lineage of vaccine escape mutants. This lineage probably evolved in commercial chicken farms where vaccination using standard LPAIV H5 strains was practiced. Recent studies have shown that new vaccines might be required to efficiently induce protective immunity against lineage 2.2.1v viruses in poultry [11]. The multiplex assay therefore may also be instrumental in decision-making regarding the type of vaccine to be used for the specific outbreak situation.

METHODS

Reference Viruses and Bacteria

A panel of 50 HPAIV H5N1 strains isolated in SPFchicken eggs in the National Laboratory for Quality Control on Poultry Production (NLQP) in Egypt was used for determination of the analytical specificity and sensitivity of the PCR assays (Additional file 1, Table S1). In addition 42 further avian influenza viruses of subtypes H1 [4], H2 [5], H5 [22], H6 [5], H7 [3], and H9 [3] from the repository of German National Reference Laboratory for Avian Influenza, Friedrich-Loeffler Institute, were analysed (Additional file 2, Table S2). Non-orthomyxoviruses and several bacterial species were used to further determine the specificity of the assay.

Primer/Probe Design

A collection of 316 near full length H5 gene segment sequences of H5N1 viruses circulating in Egypt between 2006 and 2010 was retrieved from the public GenBank data base. Sequences were aligned using MUSCLE [25] and manually edited. Primers and probes design were selected from a variable region of the HA2 gene for detection of an 85 bp fragment and a more conserved region in the HA1 gene region for detection of a 106 bp fragment (Table 1, Figure 1). Primers P1FW-Standard-EGY, P1RVEGY and probe PRO1a-Standard-EGY were used for detection of the clade 2.2.1p strains (HEX channel). Primers P1FW-Variant-EGY, P1RV-EGY and probe PRO-Variant-EGY (FAM channel) were used for detection of the clade 2.2.1v variant strains. Primers P2FW-Standard-EGY, P2FW-Variant-EGY, P2RV-Standard-EGY and P2RV-Variant-EGY and probe PRO2-EGY were used for detection of both lineages via the ROX channel.

Real Time RT-PCR Optimization

The concentration of primers and probes was optimized in separate PCRs (two primers, one probe) and re-adjusted when combined in the multiplex RT-qPCR to increase the efficiency of amplification. Likewise, different annealing temperatures ranging from 50 to 60°C and different chemistries (SuperScript III OneStep RT-PCR system with Platinum Taq DNA polymerase [Invitrogen]; Quantitect One step kit; [Qiagen]) were evaluated. Reactions were carried out in a 25-µl volume on an MX3005P real time PCR machine (Stratagene).

Quantitative Analysis

For preparation of standard controls, the cloned H5 gene segment from A/chicken/Egypt/0879-NLQP/2008 (clade 2.2.1v) and A/chicken/Egypt/NLQP-0918Q/2009 (clade 2.2.1p) was used for generating cRNA in vitro by run-off transcriptions performed as previously described [18]. Detection limit of the RT-qPCR was determined using 10-fold serial dilutions (101 -107 copies) of cRNA.

Samples from Experimentally Infected Chickens

All animal experiments were conducted following official German animal welfare regulations (LALLF M-V/ TSD/7221.3-2.1-031/09). Six weeks old SPF chickens (n = 10, each) were infected by the oculo-nasal route with a dose of $10^{6.0}$ $TCID_{50}$ of A/chicken/Egypt/0879-NLQP/ 2008 (clade 2.2.1v) or A/chicken/Egypt/NLQP-0918Q/ 2009 (clade 2.2.1p) (Grund et al., unpublished). RNA was extracted from mixed oropharyngeal and cloacal swabs collected from individual birds 2 or 7 days post infection (dpi). RNA was used in quantitative RT-qPCRs described in this study and compared to M and H5 based RT-qPCR assays as previously described [14,23].

Field Samples

Tracheal and cloacal swabs were collected both from poultry at commercial farms (n = 11) and backyard holdings (n = 39) in the frame of the national surveillance scheme in poultry sectors in Egypt from 2008 to 2010. Oropharyngeal and cloacal swabs from five birds were pooled for RNA extraction and PCR analysis.

Nucleic acid Extraction

Extraction of RNA from 140 µl of allantoic fluid (RNA viral isolates) or swab fluid (field samples, experimental infections) was carried out using the QIAamp viral RNA mini kit (Qiagen, Hilden, Germany) following the manufacturer's instructions. The RNA was eluted from the columns with 50 µl of DEPC-treated water and used immediately or after storage at -80°C. Likewise, DNA was extracted from bacterial species and DNA viruses (Additional file 2, Table S2) using the QIAamp DNA mini kit (Qiagen, Hilden, Germany).

Statistics

In addition to the descriptive evaluation of the test results, the technical sensitivity of the new multiplex RT-qPCR was investigated by comparing positive Ctvalues with those of the standard H5-specific RT-qPCR test recommended by Slomka et al. [23]. For this purpose, Fisher's exact tests were applied considering the one-sided hypothesis of achieving lower Ct-values by the multiplex RT-qPCR than by the standard test. All statistical calculations have been performed using R, Version 2.8.1 (2008-12-22) [26].

COMPETING INTERESTS

The authors declare that they have no competing interests.

AUTHORS' CONTRIBUTIONS

ESAW carried out part of the development studies, the analytical and the diagnostic evaluation; he performed the sequence alignments and helped to draft the manuscript. AME and ASA carried out parts of the analytical and diagnostic evaluation using Egyptian samples. MZ carried out the statistical tests. CG, MMA and HMH provided samples for analysis, participated in the design of the study and helped to draft the manuscript. TCH conceived and coordinated the study, and drafted the manuscript. All authors read and approved the final manuscript.

ACKNOWLEDGEMENTS

We are grateful to our co-workers and colleagues for excellent technical work in the FLI, Germany, and the NLQP, Egypt. El-Sayed is funded by a grant from the German Academic Exchange Service to the Free University of Berlin. Work by the authors on Egyptian HPAIV H5N1 is co-funded in the frame of an O.I.E. Twinning project between the Animal Health Research Institute, Giza, and the FLI, Riems.

REFERENCES

1. WHO/OIE/FAO H5N1 Evolution Working Group: Continuing progress towards a unified nomenclature for the highly pathogenic H5N1 avian influenza viruses: divergence of clade 2.2 viruses. Influenza Other Respi Viruses 2009, 3:59-62.
2. Meleigy M: Egypt battles with avian influenza. Lancet 2007, 370:553-554.
3. Fasina FO, Ifende VI, Ajibade AA: Avian influenza A(H5N1) in humans: lessons from Egypt. Euro Surveill 2009, 15(4) [http://www.eurosurveillance. org/ ViewArticle.aspx?ArticleId = 19473], pii = 19473.
4. Peiris JS, de Jong MD, Guan Y: Avian influenza virus (H5N1): a threat to human health. Clinical Microbiol Rev 2007, 20:243-267.
5. Aly MM, Arafa A, Hassan MK: Epidemiological findings of outbreaks of disease caused by highly pathogenic H5N1 avian influenza virus in poultry in Egypt during 2006. Avian Dis 2008, 52:269-277.
6. Abdelwhab EM, Selim A, Arafa A, Galal S, Kilany WH, Hassan MK, Aly MM, Hafez HM: Circulation of avian influenza H5N1 in live bird markets in Egypt. Avian Dis 2010, 54:911-914.
7. Arafa A, Suarez DL, Hassan MK, Aly MM: Phylogenetic analysis of HA and NA genes of HPAI-H5N1 Egyptian strains isolated from 2006 to 2008indicates heterogeneity with multiple distinct sublineages. Avian Dis 2010, 54:345-349.
8. Balish AL, Davis T, Saad MD, El-Sayed N, Esmat H, Tjaden JA, Earhart KC, Ahmed LE, Abd El-Halem M, Hakem M, Ali AM, Nassif SA, El-Ebiary EA, Taha M, Aly MM, Arafa A, O'Neill E, Xiyan X, Cox NJ, Donis RO, Klimov AI: Antigenic and genetic diversity of highly pathogenic avian influenza A (H5N1) viruses isolated in Egypt. Avian Dis 2010, 54:329-334.
9. van den Berg T, Lambrecht B, Marché S, Steensels M, Van Borm S, Bublot M: Influenza vaccines and vaccination strategies in birds. Comp Immunol Microbiol Infect Dis 2008, 31:121-165.
10. WHO. 2010. World Health Organization: Situation updates–avian influenza. [http://www.who.int/csr/ disease/avian_influenza/updates/en/index.html], [cited 2010, June 13;].

11. Kim J, Kayali G, Walker D, Forrest HL, Ellebedy AH, Griffin YS, Rubrum A, Bahgat MM, Kutkatd MA, Ali MAA, Aldridge JR, Negovetich NJ, Krauss S, Webby RJ, Webster RG: Puzzling inefficiency of H5N1 influenza vaccines in Egyptian poultry. Proc Natl Acad Sci 2010, 107:11044-9.

12. Terregino C, Toffan A, Cilloni F, Monne I, Bertoli E, Castellanos L, Amarin N, Mancin M, Capua I: Evaluation of the protection induced by avian influenza vaccines containing a 1994 Mexican H5N2 LPAI seed strain against a 2008 Egyptian H5N1 HPAI virus belonging to clade 2.2.1 by means of serological and in vivo tests. Avian Pathol 2010, 39:215-222.

13. Di Trani L, Bedini B, Donatelli I, Campitelli L, Chiappini B, De Marco MA, Delogu M, Buonavoglia C, Vaccari GA: Sensitive one-step real-time PCR for detection of avian influenza viruses using a MGB probe and an internal positive control. BMC Infect Dis 2006, 6:87.

14. Spackman E, Senne DA, Myers TJ, Bulaga LL, Garber LP, Perdue ML, Lohman K, Daum LT, Suarez DL: Development of a real-time reverse transcriptase PCR assay for type A influenza virus and the avian H5 and H7 haemagglutinin subtypes. J Clin Microbiol 2002, 40:3256-3260.

15. Muradrasoli S, Mohamed N, Belák S, Czifra G, Herrmann B, Berencsi G, Blomberg J: Broadly targeted triplex real-time PCR detection of influenza A, B and C viruses based on the nucleoprotein gene and a novel "MegaBeacon" probe strategy. J Virol Methods 2010, 163:313-322.

16. Suarez DL, Das A, Ellis E: Review of rapid molecular diagnostic tools for avian influenza virus. Avian Dis 2007, 51:201-208.

17. Chen W, He B, Li C, Zhang X, Wu W, Yin X, Fan B, Fan X, Wang J: Real-time RT-PCR for H5N1 avian influenza A virus detection. J Med Microbiol 2007, 5:603-607.

18. Hoffman E, Stech J, Guan Y, Webster RG, Perez D: Universal primer set for the full-length amplification of all influenza A viruses. Arch Virol 2001,146:2275-2289.

19. Li PQ, Zhang J, Muller CP, Chen JX, Yang ZF, Zhang R, Li J, He YS: Development of a multiplex real-time polymerase chain reaction for the detection of influenza virus type A including H5 and H9 subtypes. Diagn Microbiol Infect Dis 2008, 61:192-197.

20. Lu YY, Yan JY, Feng Y, Xu CP, Shi W, Mao HY: Rapid detection of H5 avian influenza virus by TaqMan-MGB real-time RT-PCR. Lett Appl Microbiol 2008, 46:20-25.

21. Ng LF, Barr I, Nguyen I, Noor SM, Tan RS, Agathe LV, Gupta S, Khalil H, To TL, Hassan SS, Ren EC: Specific detection of H5N1 avian influenza A virus in field specimens by a one-step RT-PCR assay. BMC Infect Dis 2006, 6:40.

22. Payungporn S, Chutinimitkul S, Chaisingh A, Damrongwantanapokin D, Buranathai C, Amonsin A, Theamboonlers A, Poovorawan YJ: Single step multiplex real-time RT-PCR for H5N1 influenza A virus detection. J Virol Methods 2006, 131:143-147.

23. Slomka MJ, Pavlidis T, Banks J, Shell W, McNally A, Essen S, Brown IH: Validated H5 Eurasian real-time reverse transcriptase-polymerase chain reaction and its application in H5N1 outbreaks in 2005-2006. Avian Dis 2007, 51:373-377.

24. Arafa A, Selim AA, Hassan MK, Aly MM: Genetic characterization of variant strains of highly pathogenic avian influenza H5N1 that escaped detection by real-time reverse transcriptase-PCR diagnostic tests. Avian Dis 2010, 54:673-676.

25. Edgar RC: MUSCLE: multiple sequence alignment with high accuracy and high throughput. Nucleic Acids Res 2004, 32:1792-1797.

26. R Development Core Team: R: A language and environment for statistical computing. R Foundation for Statistical Computing, Vienna, Austria 2008 [http://www.R-project.org], ISBN 3-900051-07-0.

CHAPTER 4

PASSIVE IMMUNIZATION AGAINST HIGHLY PATHOGENIC AVIAN INFLUENZA VIRUS (AIV) STRAIN H7N3 WITH ANTISERUM GENERATED FROM VIRAL POLYPEPTIDES PROTECT POULTRY BIRDS FROM LETHAL VIRAL INFECTION

Mirza Imran Shahzad, Khalid Naeem, Muhammad Mukhtar, and Azra Khanum

BACKGROUND

Avian influenza virus (AIV) besides reducing commercial production of poultry is also a causative agent for influenza among humans by cross-species infections [1]. The viral genome encodes 10 proteins and among these two surface proteins haemagglutinin and neuraminidase have main importance in viral classification [2]. AIV grouping is based on antigenic variations in haemagglutinin (H1 – H16) and neuraminidase (N1 – N9) proteins and each strain of virus is named based on respective H and N antigenicity [3]. According to virulence pattern in poultry, the AIV is mainly classified into two major groups: highly pathogenic avian influenza (HPAI) and low pathogenic avian influenza (LPAI). The HPAI strains are highly virulent and associated with bird mortality approaching 100%, whereas LPAI viruses manifest mild symptoms like decreased egg production and scruffy feathers. Throughout the world majority of avian influenza epidemics are due to HPAI viruses showing H5 and H7 antigenicity [4,5]. In Pakistan, low pathogenic H9N2 along-with high pathogenic H7N3 and H5N1 are the most predominant AIV strains and several outbreaks over the past decades are ascribed to these particular strains [6-8].

Avian influenza (AI) has emerged as a disease with significant potential to disrupt commercial poultry production, resulting in heavy losses to poultry farmers in several parts of the world. Due to fastidious viral genome, conventional antivirals against AIV are unable to control the infection and very few effective vaccines are available. Moreover, geographic strain variations have made it difficult to implement universal avian influenza vaccine strategy. As such, there has been an urgent need to develop broad spectrum antivirals against AIV or vaccines capable of coping with viral genomic changes. One of the most plausible options to control AI is development of regional immunization programs against the serotype involved in an outbreak. However, as the immunization has to be carried out prior to disease for establishing therapeutic levels of antibodies against the infection, in case of its sudden outbreak such control measures are not possible. Passive immunization has emerged as an effective therapeutic tool in the face of an outbreak; however its effectiveness in the case of AIV has not yet been investigated. During past decade, AIV, H7 serotype has caused high poultry birds mortality in different countries including Pakistan [6]. The whole virus killed AIV vaccines used against H7 has been found to be effective in reducing the clinical conditions of the birds upon subsequent field challenge [2]. However, practically it is always difficult to make use of any kind of killed vaccines during the outbreaks due to very short incubation period associated with highly pathogenic AI infection. Keeping this in view, the present study was designed to compare various viral proteins for their potentials as a vaccine candidate. According to our data nucleoprotein (NP) antiserum

Citation: Mirza Imran Shahzad, Khalid Naeem, Muhammad Mukhtar, and Azra Khanum, Passive immunization against highly pathogenic Avian Influenza Virus (AIV) strain H7N3 with antiserum generated from viral polypeptides protect poultry birds from lethal viral infection, doi:10.1186/1743-422X-5-144.

significantly protected birds from viral infection induced morbidity/mortality and lowered viral shedding compared with antiserum from other viral proteins like hemagglutinin (HA) neuraminidase (NA), viral polypeptides mix, non structural protein and viral polymerase enzyme. This proof of concept study provides initial data to rely on utilization of individual viral protein for passive immunization programs.

RESULTS

Our initial work on SDS-PAGE analysis of H7N3 viral lysate showed five major viral proteins: high molecular weight polymerase (PM), hemagglutinin (HA), nucleoprotein (NP), neuraminidase (NA) and non-structural protein (NS) as shown in Figure 1. These polypetides were further concentrated and subjected to electrophoresis on SDS-PAGE. Five obvious bands of AIV viral polypetides were cut from the gel, triturated and diluted with 1.0 ml of normal saline. This follows generation of polypeptide specific antibodies against each polypeptide and also a mixture of all was used to generate antisera. The specificity of each polypeptide antiserum was confirmed by Dot-ELISA. Intriguingly, the viral peptides mix antisera detected H7N3 viral particles at 1:4 dilution (Figure 2).

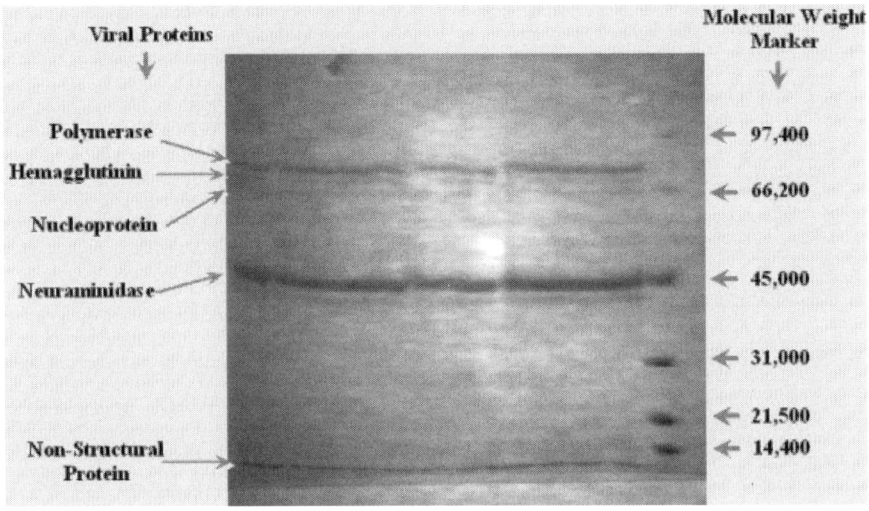

Figure 1: SDS-PAGE analysis of avian influenza Virus strain H7N3 proteins. Five major viral proteins are marked on gel corresponding to their molecular weight ascertained through protein molecular weight marker.

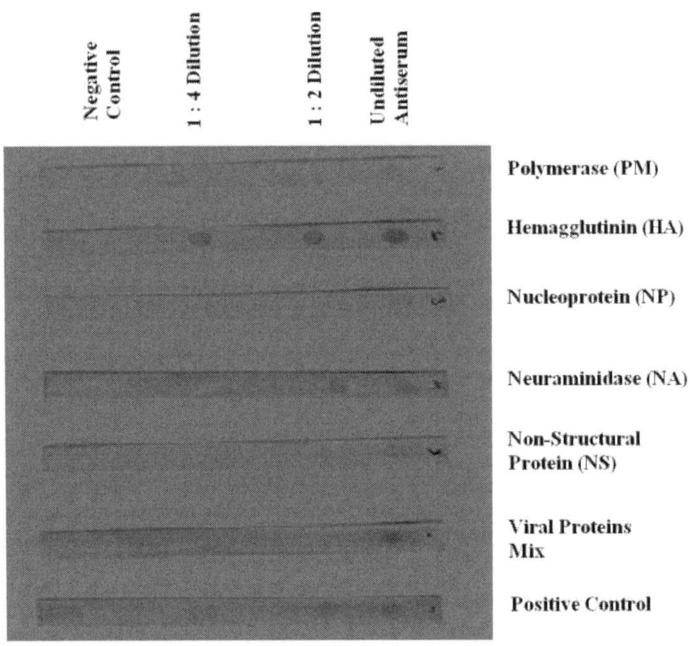

Figure 2: Dot-ELISA. confirms the antiserum specificity against respective polypetide.

All the birds used in this study were confirmed negative against AIV H7N3 antibodies by HI test. Passive immunization with individual polypeptide/protein specific antisera followed challenge with highly pathogenic AIV, H7N3. After 48 hours birds immunized with antisera and non-immunized control group were challenged with 0.2 ml of H7N3 viral strain A/Chicken/Pakistan/Murree/NARC/69/04 (H7N3). Birds' morbidity, mortality and cloacal shedding were observed over a time period of two-week. Four out of the six vaccinated group showed protection from lethal viral challenge whereas negative control group showed highest level of mortality, morbidity and cloacal shedding. The level of protection in the four groups varied and nucleoprotein antiserum vaccinated group birds showed highest protection revealed by least mortality, and low viral shedding (60%). The birds passively immunized with polymerase and non-structural protein antiserum showed no protection at all. Upon viral challenge, seven out of ten birds died in polymerase and non-structural protein antiserum vaccinated groups, whereas eight in non-vaccinated control group (Table 1). This trend continued and on day 4th all the birds in PM, NS and control (normal saline group) were dead. Mortality was associated with extensive morbidity in polypeptides groups showing less protection. One of the groups was vaccinated with antiserum generated from a mixture of all five peptides (viral polypeptides mix group). It was intriguing to note that on day 4th two birds died in this group without any further mortality thus showing 80% protection. No mortality (100% protection) was observed in birds pre-vaccinated with hemagglutinin, nucleoprotein, and neuraminidase antisera. However, morbidity and viral shedding revealed 80–100% birds infected in HA vaccinated group, 20 – 60% in NP and 80–100% in NA groups (Table 1).

Table 1: Protection of vaccinated poultry birds against highly pathogenic AIV H7N3 strain

Groups	Antisera against viral protein(s)	Post-challenge mortality at different days				Post-challenge morbidity at different days				Post-challenge cloacal shedding at different days			
		2	4	7	14	2	4	7	14	2	4	7	14
1	Viral polymerase	7/10	10/10	--	--	3/10	--	--	--	3/10	--	--	--
2	Hemagglutinin	0/10	0/10	0/10	0/10	4/10	8/10	8/10	0/10	10/10	10/10	8/10	2/10
3	Nucleoprotein	0/10	0/10	0/10	0/10	2/10	6/10	5/10	0/10	6/10	6/10	5/10	0/10
4	Neuraminidase	0/10	0/10	0/10	0/10	9/10	10/10	8/10	0/10	10/10	10/10	8/10	3/10
5	Non-structural protiens	7/10	10/10	--	--	3/10	--	--	--	3/10	--	--	--
6	Viral polypeptides mixed	0/10	2/10	0/8	0/8	10/10	8/8	7/8	0/8	10/10	8/8	7/8	2/8
7	Normal saline	8/10	10/10	--	--	2/10	--	--	--	2/10	--	--	--

Morbidity describes disease condition and prevalence of various symptoms associated with viral infection in birds. In case of bird flu outbreak, the infected birds manifest quite distinctive symptoms like ruffled feathers, excessive thirst, areas of diffuse hemorrhage between the hocks and feet, edema surrounding the eyes, watery green diarrhea progressing to white and several others. Mortality in the control (non-vaccinated) and two of the viral peptides (PM, NS) antisera manifesting least protection (0%) was associated with several disease symptoms an indicator for high morbidity (100%). In comparing the data of all protective antisera groups, the level of morbidity was higher in viral polypeptides and neuraminidase groups (100%) followed by hemagglutinin (80%) on day 4th. The nucleoprotein antiserum immunized group showed the least morbidity (maximum 60%) at day 4 along-with no mortality (0%) and lowest level of cloacal shedding makes it a potential candidature for poultry vaccine against H7N3 especially through passive immunization route.

In vaccinated groups challenged with lethal AIV, NP groups showed least cloacal shedding of virus among all the groups. All other vaccinated and non-vaccinated control manifested cloacal shedding of virus. These data are quite interesting and will help us in designing future vaccine for AIV in poultry.

DISCUSSION

Infections associated with AIV are threatening economy of several countries throughout the World. Particularly in South-East Asia viral infection has inflicted major losses to poor poultry farm holders as well as it poses a threat of cross-species infection among humans. AIV is a member of Type A group viruses and compared with its counterparts Type B and C has broad host range capable of causing infections in several birds and mammals. One of the major threats of AIV has been its capability to cross-species jumping i.e. from birds to humans [9].

According to a report from the International Federation for Animal Health (IFAH) vaccination strategies for controlling AIV infection in birds is one of the major viable options compared with other control measures [10]. Several vaccine strategies including production of vaccine from virus like particles are on horizon [11,12]. Killed vaccines have also been considered to control viral pandemic in flocks in-spite of its limitation in surveillance programs involving differentiation of infected from vaccinated animals (DIVA) test [2] particularly if killed vaccines are being used. For differentiating vaccinated birds from the naturally infected ones DIVA test strategy relies on detecting antibodies against N-type only found in infected birds and not against serotype of vaccine strain, besides general monitoring strategy of unvaccinated sentinels.

Passive immunization with antiserum generated from viral polypeptides antigenic determinants has shown significant protection in mammals [13,14] and also in birds [15]. We employed a passive immunization strategy by utilizing various proteins of AIV to ascertain which one of these could be comparatively a better candidate for the generation of antisera to be used for passive immunization. The viral polypeptides used in this study were from a highly pathogenic avian influenza virus serotype H7N3 that has been previously reported in Pakistan and several other parts of the world [6,7,16]. Our proof of concept studies reveal that it is possible to develop passive immunization strategies against AIV subtype by using viral proteins and among the five viral proteins (hemagglutinin, neuraminidase, nucleoproteins, non-structural protein, polymerase, and a mixture of all these) nucleoprotein generated antiserum provided better protection in birds upon challenge with highly pathogenic avian influenza virus.

Four out of six vaccines have given protection in decreasing order NP>HA>NA>viral polypeptides mix. In case of HA, NA and viral polypeptides mix, the level of infection increased from day 0 to day 4 and then it decreased till the end of experiment i.e. day 14. NP antiserum besides providing 100% protection also boosted chick's immunity manifested as sustained resistance against infection (low level of morbidity and viral shedding) as compared to other vaccine groups. These data suggest that passively transfused anti-NP antibodies have a better antiviral neutralizing effect and overall protection from AIV. Overall, a better protection provided during days 7–14 is due to immune regulation.

Considering the situation of developing nations like Pakistan passive immunization strategy will be economical and targeted. Avian Influenza is capable of changing antigenic determinants that leads to inefficacy of vaccines. A locally produced economical vaccine will provide effective and long lasting solution to this pandemic especially the non-variant parts (nucleoproteins) that hold the promising future of AIV vaccines.

Materials and Methods

Prior to beginning this study the protocol was reviewed and approved by the animal biosafety committee of the Pir Mehr Ali Shah Arid Agriculture University Rawalpindi, and all the viral challenges and preparations were conducted at the biosecure facilities of the National Reference Laboratory for Poultry Diseases (NRLPD) at the Animal Sciences Institute, National Agriculture Research Center (NARC), Islamabad, Pakistan.

Viral Stocks

A previously isolated highly pathogenic AIV serotype H7N3 A/Chicken/Pakistan/Murree/NARC/69/04 (H7N3) [17] was obtained from the repository of the NRLPD at Animal Sciences Institute, National Agricultural Research Center (NARC), Islamabad. The viral stock was reactivated in the allantoic cavity of embryonated hen's eggs as described previously [18]. Agar gel precipitation test was used to confirm the presence of AIV in the allantoic fluid [19] and HA test was performed to calculate the viral titer, whereas embryo lethal dose 50 (ELD_{50}) titer of the fresh viral stock was determined by classical Reed and Muench [20] methodology. In brief, this involves 10 fold serial dilutions of stock virus in normal saline (10^1 to 10^{12}) followed by injecting 0.2 ml of each dilution into the chorioallantoic region of embryonated eggs. The mortality of eggs is recorded and ELD_{50} calculated as described previously[20].

Preparation of Viral Polypeptides and Production of Monovalent Hyperimmune Antisera

Purified fresh stock of H7N3 AIV was lysed with 4% Triton X-100 using 0.01 M Tris buffer (pH 7.2) in the presence of 1 mM KCl. Viral lysate was stirred for 45 minutes at room temperature followed by centrifugation at 10,000 × g to get the supernatant containing HA, NA and matrix (M) proteins. The pellet containing NP protein was washed with phosphate buffer saline (PBS), by re-centrifuging at 10,000 × g for 1 hour at 4°C. To remove viral DNA/viral particles the supernatant was centrifuged at 200,000 × g by using Beckman ultracentrifuge L8-80 on 50 Ti rotor (Beckman, USA) for 1 hour to remove the viral DNA and viral particles. The supernatant was collected and dialyzed against 0.01 M PBS for 48 hours. It was again centrifuged at 10,000 × g for 10 minutes to separate M protein out of these preparations and the resulting pellet was suspended in PBS. The supernatant containing HA, NA, polymerase (PM) and non-structural (NS) proteins was collected by centrifuging three times repeatedly at 10,000 × g for 10 minutes at 4°C. The supernatants were dialyzed and the resultant collections were analyzed on 12% polyacrylamide gel. Five bands of AIV proteins separated on the gel were cut, triturated and diluted with 1 ml of normal saline solution (NSS). The material was centrifuged at 1000 × g for 10 min and supernatant was quantified by Lowry's method [21]. Each polypeptide was emulsified with complete Freund's adjuvant and injected @ 4 µg/bird/injection via subcutaneous route in six groups of four birds each (fourth bird was a negative control), twice at two weeks interval, respectively.

Dot-ELISA

Dot-ELISA was standardized and performed to check the specificity of each polypeptide specific antisera against AI H7N3 virus. Antigen dots were used in different dilutions ranging from Neat virus to 1:4 dilutions with NSS along with a dot containing BSA as a negative control.

Passive Immunization with Polypeptides Specific Antisera

Broiler chicks tested negative for AIV were divided equally into seven group of ten each. These birds were reared under strict isolation and high security conditions in chicken isolators. At the age of two weeks, birds were passively immunized with 4 ml each of the polypeptide specific antisera. Birds were challenged while rearing in chicken isolators at 24 hours post inoculation (PI) with live virus of AI serotype H7N3 at a dose 10^5 ELD_{50}. The birds were examined for clinical signs, mortality and cloacal shedding, up to 14 days post-challenge (PC).

ACKNOWLEDGEMENTS

This work was supported by the Agricultural Linkage Program (ALP) grant from the Pakistan Agricultural Research Council to KN, Department of Biochemistry, PMAS Arid Agricultural University Rawalpindi research funds to AK and Foreign Faculty Hiring Program of the Higher Education Commission Pakistan support to MM.

REFERENCES

1. Alexander DJ. Avian influenza viruses and human health. Developments in biologicals. 2006;124:77–84.
2. Suarez DL. Overview of avian influenza DIVA test strategies. Biologicals. 2005;33:221–226. doi: 10.1016/j.biologicals.2005.08.003.
3. World Health Organization Expert Committee report on "A revision of the system of nomenclature for influenza viruses: a WHO Memorandum". Bull WHO. 1980;58:585–591.
4. Tweed SA, Skowronski DM, David ST, Larder A, Petric M, Lees W, Li Y, Katz J, Krajden M, Tellier R, Halpert C, Hirst M, Astell C, Lawrence D, Mak A. Human illness from avian influenza H7N3, British Columbia. Emerging infectious diseases. 2004;10:2196–2199.
5. Tran TH, Nguyen TL, Nguyen TD, Luong TS, Pham PM, Nguyen VC, Pham TS, Vo CD, Le TQ, Ngo TT, Dao BK, Le PP, Nguyen TT, Hoang TL, Cao VT, Le TG, Nguyen DT, Le HN, Nguyen KT, Le HS, Le VT, Christiane D, Tran TT, Menno de J, Schultsz C, Cheng P, Lim W, Horby P, Farrar J. Avian influenza A (H5N1) in 10 patients in Vietnam. The New England journal of medicine.2004;350:1179–1188. doi: 10.1056/NEJMoa040419.
6. Naeem K, Siddique N. Use of strategic vaccination for the control of avian influenza in Pakistan.Developments in biologicals. 2006;124:145–150.
7. Naeem K, Siddique N, Ayaz M, Jalalee MA. Avian influenza in Pakistan: outbreaks of low- and high-pathogenicity avian influenza in Pakistan during 2003–2006. Avian diseases. 2007;51:189–193. doi: 10.1637/7617-042506R.1.
8. Naeem K, Ullah A, Manvell RJ, Alexander DJ. Avian influenza A subtype H9N2 in poultry in Pakistan. The Veterinary record. 1999;145:560.
9. Ernst WA, Kim HJ, Tumpey TM, Jansen AD, Tai W, Cramer DV, Adler-Moore JP, Fujii G. Protection against H1, H5, H6 and H9 influenza A infection with liposomal matrix 2 epitope vaccines. Vaccine.2006;24:5158–5168. doi: 10.1016/j.vaccine.2006.04.008.
10. IFAH IFAH background paper on Avian Influenza Vaccination. 2006. http://www.ifahsec.org
11. Quan FS, Huang C, Compans RW, Kang SM. Virus-like particle vaccine induces protective immunity against homologous and heterologous strains of influenza virus. Journal of virology. 2007;81:3514–3524. doi: 10.1128/JVI.02052-06.
12. Swayne DE, Suarez DL. Current developments in avian influenza vaccines, including safety of vaccinated birds as food. Developments in biologicals. 2007;130:123–133.
13. Foo DG, Alonso S, Chow VT, Poh CL. Passive protection against lethal enterovirus 71 infection in newborn mice by neutralizing antibodies elicited by a synthetic peptide. Microbes and infection.2007;9:1299–1306. doi: 10.1016/j.micinf.2007.06.002.
14. Ledizet M, Kar K, Foellmer HG, Bonafe N, Anthony KG, Gould LH, Bushmich SL, Fikrig E, Koski RA. Antibodies targeting linear determinants of the envelope protein protect mice against West Nile virus. The Journal of infectious diseases. 2007;196:1741–1748. doi: 10.1086/523654.
15. Chambers TM, Kawaoka Y, Webster RG. Protection of chickens from lethal influenza infection by vaccinia-expressed hemagglutinin. Virology. 1988;167:414–421.
16. Alexander DJ. Summary of avian influenza activity in Europe, Asia, Africa, and Australasia, 2002–2006. Avian diseases. 2007;51:161–166. doi: 10.1637/7602-041306R.1.
17. Sattar S, Naeem K, Ahmed Z, Malik SA. Influence of virus strain on the efficacy of vaccine against avian influenza virus subtype H7N3. International Journal of poultry science. 2007;6:989–993.
18. Lu J, Guo Z, Pan X, Wang G, Zhang D, Li Y, Tan B, Ouyang L, Yu X. Passive immunotherapy for influenza A H5N1 virus infection with equine hyperimmune globulin F(ab')2 in mice. Respiratory research. 2006;7:43. doi: 10.1186/1465-9921-7-43.
19. Sakai K, Yada K, Sakabe G, Tani O, Miyaji K, Nakamura M, Takehara K. Serological and virological studies of Newcastle disease and avian influenza in slaughter-age ostriches (*Struthio camelus*) in Japan.The Journal of veterinary medical science/the Japanese Society of Veterinary Science. 2006;68:491–494.
20. Reed LJ, Muench H. A simple method of estimating fifty percent end points. American journal of hygiene. 1938;27:493–498.
21. Lowry OH, Rosbrough NJ, Farr AL, Randall RJ. Protein measurement with the Folin phenol reagent.J Biol Chem. 1951;193:265–275.

CHAPTER 5

PREVIOUS INFECTION WITH VIRULENT STRAINS OF NEWCASTLE DISEASE VIRUS REDUCES HIGHLY PATHOGENIC AVIAN INFLUENZA VIRUS REPLICATION, DISEASE, AND MORTALITY IN CHICKENS

Mar Costa-Hurtado, Claudio L. Afonso, Patti J. Miller, Eric Shepherd, Ra Mi Cha, Diane Smith, Erica Spackman, Darrell R. Kapczynski, David L. Suarez, David E. Swayne and Mary J. Pantin-Jackwood

INTRODUCTION

Avian influenza virus (AIV) and Newcastle disease virus (NDV) affect poultry worldwide and cause important economic losses [1]. Lower virulence viruses produce subclinical infections and occasionally upper respiratory disease and drops in egg production; however, more virulent forms of these viruses cause high mortality in poultry. AIV and NDV are single-stranded, negativesense RNA viruses. AIVs are type A Orthomyxoviruses and are classified as low pathogenicity (LP) or high pathogenicity (HP) viruses based on their virulence in chickens and the presence of multiple basic amino acids at the cleavage site of the hemagglutinin precursor protein [2]. NDV's, also known as avian Paramyxovirus type-1 viruses (APMV-1), are members of the genus Avulavirus in the Paramyxoviridae family. NDVs also vary in the type and severity of the disease they produce, and different pathotypes, based on their virulence in chickens and principally determined by the sequences surrounding the protease cleavage site of the fusion protein, have been described in poultry [3]. The original classification of NDV isolates into 1 of 3 virulence groups by chicken embryo inoculation as virulent (velogenic), moderately virulent (mesogenic), or as low virulence (lentogenic) has been recently modified for regulatory purposes. Velogenic and mesogenic viruses are now classified as virulent NDV, the cause of Newcastle disease (ND), whereas lentogenic viruses are the low virulence NDV widely used as vaccines [3]. The intracerebral pathogenicity index (ICPI) in day-old chicks is used to differentiate lentogenic viruses with ICPI values of less than 0.7, from virulent mesogenic strains with ICPI values equal to or greater than 0.7 and less than 1.5, and velogenic viruses with ICPI values greater than 1.5. Variant APMV-1 isolates from pigeons, referred to as pigeon paramyxovirus type 1(PPMV-1), by containing multiple amino acids in the fusion cleavage site are considered virulent NDV [3]. Virulent NDV isolates, the cause of ND, can spread rapidly leading to high mortality rates in poultry; the latter is an especially prominent feature of velogenic (v) NDV. In the US, and in many countries worldwide, ND is not endemic and prevention is focused on biosecurity and the vaccination of poultry with both live and inactivated NDV vaccines [3]. In order to control ND, intensive vaccination programs have been established in different countries around the world. The most widely used vaccines are formulated with NDV strains of low virulence, such as B1 and LaSota [3]. Some countries, usually those with endemic virulent NDV, use live mesogenic (m) NDV vaccines as booster vaccines [4-6]; but these mNDV are considered by the World Organization for Animal Health (OIE) to be virulent and therefore reportable for the purposes of trade if isolated from poultry [3]. HPAI is also a devastating disease for poultry and strategies for its control differ depending on the country, resources, subtype of the virus, and risk to

Citation: Mar Costa-Hurtado, Claudio L. Afonso, Patti J. Miller, Eric Shepherd, Ra Mi Cha, Diane Smith, Erica Spackman, Darrell R. Kapczynski, David L. Suarez, David E. Swayne and Mary J. Pantin-Jackwood, Previous infection with virulent strains of Newcastle disease virus reduces highly pathogenic avian influenza virus replication, disease, and mortality in chickens, DOI 10.1186/s13567-015-0237-5.

public health. Eradication is the main goal, but has not been a realistic option for all countries, where vaccination may be considered as the only feasible option [7]. The current vaccines most commonly used are inactivated whole virus vaccines, but vectored vaccine use is increasing [8]. Little is known on the interactions between NDV and AIV when simultaneously infecting poultry. Co-infection of poultry with more than one etiologic agent is common and has resulted in increased clinical signs when compared to single agent infections [9-14]. Conversely, infection of a host with one virus may affect infection by a second virus, a phenomenon known as viral interference [15]. Exposure to NDV, either live vaccines or field strains, is almost unavoidable for commercial and non-commercial poultry worldwide, so co-infections with HPAIV are expected to occur in outbreaks and endemic situations. It is not clear if co-infections will exacerbate clinical signs of disease or if viral interference might occur and consequently mask or affect infections by one or other virus. AIV and NDV coinfections have been studied using cell cultures or chicken embryos, with interference between these viruses being reported, one virus inhibiting the growth of the other [16-19]; however, the mechanisms involved in such interference remains to be elucidated. In vivo studies examining AIV and NDV co-infections are scarce. A study using mallard ducks showed that co-infection with lentogenic (l) NDV and LPAIV did not affect the ability of ducks to become infected with either virus and had a minimal effect on viral shedding [20]. On the other hand, we found that coinfection of chickens and turkeys with a lNDV and a LPAIV affected the replication dynamics of these viruses, especially if given sequentially [21]. Similarly, a reduction of virus shedding and transmission was found in Pekin ducks coinfected with a vNDV and a LPAIV, and an increase in mean death time was observed in ducks infected with vNDV followed by a HPAIV compared to ducks infected only with HPAIV [22]. AIV's are a continuous threat to poultry and public health worldwide. In order to control AI, it's important to understand the pathogenesis of AIV's in field conditions, and this includes co-infections with other viruses. The objective of this study was to determine if co-infection of chickens with NDV strains of different virulence could affect the outcome of infection with HPAIV's. For this, we conducted experiments to examine the effect of simultaneous or sequential infection of chickens with lentogenic, and virulent (mesogenic and velogenic) NDV strains on HPAIV infections. Pathogenesis (clinical signs, lesions), presence of the viruses in tissues, and virus shedding were evaluated.

MATERIALS AND METHODS

Viruses

The following viruses were obtained from the Southeast Poultry Research Laboratory (SEPRL) repository: low virulent (lentogenic) NDV LaSota/1946 (vaccine strain, ICPI = 0.4), virulent (mesogenic) NDV Pigeon/1984 (ICPI = 1.45), virulent (velogenic) NDV CA/2002 (ICPI = 1.85); and HPAIVs viruses A/chicken/Queretaro/14588-1988 (H5N2) and A/chicken/Jalisco/CPA-12283-12/2012 (H7N3). The viruses were propagated in specific pathogen free (SPF) embryonating chicken eggs (ECE), as previously described [23]. Virus-infected allantoic fluid was diluted in brain heart infusion (BHI) medium (BD Bioscience, Sparks, MD) in order to obtain an inoculum with titers of 104 to 107 50% egg infectious dose (EID50) per 0.1 mL/bird. Sham inoculum was made using non-infected allantoic fluid from SPF ECE diluted 1:300 in brain heart infusion (BHI) medium (BD Bioscience, Sparks, MD, USA). The experiments were performed in biosecurity level-3 enhanced (BSL-3E) facilities in accordance with procedures approved by the SEPRL's Institutional Biosecurity Committee.

Birds

Specific pathogen free (SPF) white leghorn chickens were obtained from SEPRL's in-house flocks. The birds were housed in self-contained isolation units that were ventilated under negative pressure with inlet and exhaust HEPA-filtered air and maintained under continuous lighting. Feed and water were provided with ad libitum access. Birds were cared for in accordance to an SEPRL's Institutional Animal Care and Use Committee approved animal use protocol.

Experimental Design

Three similar experiments were conducted. The first experiment examined the effect of simultaneous or previous inoculation of chickens with low virulence (lentogenic) (l), virulent (mesogenic) (m), and virulent (velogenic) (v) strains of NDV on HPAIV infection. A second experiment examined the effect of co-infection of chickens with the same mNDV and HPAIV strains, but the timing and dose of the HPAIV inoculation were modified. In order to corroborate the results of the second study, a third study further examined the effect of co-infection by using the same mNDV strain and a different HPAIV isolate in chickens at two different ages.

Study 1

Five-week-old SPF chickens were separated into a control group and virus-inoculated groups. The control group contained 12 birds, which were intraocularly (IO) (conjunctival sac of the right eye) and intranasally (IN) (choanal cleft) inoculated with 0.1 mL total of sham inoculum (group 1). The virus-inoculated groups, each also containing 12 birds, were inoculated IO and IN with $10^{4.7}$–6.9EID_{50} in 0.1 mL of the following viruses: group 2, lNDV strain: LaSota/1946 ($10^{6.9}EID_{50}$); group 3, mNDV strain: Pigeon/1984 ($10^{6.3}$ EID_{50}); groups 4 and 5, vNDV strain: CA/2002 (low and high dose: $10^{4.7}$ or 6.3 EID_{50}); and group 6, HPAIV: A/chicken/Queretaro/ 14588-1988 H5N2 (106.9 EID50). When birds were challenged with two viruses, both viruses were either administered at the same time (day 0; groups 7–10), or sequentially (HPAIV given 2 days after the NDVs; groups 11–14). Control groups were only exposed to one virus.

Study 2

Three groups of 12 3-week-old SPF chickens were inoculated IO and IN with the following: Group 1: 0.1 mL of sham inoculum given at day 0; Group 2: mNDV (Pigeon/1984) (10^{6} EID_{50} in 0.1 mL) given at day 0 followed by A/chicken/Queretaro/14588-1988 (H5N2) ($10^{5.3}$ EID_{50} in 0.1 mL) given at day 3; and Group 3: A/ chicken/Queretaro/14588-1988 H5N2 ($10^{5.3}$ EID_{50} in 0.1 mL) given at day 3.

Study 3

Two and 4-week-old SPF chickens were separated into control groups and virus-inoculated groups. The control groups contained 6–8 birds, which were IO and IN inoculated with 0.1 mL of a sham inoculum (groups 1 and 6). The virus-inoculated groups, each containing 12 birds, were inoculated IO and IN with the following viruses: mNDV (Pigeon/1984) (10^6 EID_{50} in 0.1 mL; groups 2 and 7) and HPAIV A/chicken/Jalisco/CPA-12283-12/2012 (H7N3) (10^5 EID_{50} in 0.1 mL; groups 3 and 8). When birds were challenged with two viruses, the viruses were either administered at the same time (day 0; groups 4, 9) or sequentially (HPAIV given 3 days after the mNDV; groups 5, 10). In all studies, birds were observed for clinical signs of disease over a 10-day period. Oropharyngeal (OP) and cloacal (CL) swabs were collected from all birds at days 1 and 2 (Study 1 and 2) or 1 through 7 (Study 3) days post-inoculation (dpi) to determine virus shedding. Two birds from each group in Studies 1 and 3 were euthanized at 2 dpi in single and simultaneously inoculated groups, and 2 days after HPAIV inoculation in sequentially infected groups. Gross lesions were recorded and tissues were collected in 10% neutral buffered formalin solution to evaluate microscopic lesions and the extent of virus replication in tissues by immunohistochemistry as described previously [24,25]. Portions of lung and spleen were also stored at −70°C for virus detection. Birds that stopped eating or drinking, had severe neurological signs, or remained recumbent were euthanized and counted dead as for the next day. Birds euthanized for necropsy, moribund birds, and all birds remaining at the end of the experiments were euthanized by the intravenous (IV) administration of sodium pentobarbital (100 mg/kg body weight).

Virus Titrations

OP and CL swabs were collected in 2 mL of BHI broth with a final concentration of 10 µg/mL of gentamicin, 100 units/mL of penicillin G, and 56 µg/mL of amphotericin B, and kept frozen at −70°C until processed. RNA was extracted using the MagMax AI/ND RNA isolation kit (Ambion, Inc. Austin TX, USA). Quantitative real time RRT-PCR (qRRT-PCR) for AIV and Newcastle disease virus (NDV) detection was performed as previously described [26,27] with modifications [21]. qRRTPCR reactions targeting the influenza virus M gene [28] and the NDV M gene [29] were conducted using AgPath-ID one-step RT-PCR Kit (Ambion, Austin, TX) and the ABI 7500 Fast Real-Time PCR system (Applied Biosystem, Calsbad, CA). The RT step conditions for reactions were 10 min at 45°C and 95°C for 10 min. The cycling conditions for AIV were 45 cycles of 15 s, 95°C; 45 s, 60°C; and for NDV were 40 cycles of 10 s, 94°C; 30 s, 56°C; 10 s, 72°C. Virus titers in frozen lung and spleen samples were determined by weighing, homogenizing tissues, and diluting in BHI to a 10% (wt/vol) concentration. Equal amounts of RNA extracted from the tissue samples were used in the qRRT-PCR assay (50 ng/µL). For virus quantification, a standard curve was established with RNA extracted from dilutions of the same titrated stock of the challenge virus. Ct (cycle threshold) values of each viral dilution were plotted against viral titers. The resulting standard curve had a high correlation coefficient ($r^2 > 0.99$), and it was used to convert Ct values to EID_{50}. Results were reported as EID_{50}/mL or EID_{50}/g equivalents and the lower limit of detection was was $10^{1.5}$ EID_{50}/mL for AIV and $10^{1.7}$ EID_{50}/mL for NDV.

Serology

In study 3, serology was conducted on serum from birds that survived. Hemagglutination inhibition (HI) assays were used to quantify antibody responses to virus infection as previously described [30]. Serum was collected from birds at 10 dpi (7 dpi from the second virus given in groups exposed to the viruses sequentially). Titers were calculated as the highest reciprocal serum dilution providing complete hemagglutination inhibition. Serum titers of Log2 3 or lower were considered negative for antibodies against AIV or NDV.

Statistical Analyses

Data were analyzed using Prism v.5.01 software (GraphPad Software Inc. La Jolla, CA, USA). The survival rate data was analyzed using the Mantel-Cox Log-Rank test. Differences in the number of chickens positive for virus shed in co-infected groups compared to singleinfected were analyzed with Fisher's exact test (two-tailed). One-way ANOVA with Bonferroni multiple comparison analysis was used to evaluate virus titers in swabs. For statistical purposes, all OP and CL swabs and tissues from which virus was not detected were given a numeric value of $10^{1.4}$ EID_{50}/mL for AIV and $10^{1.6}$ EID_{50}/mL for NDV. These values represent the lowest detectable level of virus in these samples based on the methods used. Statistical significance was set at $P < 0.05$.

RESULTS

Study 1

This study examined the effect of simultaneous or previous infection of 5-week-old chickens with lNDV, mNDV and vNDV on HPAIV infection.

Clinical Signs and Survival

None of the chickens inoculated with sham inoculum showed clinical signs. Chickens inoculated with lNDV and mNDV had mild conjunctivitis. All the vNDV and HPAIV-inoculated birds became severely sick and died with mean death times (MDT's) between 1.9 and 5.2 days (Table 1 and Figure 1). Most of the birds died without previous overt clinical signs (peracute disease), but some showed non-specific clinical signs including conjunctivitis, ruffled feathers, lethargy, anorexia, swelling of the head and prostration, especially those birds infected only with vNDV, which also survived for longer. Some birds in the groups that received HPAIV also presented petechial-to-ecchymotic subcutaneous hemorrhages in leg shanks, feet and combs. No difference in the presence of clinical signs was observed between singleinfected and co-infected birds. Bird survival was compared among groups after single or simultaneous inoculation of the viruses, and after inoculation with HPAIV in groups sequentially infected with the viruses. All birds inoculated only with lNDV or mNDV, survived and showed significant differences in survival when compared with the rest of the groups. Differences in survival were also found between singleinfected vNDV groups (CA/2002) depending on the dose, $10^{4.7}$or $10^{6.3}$ EID_{50}, ($P < 0.01$); MDT was lower with the higher dose (MDT's of 4.4 days vs. 5.2 days). Differences in survival were observed between groups infected only with the vNDV (low or high dose) and birds inoculated simultaneously or 2 days later with the HPAIV ($P < 0.0001$). Low dose of vNDV increased the survival of the birds when given HPAIV either simultaneously or sequentially, when compared to the group inoculated only with HPAIV ($P < 0.05$), contrary to the high dose vNDV which did not have that effect. Simultaneous co-infection with lNDV or mNDV and HPAIV did not significantly increase the survival of the birds; however, a significant difference in mean death time was found when comparing birds sequentially infected with these viruses and HPAIV and birds that received only HPAIV ($P < 0.01$), MDT's of 2.6 and 3.1 versus 2.0 days.

Viral Shedding

OP and CL viral shedding were examined at 1 and 2 dpi by qRRT-PCR and results are shown in Table 1. To compare the effect of co-infection with NDV on HPAIV replication, birds sequentially inoculated with NDV and HPAIV were evaluated at days 3 and 4 after NDV inoculation), corresponding to days 1 and 2 after HPAIV inoculation. NDV was detected in most OP swabs from birds inoculated only with NDV (l, m or v) or from birds sequentially inoculated with the HPAIV, these later groups corresponding to 3 and 4 days after NDV inoculation. Fewer birds were positive for lNDV at 2 dpi (OP and CL swabs) ($P < 0.01$), and lower OP NDV titers ($P < 0.05$) at 1 and 2 dpi, were observed from the group inoculated simultaneously with HPAIV when compared to the lNDV single-inoculated group. Similarly, lower numbers of birds shed mNDV or vNDV (low dose) by the OP route ($P < 0.01$) and lower viral titers ($P < 0.05$) at 2 dpi, were observed from the groups inoculated simultaneously with HPAIV when compared to the singleinoculated groups. None of the birds inoculated simultaneously with the vNDV (low dose) and HPAIV shed vNDV by the CL route at 2 dpi, different to the NDV single-inoculated group where 8 of 10 birds shed virus at this time point. All birds simultaneously inoculated with the high-dose vNDV and HPAIV shed virus by the OP route at 1 and 2 dpi, and higher OP titers were found at 1 dpi in the co-inoculated groups versus the single vNDV inoculated group ($P < 0.05$). However, birds in the simultaneously inoculated group did not shed virus by the CL route at 2 dpi compared to 10 of 10 birds shedding virus by this route in the single virus inoculated group. The number of birds shedding HPAIV by the OP and CL route, and the titers of virus shed, were similar at both 1 and 2 dpi among groups inoculated only with HPAIV and groups simultaneously co-infected with NDV, with the exception of birds co-inoculated with lNDV which shed higher titers of HPAIV by the CL route ($P < 0.05$). Similarly, no difference in the number of birds shedding virus by the OP route was found between birds sequentially inoculated with lNDV or mNDV and HPAIV compared to HPAIV only inoculated birds. Strikingly, all birds inoculated with vNDV (low and high dose) then sequentially inoculated with HPAIV, showed a reduction in the number of birds shedding HPAIV by the OP and CL routes ($P < 0.001$; $P < 0.0001$), and a reduction of viral titers ($P < 0.05$) when examined at 1 dpi, indicating that HPAIV replication was hampered by vNDV. However, this effect was short-lived since no difference between sequentially inoculated and single-virus-inoculated groups was observed as of 2 dpi and beyond.

Table 1: Study 1: Mortality, mean death time (MDT), oropharyngeal (OP) and cloacal (CL) virus shedding in chickens inoculated with NDV and HPAIV

Virus	Number of dead/ total (MDT)	OP shedding: number of birds shedding/total (Log_{10} EID_{50}/mL)^{a}				CL shedding: number of birds shedding/total (Log_{10} EID_{50}/mL)^{a}			
		NDV		HPAIV		NDV		HPAIV	
		1 dpi	2 dpi	1 dpi	2 dpi	1 dpi	2 dpi	1 dpi	2 dpi
lNDV	0/10	10/10 (5.3 ± 0.1)^{A}	10/10 (6.1 ± 0.1)^{A}	nd	nd	1/10 (1.6)^{A}	5/10 (3.3 ± 0.08)^{A}	nd	nd
mNDV	0/10	7/10 (2.9 ± 0.4)^{A}	10/10 (3.2 ± 0.1)^{A}	nd	nd	0/10 (0)^{A}	0/10^{A}	nd	nd
vNDV low dose	10/10 (5.2)	10/10 (4.4 ± 0.1)^{A}	10/10 (5.9 ± 0.1)^{A}	nd	nd	0/10 (0)^{A}	8/10 (3.8 ± 0.2)^{A}	nd	nd
vNDV high dose	10/10 (4.4)	10/10 (3.6 ± 0.1)^{A}	10/10 (6.4 ± 0.1)^{A}	nd	nd	0/10 (0)^{A}	10/10 (5.2 ± 0.1)^{A}	nd	nd
HPAIV	10/10 (2.0)	nd	nd	10/10 (4.6 ± 0.1)^{A}	2/2 (5.7 ± 0.1)^{A}	nd	nd	8/10 (3.1 ± 0.3)^{A}	2/2 (4.9 ± 0.3)^{A}
lNDV + HPAIV	10/10 (2.3)	9/10 (4 ± 0.1)^{B}	5/10 (4.5 ± 0.1)^{*B}	9/10 (4.6 ± 0.1)^{A}	5/5 (6 ± 0.1)^{A}	0/10 (0)^{A}	0/5^{B}	8/10 (2.7 ± 0.3)^{A}	5/5 5.6 ± 0.2)^{B}
mNDV + HPAIV	10/10 (2.2)	2/10 (1.8 ± 0.04)^{*A}	1/3 (2.2)^{*B}	10/10 (4.2 ± 0.1)^{A}	3/3 (5.9 ± 0.3)^{A}	0/10 (0)^{A}	0/3^{A}	8/10 (2.4 ± 0.2)^{A}	3/3 (5.4 ± 0.03)^{A}
vNDV low dose + HPAIV	10/10 (2.4)	5/10 (4.2 ± 0.2)^{*A}	5/5 (4.7 ± 0.1)^{B}	10/10 (4.5 ± 0.1)^{A}	5/5 (5.8 ± 0.2)^{A}	0/10 (0)^{A}	0/5^{**B}	7/10 (2.5 ± 0.4)^{A}	5/5 (5.5 ± 0.2)^{A}
vNDV high dose + HPAIV	10/10 (2.1)	10/10 (4.6 ± 0.3)^{B}	3/3 (6.3 ± 0.1)^{A}	10/10 (4.3 ± 0.1)^{A}	3/3 (6.4 ± 0.3)^{A}	0/10 (0)^{A}	0/3^{**B}	7/10 (2.5 ± 0.4)^{A}	3/3 (5.5 ± 0.2)^{A}
lNDV + HPAIV 2 days later	10/10 (2.6)	10/10 (3.9 ± 0.1)^{A}	7/7 (6.6 ± 0.2)^{A}	10/10 (4.5 ± 0.2)^{A}	7/7 (5.9 ± 0.2)^{A}	2/10 (2.1 ± 0.01)^{A}	0/7^{*A}	7/10 (2.3 ± 0.4)^{A}	7/7 (5.3 ± 0.04)^{A}
mNDV + HPAIV 2 days later	10/10 (3.1)	7/10 (2.4 ± 0.2)^{A}	7/9 (3.1 ± 0.1)^{A}	10/10 (3.8 ± 0.2)^{A}	9/9 (5.8 ± 0.1)^{A}	4/10 (1.8 ± 0.1)^{B}	5/9 (3.2 ± 0.2)^{*B}	10/10 (3.1)^{A}	9/9 (5.2 ± 0.04)^{A}
vNDV low + HPAIV 2 days later	10/10 (2.5)	10/10 (6.4 ± 0.1)^{B}	7/7 (7.6 ± 0.1)^{B}	3/10 (1.5 ± 0.7)^{**B}	7/7 (5.3 ± 0.1)^{A}	10/10 (5.5 ± 0.1)^{***B}	7/7 (7.1 ± 0.1)^{B}	0/10^{***B}	7/7 (5.3 ± 0.05)^{A}
vNDV high + HPAIV 2 days later	10/10 (1.9)	10/10 (6.8 ± 0.1)^{B}	2/2 (8.0 ± 0.3)^{B}	3/10 (2.1 ± 0.1)^{**B}	2/2 (5.3 ± 0.1)^{A}	10/10 (6.3 ± 0.1)^{***B}	2/2 (7.6 ± 0.3)^{B}	0/10^{***B}	2/2 (5.3 ± 0.05)^{A}

For groups inoculated sequentially with the viruses, HPAIV titers are from 1 or 2 days after HPAIV inoculation, which corresponds to days 3 and 4 after NDV inoculation.

^{a}Viral titers average ± SEM (standard error mean) from the positive birds; Log_{10} EID_{50} equivalents were determined by qRRT-PCR. nd = not done. *Significant difference in number of chickens virus positive compared to single virus infected groups ($^{*}P < 0.01$; $^{**}P < 0.001$; $^{***}P < 0.0001$).

^{A,B}Different superscript uppercase denote significant difference in virus titers compared to single virus-infected groups ($P < 0.05$).

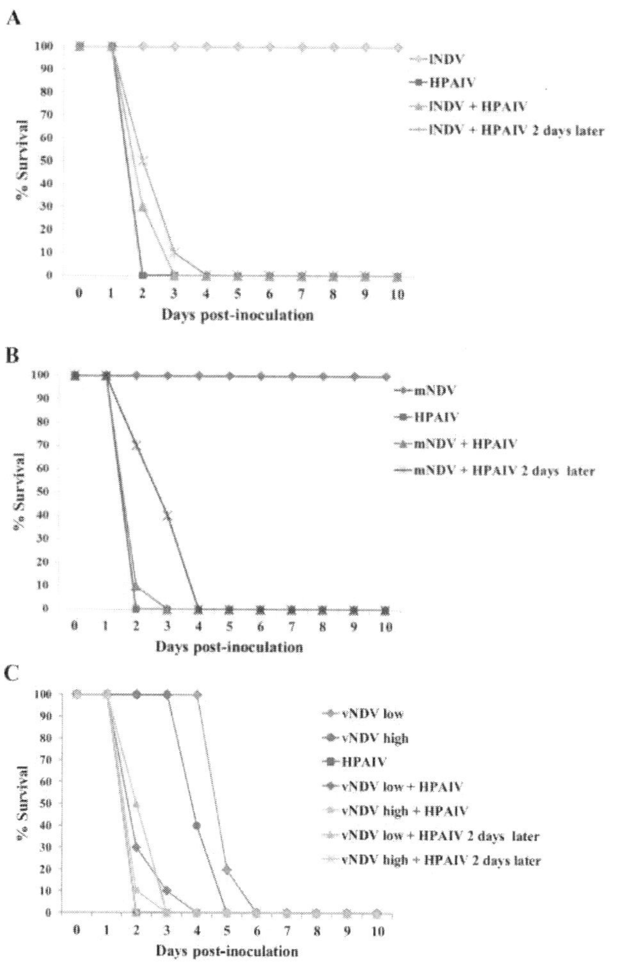

Figure 1: Survival curves after inoculation of chickens with NDV and HPAI viruses (Study 1). Chickens were inoculated simultaneously or sequentially with different strains of NDV and with HPAIV. Survival curves for chickens inoculated with lNDV and HPAIV (A), mNDV and HPAIV, and (B) vNDV and HPAIV (C).

Gross, Microscopic Lesions and Viral Antigen Staining in Tissues

Two birds from each group were necropsied 2 dpi (or 2 days after inoculation with HPAIV in the groups that received the viruses sequentially). No gross lesions were observed at 2 dpi in chickens inoculated with the sham inoculum. Mild sinusitis and conjunctivitis were present in all birds inoculated with lNDV or the mNDV. Chickens inoculated with the vNDV had severe sinusitis, edema and hemorrhages in eyelids, and enlarged mottled spleens. Birds infected with HPAIV, regardless of exposure to NDV, had sinusitis, hemorrhages on serosal surfaces of internal organs, especially in the coronary fat and on the epicardium, within the pectoral muscles and in the cecal tonsils and Meckel's diverticulum; swollen kidneys; enlarged mottled spleens; and malacic brains. Lesions were slightly more severe in birds simultaneously inoculated with vNDV and HPAIV. Microscopic lesions in chickens inoculated with lNDV and mNDV were confined to the sites of virus inoculation and included mild catarrhal rhinitis, sinusitis, and mild edema of the eyelid. By contrast, more severe and widespread histopathological findings were seen with vNDV and HPAIV infections, consistent with previous descriptions [24,31]. Viral antigen was present in several organs of birds inoculated with mNDV, vNDV or HPAIV, suggesting systemic infection with all three viruses (Additional files 1 and 2, Figure 2). In birds inoculated with NDV, NDV-nucleoprotein (NDV-NP) antigen staining was intracytoplasmic. The tissues with the strongest staining were; the nasal cavity, eyelids, and lymphoid organs, similar to previously described [24]. In birds inoculated with the HPAIV, AIV-nucleoprotein (AIVNP) antigen was detected intranuclear and intracytoplasmic in blood vessel endothelial cells throughout the body, and in various cell types within areas of necrosis and inflammation in many tissues including nasal cavity, lymphoid tissues, lung, brain, liver, and spleen, similar lo previously described [31]. Specifically, virus antigen was present in parenchymal cells of some organs including nasal epithelium, cardiac myocytes, Kupffer cells, hepatocytes, microglial cells and neurons, respiratory epithelial cells in the lung, and tubular epithelial and glomerular cells of the kidney. Compared to single-virus infected birds, tissues from birds simultaneously infected with lNDV and HPAIV showed more widespread AIV-NP antigen staining. Simultaneous co-infection with mNDV and HPAIV and low dose vNDV and HPAIV showed no differences in HPAIV-NP antigen staining compared with single-virus infected birds; however, in birds simultaneously coinfected with high dose vNDV and HPAIV there was widespread NDV-NP and AIV-NP antigen staining in tissues. Regarding birds sequentially infected with the lNDV and HPAIV, and taking into account that tissues were examined 2 dpi after HPAIV inoculation (4 dpi for NDV), an enhanced AIV-NP antigen staining was observed when compared to birds infected only with HPAIV. Widespread AIV-NP antigen staining was found both in spleen and lung (Figure 2). In birds simultaneously infected with mNDV and HPAIV, AIV-NP antigen staining was less widespread than in birds infected only with HPAIV alone. Interestingly, in birds sequentially infected with vNDV (low and high dose) and HPAIV, AIV-NP antigen staining in tissues was minimal and NDV-NP widespread (Figure 2).

Virus Titers in Lung and Spleen

Viral titers were determined in lungs and spleens collected from 2 birds per group at 2 dpi (Additional file 3). NDV viral titers were under the limit of detection in tissues from lNDV and mNDV single-inoculated and simultaneously inoculated birds. Low to moderate NDV virus titers were present in the lungs and spleens of chickens inoculated with vNDV given single or coinfected simultaneously with HPAIV (2–4 log10 EID_{50}/ gr); but NDV titers in spleen and lung were under the limit of detection in birds simultaneously infected with the low dose of vNDV and HPAIV. NDV titers were higher in the sequentially inoculated groups because it corresponded to 4 dpi after NDV inoculation. At this time, lNDV and mNDV were detected in tissues. High HPAIV titers were present in both lung and spleen collected from all birds inoculated with HPAIV; however birds previously infected with vNDV (low or high dose) had lower HPAIV titers (log_{10} 3.3-4.2 EID_{50}/gr) than the other HPAIV-inoculated groups (log_{10} 4.4-8.2 EID_{50}/gr).

Study 2

Based on the slight delay of mortality time observed in the group sequentially infected with mNDV and HPAIV, and to accentuate the effects of NDV on HPAIV replication, a study was performed in which 3-week-old chickens were inoculated with the same mNDV (Pigeon 1984) followed by the same HPAIV (A/chicken/Queretaro/ 14588-1988 H5N2) three days later. The one day delay in the HPAIV challenge allows mNDV pigeon isolate to achieve more replication as the viral titers typically don't reach their peak until 4 dpi. The titer of the HPAIV inoculum was also reduced to one log lower ($10^{5.3}$ EID_{50}) than in the previous experiment to more likely observe differences between groups. Most birds inoculated with mNDV were protected against HPAIV disease, with only 1 of 12 birds dying at 8 dpi in the sequential co-infected group, compared to 11 of 12 birds dying in the group that received the HPAIV alone (MDT of 3.5 days). In addition, while all the chickens in the HPAIV-inoculated group were shedding virus at 2 dpi, only 1 bird out of 12 was positive for HPAIV in the sequentially co-infected group (data not shown). These results suggested that the timing of exposure and the titers of the viruses might affect the outcome of infection.

Study 3

A third experiment was performed to determine if the reduction in severity of HPAIV infection in birds previously infected with mNDV occurs independently of age and virus strain. Different age chickens (2 and 4-week-old) were used to compare the effect of immune system maturity on the results of the co-infections. Furthermore a different HPAIV isolate (A/chicken/Jalisco/CPA-12283-12/ 2012 H7N3) was used to determine if the protective effect of mNDV applies to other HPAIV strains. Similar to Study 2, a lower dose of HPAIV was used, and the inoculation of HPAIV was performed 3 days after inoculation with the mNDV.

Clinical Signs

None of the birds inoculated with sham inoculum showed clinical signs. Chickens inoculated with the mNDV had mild conjunctivitis. Six to 8 out of 10 birds (2 and 4 week-old birds respectively) inoculated only with HPAIV died, with a MDT of 1.6 days for the 2-week-old chickens and 3.5 days for the 4-week-old birds (Tables 2 and 3, Figure 3). Eight of 10 birds died in the groups that were inoculated simultaneously with the mNDV and the HPAIV, with MDT's of 2 or 2.1 days. In general, chickens inoculated with the HPAIV virus in these groups in both experiments had similar clinical signs, including ruffled feathers, lethargy, anorexia and prostration and some birds presented respiratory distress, swollen head, and cyanotic comb, wattles and legs. Most birds died without showing previous clinical signs (peracute disease). Only 1 or 2 of 10 birds 2 and 4-week-old birds respectively) died in the groups sequentially inoculated with the viruses, with a MDT of 7 days for the 2-week-old birds and 9 for the 4-week-old birds. The surviving birds in these groups showed no clinical signs. No significant difference in survival among groups was found regarding the age of the birds and both behaved similarly. None of the 2-week and 4-week-old chickens inoculated only with mNDV died; thus, they showed statistically significant differences in survival when compared to the HPAIV-single-infected birds (2-week-old, $P < 0.05$; 4- week old $P < 0.001$), and those simultaneously co-infected with HPAIV (2 and 4-week-old, $P < 0.001$). When HPAIV was given 3 days after the mNDV, almost all birds survived and the birds that died had a longer mean death time when compared to the group that was only inoculated with the HPAIV (2 and 4-week-old; $P < 0.001$).

Viral Shedding

The number of birds shedding virus by the OP and CL routes and virus titers at 2, 3, 4, and 7 dpi are shown in Tables 2 and 3. Birds sequentially co-infected with mNDV and HPAIV were evaluated at days 5, 6, 7 and 10 of the experiment, corresponding to days 2, 3, 4, and 7 after HPAIV inoculation. All 2-week-old chickens inoculated with mNDV and co-infected or not with HPAIV, were shedding NDV by the OP route at the time points examined, and most CL swabs were positive by 3 dpi. No significant difference in mNDV titers was found in OP and CL swabs when comparing groups. Some of the 4-week-old birds co-infected with mNDV and HPAIV were not shedding mNDV but the number of birds shedding was not different than mNDV only-inoculated birds, with the exception of the group sequentially inoculated with the viruses, in which more birds shed virus by the CL route at 4 dpi and at higher titers. This difference can be attributed to the fact that this time point corresponds to day 7 after mNDV inoculation. At 2 dpi, both 2 and 4-week-old chickens in the groups sequentially infected with mNDV and HPAIV lacked HPAIV OP and CL shedding, significantly lower than to the groups that only received HPAIV ($P < 0.001$, $P < 0.0001$). Furthermore, at 3, 4 and 7 dpi, only 0 to 3 out of 9 or 10 birds from these groups (2 and 4-week-old birds respectively) were shedding HPAIV by the OP or CL route. In the groups simultaneously infected with mNDV and HPAIV, there was no significant reduction in the number of surviving birds shedding HPAIV or the viral titers shed at 2 dpi compared to the groups that only received HPAIV. However, 4 of 4 4- week-old birds inoculated only with the HPAIV were still shedding by the OP route at 4 dpi while none of the coinfected birds were.

Table 2: Study 3: Mortality, mean death time (MDT), and OP virus shedding in 2 and 4 week-old chickens inoculated with mNDV and HPAIV

Age	Virus	Number of dead/ total (MDT)	OP shedding: number of birds shedding/total (Log_{10} EID_{50}/mL)^{a}							
			NDV				HPAIV			
			2 dpi	3 dpi	4 dpi	7 dpi	2 dpi	3 dpi	4 dpi	7 dpi
2 weeks old	HPAIV	6/10 (1.6)	nd	nd	nd	nd	5/7 (7.9 ± 3)^{A}	3/4 (1.9 ± 0.4)^{A}	0/4^{A}	0/4^{A}
	mNDV	-	9/10 (3.7 ± 0.3)^{A}	9/10 (3.7 ± 0.1)^{A}	10/10 (4.3 ± 0.3)^{A}	10/10 (4.8 ± 0.2)^{A}	nd	nd	nd	nd
	mNDV + HPAIV	8/10 (2)	4/4 (3.3 ± 0.6)^{A}	2/2 (4.1 ± 0.1)^{A}	2/2 (4.5 ± 0.4)^{A}	2/2 (3.8 ± 0.2)^{A}	3/4 (7.3 ± 2.9)^{A}	0/2^{A}	0/2^{A}	0/2^{A}
	mNDV + HPAIV 3 days later	2/10 (7.0)	10/10 (3.4 ± 0.3)^{A}	9/10 (3.6 ± 0.3)^{A}	9/9 (3.6 ± 0.3)^{A}	9/9 (4.8 ± 0.2)^{A}	0/10^{**B}	0/10^{*A}	1/9 (4.5)^{A}	1/9 (9.3)^{A}
4 weeks old	HPAIV	8/10 (3.5)	nd	nd	nd	nd	8/9 (7.1 ± 0.9)^{A}	0/5^{A}	4/4 (3.3 ± 0.8)^{A}	0/2^{A}
	mNDV	-	10/10 (3.5 ± 0.2)^{A}	8/8 (3.5 ± 0.2)^{A}	8/8 (3.8 ± 0.3)^{A}	8/8 (4.7 ± 0.2)^{A}	nd	nd	nd	nd
	mNDV + HPAIV	8/10 (2.1)	3/5 (2.7 ± 0.5)^{A}	2/2 (3 ± 0.9)^{A}	2/2 (2.9 ± 0.7)^{A}	2/2 (3.6 ± 0.8)^{A}	3/5 (6.6 ± 1.7)^{A}	0/2^{A}	0/2^{A}	0/2^{A}
	mNDV + HPAIV 3 days later	1/10 (9.0)	9/10 (3.9 ± 0.3)^{A}	8/10 (3.4 ± 0.4)^{A}	9/10 (3.5 ± 0.3)^{A}	10/10 (4.5 ± 2.1)^{A}	0/10^{***B}	0/10 ^{A}	0/10^{***A}	1/10 (10)^{A}

For groups inoculated sequentially with the viruses, the HPAIV results are 2, 3, 4, and 7 days after H7N3 HPAIV inoculation which corresponds to days 5, 6, 7 and 10 after NDV inoculations.
^{a}Viral titers average ± SEM from the positive birds; Log_{10} EID_{50} equivalents were determined by qRRT-PCR.
nd = not done.
Significant difference in number of chickens virus positive compared to single virus infected groups ($P < 0.05$; **$P < 0.01$; ***$P < 0.001$).
$^{A;\ B}$Different superscript uppercase denote significant difference in virus titers compared to single virus infected groups ($P < 0.05$).

Table 3: Study 3: CL virus shedding in 2 and 4 week-old chickens inoculated with mNDV and HPAIV

Age	Virus	CL shedding: number of birds shedding/total (Log_{10} EID_{50}/mL)^{a}							
		NDV				HPAIV			
		2 dpi	3 dpi	4 dpi	7 dpi	2 dpi	3 dpi	4 dpi	7 dpi
2 weeks old	HPAIV	nd	nd	nd	nd	3/7 (9.8 ± 0.6)^{A}	0/4^{A}	0/4^{A}	0/4^{A}
	mNDV	0/10^{A}	8/10 (2.8 ± 0.3)^{A}	8/8 (3.6 ± 0.2)^{A}	10/10 (3.6 ± 0.2)^{A}	nd	nd	nd	nd
	mNDV + HPAIV	0/4^{A}	2/2 (2.1 ± 0.4)^{A}	2/2 (3.7 ± 0.2)^{A}	2/2 (4.5 ± 0.15)^{A}	2/4 (8.5 ± 2.8)^{A}	0/2^{A}	0/2^{A}	0/2^{A}
	mNDV + HPAIV 3 days later	1/9 (2.1)^{A}	8/10 (2.9 ± 0.1)^{A}	10/10 (3.9 ± 0.2)^{A}	9/9 (4.3 ± 0.2)^{A}	0/10^{B}	0/10^{A}	3/9 (2.2 ± 0.5)^{A}	1/9 (6.9)^{A}
4 weeks old	HPAIV	nd	nd	nd	nd	4/9 (8.1 ± 0.8)^{A}	1/5 (5.6)^{A}	1/4 (5.4)^{A}	0/2^{A}
	mNDV	0/10	4/8 (1.9 ± 0.1)^{A}	8/8 (2.9 ± 0.1)^{A}	8/8 (3.9 ± 0.1)^{A}	nd	nd	nd	nd
	mNDV + HPAIV	1/5 (1.8)	1/2 (2.5)^{A}	1/2 (3.2)^{A}	2/2 (3.7 ± 0.3)^{A}	3/5 (8.1 ± 2.5)^{A}	0/2^{A}	0/2^{A}	0/2^{A}
	mNDV + HPAIV 3 days later	0/10	5/10 (2.3 ± 0.2)^{A}	10/10 (3.7 ± 0.2)^{*B}	10/10 (4.2 ± 0.2)^{A}	0/10^{*B}	1/10 (1.6)^{A}	0/10^{A}	1/10 (4)^{A}

For groups inoculated sequentially with the viruses, the HPAIV results are 2, 3, 4, and 7 days after HPAIV inoculation which corresponds to days 5, 6, 7 and 10 after NDV inoculations.
^{a}Viral titers average ± SEM from the positive birds; Log_{10} EID_{50} equivalents were determined by qRRT-PCR.
nd = not done.
Significant difference in number of chickens virus positive compared to single virus infected groups ($P < 0.05$).
$^{A;\ B}$Different superscript uppercase denote significant difference in virus titers compared to single virus infected groups ($P < 0.05$).

Gross, Microscopic Lesions and Viral Antigen Staining in Tissues

Two birds from each group were necropsied at day 2 or 2 days after inoculated with the HPAIV in the groups that got the viruses sequentially. No gross lesions were observed in chickens inoculated with the sham inoculum, and mild conjunctivitis and sinusitis was present in the birds inoculated with mNDV or sequentially co-infected with mNDV and HPAIV. Chickens inoculated only with HPAIV or co-infected with mNDV and HPAIV simultaneously, presented lesions typical of HPAIV infection as described in Study 1. Histopathological findings were consistent with mNDV and HPAIV infections [3,25]. No or less severe lesions were observed in the birds sequentially inoculated with the viruses compared to birds that received only HPAIV or were inoculated simultaneously with the viruses. Mild conjunctivitis and sinusitis was present in birds inoculated only with mNDV. Mild to severe, diffuse catarrhal rhinitis and mild to moderate tracheitis was present in all birds inoculated only with HPAIV or simultaneously inoculated with mNDV. The rest of the lesions were similar to those observed in study 1. By immunohistochemistry, NDV-NP antigen was detected mostly in the nasal cavity, lung, spleen and cecal tonsils, but also in liver and intestine in birds inoculated only with mNDV. NDV-NP antigen staining was more common in tissues from birds co-infected with the HPAIV, including viral staining in eyelid and trachea (Additional file 4). Widespread AIV-NP antigen staining was present in tissues from birds inoculated with the HPAIV given alone or given simultaneously with mNDV (Additional file 5). In contrast, AIV-NP staining was limited to nasal cavity, lung and spleen in birds receiving the viruses sequentially.

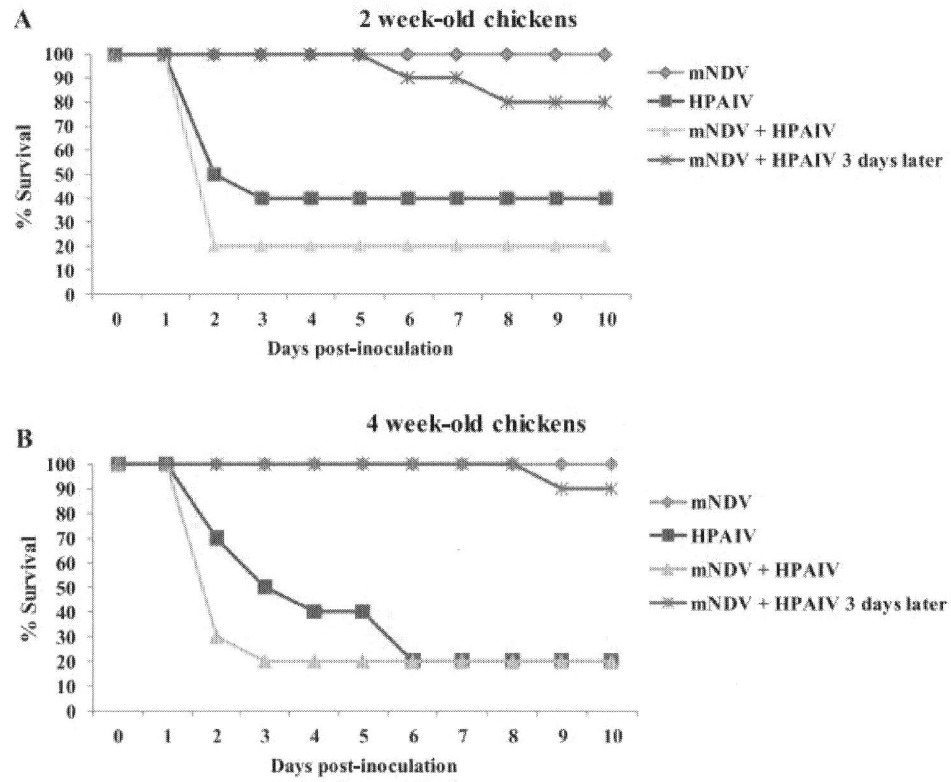

Figure 3: Survival curves after inoculation of chickens with mNDV and HPAIV (Study 3). A. 2-week-old chickens. B. 4-week-old chickens. Chickens inoculated simultaneously or sequentially with mNDV and with HPAIV.

Virus Titers in Lung and Spleen

Viral titers were determined in lungs and spleens collected from the 2 birds per group necropsied at 2 dpi (Additional file 6). Similar to study 1, HPAIV titers were generally lower in both lung and spleen in birds sequentially inoculated with mNDV and HPAIV (3.3-5.2 log10 EID50/gr) when compared to groups only receiving HPAIV (6.2-8.1 log10 EID50/gr).

Serology

HI assays were used to test for antibodies against AIV and NDV in surviving birds in study 3. Because serum samples were not taken the same day for single-virus, simultaneously and sequentially exposed birds (10 dpi for the first two and 7 dpi for the second), the level of the HI titers couldn't be strictly compared, but antibody titers provided an alternative method for determining viral infection. All the surviving chickens seroconverted to mNDV, but not to HPAIV when exposed, with no differences in titers among the treatment groups (data not shown).

DISCUSSION

The goal of these studies was to evaluate the effect of NDV on HPAIV infection in chickens. Our results showed that the severity of the disease caused by HPAIV, the number of birds shedding virus and the titers of virus shed can be reduced by previous infection with virulent strains of NDV (mNDV and vNDV) if given 2–3 days before HPAIV challenge. As expected, single-virus infection with lNDV and mNDV did not cause disease or death in chickens. The mNDV isolate used in this study was a pigeon strain (PPMV-1), which typically cause high mortality in pigeons but not in chickens. These viruses can acquire virulence after multiple passages in chickens [32]. In contrast, infection with vNDV or HPAIV caused severe disease and high mortality (60-100%). In study 1, there were no differences in mortality rates in groups that received HPAIV alone and sequentially and simultaneously infected groups, but increased MDT's were observed in groups co-infected with lNDV, mNDV and low dose of vNDV. In addition, OP and CL HPAIV shedding at 1 dpi were significantly lower in the sequential vNDV (low or high dose) and HPAIV-inoculated groups when compared to the group inoculated only with HPAIV or simultaneously with vNDV and HPAIV. By 2 dpi, the OP and CL viral shedding were similar in all groups, but IHC on tissues from sequentially infected groups showed little or no HPAIV antigen as compared to high levels of HPAIV antigen in the other groups. This indicates that although vNDV interfered with initial HPAIV replication, the HPAIV challenge was sufficiently high to overcome vNDV interference and killed the chickens, the vNDV itself killed the chickens, or a combination of both. In studies 2 and 3, we found that when HPAIV was given at a lower dose three days after inoculation with the less virulent NDV (mNDV), bird survival increased and virus shedding decreased. By using this less virulent NDV strain that also replicates systemically, the replication of HPAIV was reduced to a point that allowed the host immune response to control the infection and prevent death. These results agree with our previous studies showing viral interference between less pathogenic forms of NDV and AIV in chickens and turkeys [21]. In this previous study, although no differences in clinical signs were observed in chickens, prior exposure to lNDV modified viral shedding patterns by delaying the LPAIV shedding. In the present study, lNDV had little effect on HPAIV replication, failing to stop viral systemic spread and replication leading to death. Nevertheless, the presence of

HPAIV affected lNDV replication when given simultaneously. In studies 2 and 3, mNDV given 3 days prior to a lower dose HPAIV challenge interfered with HPAIV replication protecting chickens against death. mNDV and vNDV strains have a multi-basic amino acid motif at the fusion cleavage site and can be cleaved intracellularly by ubiquitous furin-like proteases found in most host tissues [33]. This results in a systemic infection that often is fatal with vNDV's. On the other hand, mNDV's are less pathogenic, rarely producing neurological disease, and death only seen in young birds [3]. Both NDV and AIV have mechanisms to interfere with the host innate immune responses including reduction of the interferon (IFN) response. The V protein of NDV is an IFN antagonist [34], but this effect appears to be strain dependent. For instance, the V protein of the mesogenic Beaudette C strain exhibits a greater antagonistic effect on IFN induction in vitro than that of the lentogenic LaSota strain [35]. Interestingly, in the case of the velogenic strain CA02, a block of IFN pathways did not occur at the mRNA level when the chicken immune response was evaluated by microarray, in fact CA02 elicited a strong immune response in chickens suggesting that the host response itself may contribute to the pathogenesis of vNDV [36]. The most likely mechanism for the viral interference observed is that the mNDV produced a robust IFN response that resulted in blocking or greatly reducing HPAIV viral replication in the sequentially infected birds. In the simultaneously infected birds, the IFN response did not develop in time to prevent infection, resulting in death of the birds. In the present study, mNDV might be eliciting a more balanced innate immune response, not as strong as vNDV but strong enough to activate the IFN pathway and protect against a second virus infection. In this scenario, although it is known that HPAIV can delay IFN-induced antiviral responses [37], the overall outcome of HPAIV infection depends on the early innate immune protection induced by the mNDV, therefore, the timing of the previous infection with mNDV might be crucial for protection against HPAIV infection. Because of strain differences, the observations presented cannot be generalized to all mNDV's and the interfering potential of other mNDV strains needs to be evaluated. Age-related susceptibility to disease in birds may be associated with a maturing immune system. For instance, the expression levels of components involved in the IFN system appear to increase with age [38]. However, in this study, 2 and 4-week-old chickens showed similar survival rates and virus shedding patterns suggesting no difference in immune competency. These results, although insufficient to completely characterize the mechanism involved in suppression of viral replication, might also suggest that, apart from non-specific innate immune responses induced by the earlier infection, viral competition for target host cells could also be involved, this second alternative also needing further study. The existence of previous viral-host interactions capable of modifying the outcomes of a sequential HPAIV infection indicates that it may be possible to develop novel strategies to prevent or reduce mortality in HPAIV-infected naïve birds. In some countries where ND is endemic, mNDV live vaccine strains are used to control the disease in poultry [3,39]. Some of these countries report outbreaks of HPAIV; therefore it's possible that co-infections with mNDV occur. If this was the case, disease and mortality caused by HPAIV could be curbed by previous infection with mNDV, affecting the clinical diagnosis and the control of the virus; but, as mentioned, the effect of mNDV on HPAIV infection is most likely strain dependent. Vaccination with mNDV strains to protect against HPAI could be a possible option in countries where ND is endemic. However, since virulent strains of NDV are reportable to the OIE, the use of mNDV vaccines would not be an option in countries free of ND. Further studies are needed to explore this unconventional use of live NDV vaccines. In conclusion, co-infection with NDV and HPAIV can affect the viral replication dynamics and the disease caused by these viruses in chickens, but this effect will depend on the virulence of the viruses involved, the challenge titer of the viruses and the timing of the coinfections. The identification of factors that influence a delay of infection of one virus by another will provide new insights in the pathogenesis of these viruses, allowing the development of novel ways to control viruses.

COMPETING INTERESTS

The authors declare that they have no competing interests.

ACKNOWLEDGEMENTS

The authors appreciate the assistance provided by Aniko Zsak, Tim Olivier, Dawn Williams-Coplin, Ronald Graham, and Roger Brock in conducting these studies. This research was supported by the U.S Department of Agriculture, ARS CRIS Project 6612-32000-063-00D and with federal funds from the National Institute of Allergy and Infectious Diseases, National Institutes of Health, under IAA no. AAI12004-001-00001. Its contents are solely the responsibility of the authors and do not necessarily represent the official views of the NIH. Mention of trade names or commercial products in this publication is solely for the purpose of providing specific information and does not imply recommendation or endorsement by the U.S. Department of Agriculture. USDA is an equal opportunity provider and employer.

REFERENCES

1. Alexander DJ (1995) The epidemiology and control of avian influenza and Newcastle disease. J Comp Pathol 112:105–126
2. Swayne DE, Suarez DL, Sims LD (2013) Influenza. In: Swayne DE, Glisson JR, McDougald LR, Nolan LK, Suarez DL, Venugopal N (eds) Diseases of Poultry 13th edn. Wiley-Blackwell, Ames, IA, pp 181–218
3. Miller PJ, Koch G (2013) Newcastle disease, Other Avian Paramyxoviruses and Avian Metapneumovirus Infections. In: Swayne DE, Glisson JR, McDougald LR, Nolan LK, Suarez DL and Venugopal N (eds) Diseases of Poultry 13 edn. Wiley-Blackwell, Ames, IA, pp. 89–138
4. Roy P, Venugopalan AT, Koteeswaran A (1999) Efficacy of live adjuvanted mesogenic Newcastle disease vaccine in chickens. Vaccine 17:2674–2676
5. Khan TA, Rue CA, Rehmani SF, Ahmed A, Wasilenko JL, Miller PJ, Afonso CL (2010) Phylogenetic and biological characterization of Newcastle disease virus isolates from Pakistan. J Clin Microbiol 48:1892–1894
6. Rehmani SF (1996) Newcastle disease vaccination: a comparison of vaccinesand routes of administration in Pakistan. Prev Vet Med 25:241–248

7. Swayne DE, Pavade G, Hamilton K, Vallat B, Miyagishima K (2011) Assessment of national strategies for control of high-pathogenicity avian influenza and low-pathogenicity notifiable avian influenza in poultry, withemphasis on vaccines and vaccination. Rev Sci Tech 30:839–870
8. Spackman E, Swayne DE (2013) Vaccination of gallinaceous poultry for H5N1 highly pathogenic avian influenza: current questions and newtechnology. Virus Res 178:121–132
9. Bano S, Naeem K, Malik SA (2003) Evaluation of pathogenic potential ofavian influenza virus serotype H9N2 in chickens. Avian Dis 47:817–82210. Charles SD, Nagaraja KV, Halvorson DA, Barnes DM (1993) Influence ofNewcastle disease virus on the severity of Pasteurella anatipestifer infection in turkeys. Res Vet Sci 55:209–214
10. Haghighat-Jahromi M, Asasi K, Nili H, Dadras H, Shooshtari AH (2008) Coinfection of avian influenza virus (H9N2 subtype) with infectious bronchitis live vaccine. Arch Virol 153:651–655
11. Karimi-Madab M, Ansari-Lari M, Asasi K, Nili H (2010) Risk factors for detection of bronchial casts, most frequently seen in endemic H9N2 avian influenza infection, in poultry flocks in Iran. Prev Vet Med 95:275–280
12. Pan Q, Liu A, Zhang F, Ling Y, Ou C, Hou N, He C (2012) Co-infection of broilers with Ornithobacterium rhinotracheale and H9N2 avian influenza virus. BMC Vet Res 8:104
13. Stipkovits L, Egyed L, Palfi V, Beres A, Pitlik E, Somogyi M, Szathmary S, Denes B (2012) Effect of low-pathogenicity influenza virus H3N8 infection on Mycoplasma gallisepticum infection of chickens. Avian Pathol 41:51–57
14. Kimura Y, Norrby E, Nagata I, Ito Y, Shimokata K (1976) Homologous interference induced by a temperature-sensitive mutant derived from an HVJ (Sendai virus) carrier culture. J Gen Virol 33:333–343
15. Burnet FM (1945) Experiments on the elution of Newcastle disease virus and influenza virus from fowl red cells. Aust J Exp Biol Med Sci 23:180–183
16. Carr JH (1960) Inoculation time differentials for expression of interference of Newcastle disease virus by swine influenza virus in chick embryos. Trans Kans Acad Sci 63:141–146
17. Shortridge KF, King AP (1983) Cocultivation of avian orthomyxoviruses and paramyxoviruses in embryonated eggs: implications for surveillance studies. Appl Environ Microbiol 45:463–467
18. Ge S, Zheng D, Zhao Y, Liu H, Liu W, Sun Q, Li J, Yu S, Zuo Y, Han X, Li L, Lv Y, Wang Y, Liu X, Wang Z (2012) Evaluating viral interference between Influenza virus and Newcastle disease virus using real-time reverse transcriptionpolymerase chain reaction in chicken eggs. Virol J 9:128
19. Franca M, Howerth EW, Carter D, Byas A, Poulson R, Afonso CL, Stallknecht DE(2014) Co-infection of mallards with low-virulence Newcastle disease virus and low-pathogenic avian influenza virus. Avian Pathol 43:96–104
20. Costa-Hurtado M, Afonso CL, Miller PJ, Spackman E, Kapczynski DR, Swayne DE, Shepherd E, Smith D, Zsak A, Pantin-Jackwood M (2014) Virus interference between H7N2 low pathogenic avian influenza virusand lentogenic Newcastle disease virus in experimental co-infections in chickens and turkeys. Vet Res 45:1
21. Pantin-Jackwood M, Costa-Hurtado M, Miller PJ, Afonso CL, Spackman E, Kapczynski DR, Shepherd E, Smith D, Swayne DE (2015) Experimental co-infections of domestic ducks with a virulent Newcastle disease virus and low or highly pathogenic avian influenza viruses. Vet Microbiol 177:7–17
22. Senne D (2008) Virus propagation in embryonating eggs. In L. DofourZavala, Swayne DE, Gilsson JR, Pearson JE, Jackwood MW, Reed W, and Woolcock PR (ed.), A laboratory manual for the isolation, identification and characterization of avian pathogens, 5th edn. American Association of Avian Pathologists, Jacksonville, FL, pp 204–208
23. Susta L, Miller PJ, Afonso CL, Brown CC (2011) Clinicopathological characterization in poultry of three strains of Newcastle disease virus isolated from recent outbreaks. Vet Pathol 48:349–360
24. Pantin-Jackwood MJ, Swayne DE (2007) Pathobiology of Asian highly pathogenic avian influenza H5N1 virus infections in ducks. Avian Dis 51:250–259 26. Pantin-Jackwood M, Swayne DE, Smith D, Shepherd E (2013) Effect of species, breed and route of virus inoculation on the pathogenicity of H5N1 highly pathogenic influenza (HPAI) viruses in domestic ducks. Vet Res 44:62
25. Pedersen J (2009) Programming and Result Interpretation for Avian Influenza and Avian Paramyxovirus-1 Real-Time RT-PCR Protocols withApplied Biosystems 7500 Fast Real-Time Instrumentation, SOP-AV-0003. National Veterinary Services Laboratory, Ames, IA
26. Spackman E, Senne DA, Myers TJ, Bulaga LL, Garber LP, Perdue ML, Lohman K, Daum LT, Suarez DL (2002) Development of a real-time reverse transcriptase PCR assay for type A influenza virus and the avian H5 and H7 hemagglutinin subtypes. J Clin Microbiol 40:3256–3260
27. Wise MG, Suarez DL, Seal BS, Pedersen JC, Senne DA, King DJ, Kapczynski DR, Spackman E (2004) Development of a real-time reverse-transcription PCR for detection of newcastle disease virus RNA in clinical samples. J Clin Microbiol 42:329–338
28. OIE (2015) Avian Influenza. In: Manual for Diagnostic Tests and Vaccines for Terrestrial Animals, chapter 2.3.4., http://www.oie.int/international-standardsetting/terrestrial-manual/access-online/
29. Pantin-Jackwood MJ, Swayne DE (2009) Pathogenesis and pathobiology of avian influenza virus infection in birds. Rev Sci Tech 28:113–136
30. Kommers GD, King DJ, Seal BS, Brown CC (2001) Virulence of pigeon-O/ origin Newcastle disease virus Isolates for domestic chickens. Avian Dis 45:906–921
31. Nagai Y, Klenk HD, Rott R (1976) Proteolytic cleavage of the viral glycoproteins and its significance for the

virulence of Newcastle disease irus. Virology 72:494–508

32. Park MS, Shaw ML, Munoz-Jordan J, Cros JF, Nakaya T, Bouvier N, Palese P, Garcia-Sastre A, Basler CF (2003) Newcastle disease virus (NDV)-based assay demonstrates interferon-antagonist activity for the NDV V protein and the Nipah virus V, W, and C proteins. J Virol 77:1501–1511

33. Alamares JG, Elankumaran S, Samal SK, Iorio RM (2010) The interferon antagonistic activities of the V proteins from two strains of Newcastle disease virus correlate with their known virulence properties. Virus Res 147:153–157

34. Rue CA, Susta L, Cornax I, Brown CC, Kapczynski DR, Suarez DL, King DJ, Miller PJ, Afonso CL (2011) Virulent Newcastle disease virus elicits a strong innate immune response in chickens. J Gen Virol 92:931–939

35. Cornelissen JB, Vervelde L, Post J, Rebel JM (2013) Differences in highly pathogenic avian influenza viral pathogenesis and associated early inflammatory response in chickens and ducks. Avian Pathol 42:347–364

36. Karpala AJ, Bagnaud-Baule A, Goossens KE, Lowenthal JW, Bean AG (2012) Ontogeny of the interferon system in chickens. J Reprod Immunol 94:169–174

37. Chellappa MM, Dey S, Gaikwad S, Kataria JM, Vakharia VN (2012) Complete genome sequence of Newcastle disease virus mesogenic vaccine strain R2B from India. J Virol 86:13814–13815

CHAPTER 6 MODELLING MAREK'S DISEASE VIRUS (MDV) INFECTION: PARAMETER ESTIMATES FOR MORTALITY RATE AND INFECTIOUSNESS

Katherine E Atkins, Andrew F Read, Nicholas J Savill, Katrin G Renz, Stephen W Walkden-Brown and Mark EJ Woolhouse

BACKGROUND

Marek's Disease Virus (MDV) is an oncogenic poultry herpesvirus of considerable economic importance to the poultry industry. Virus strains have become increasingly virulent since the 1960s [1, 2]. The causes of this viral evolution are unclear [3]. Quantification of patterns of viral shedding and virus-induced host mortality are necessary for a rigorous understanding of the epidemiology of a disease, not least to identify increases in virulence. Here we develop methods to do this.

MDV is an airborne pathogen with infection occurring via inhalation [4]. Virus shedding occurs by infected feather follicle epithelium [5]. The resulting dust and dander from dead stratified cells and moulted feathers can then remain in the environment and act as a reservoir for chicken infection. Clinical signs are varied and result in significant morbidity and mortality depending on host genetic susceptibility and virulence of the MDV strain [6]. Symptoms include polyneuritis (an enlargement of multiple peripheral nerves), visceral lymphoma (tumours affecting organs such as the heart, liver, spleen etc.), acute transient paralysis, immunosuppression, brain oedema and acute rash. There has been a change in the types of clinical signs since the disease was first noted [7, 8], when chronic polyneuritis was the only sign. Since then, the list of clinical signs described above expanded gradually over the decades [2].

Almost all industrialised countries have experienced MD losses in their poultry industry and a crude estimate of the cost of Marek's disease is said to be in the range of US $1-2 billion annually [7]. Control of Marek's disease is predominantly via vaccination of chickens. Crucially the MDV vaccine was the first vaccine to be developed against any cancer [9]. Three types of vaccine have been developed for use against MD. These are herpesvirus of turkeys (HVT), non-pathogenic serotype 2 MDV and non-pathogenic serotype 1 [10]. These vaccines have been used in different sequences in different countries and the vaccine types have also been commonly combined in bivalent or trivalent vaccines.

Definitions of virulence are numerous and varied [11]. Microbiologists equate virulence with the notions of both infectivity and severity of disease, whereas evolutionary biologists focus on evolutionary fitness of either the pathogen or the host [12]. Zoologists tend to focus on host fitness [13, 14] with most mathematical models describing this specifically as host mortality [14, 15].

There have been various attempts to define virulence in the context of MDV [16]. For example, Witter [1] relates the percentage of HVT or Bivalent (HVT + serotype 2) vaccinated chickens, that when infected with a particular strain develop gross lesions or die of Marek's disease within eight weeks

Citation: Katherine E Atkins, Andrew F Read, Nicholas J Savill, Katrin G Renz, Stephen W Walkden-Brown and Mark EJ Woolhouse, Modelling Marek's Disease Virus (MDV) infection: parameter estimates for mortality rate and infectiousness, DOI: 10.1186/1746-6148-7-70.

to the figure in unvaccinated chickens challenged with the same strain. This percentage score uses the protective effect of vaccines as its metric for virulence. In addition, [17] use neurovirulence as a tool for pathotyping MDV strains, noting that many of the very virulent strains are synonymous with high levels of neuropathology. This method circumvents the need for both vaccine-based definitions and therefore high numbers of birds, and also may cut the experiment time. In the first pathotyping regime, there are four recognised MDV serotype 1 pathotypes, each occupying part of the 'continuum of virulence' [1] defined as: mild, m; virulent, v; very virulent, vv; very virulent plus, vv+. This grouping correlates very well with the second pathotyping regime [18].

Control strategies for MDV require an understanding of the epidemiology of the disease, in particular how virulence relates to key parameters such as viral shedding rates and duration of infectious period. There have been efforts to find correlates with virulence of MDV isolates, most notably with viral load (virus within a bird tissue). This has been achieved by cell culture techniques [19] and intra-cellular detection during the first 10 days of infection [20]. However, the relationship between viral load (or replication) and virulence has not been convincingly tested [16]. PCR testing methods have been developed in order to quantify viral loads [20, 21, 22,23, 24, 25, 26], which have enabled the viral loads in shed dust to be directly measured [27, 28]. The infectious period is defined as the total time in which an infected bird sheds virus [29]. This is determined by the clearance rate and the lifespan of the individual host. Since there is no recovery from MDV infection, the infectious period is defined by a total of four parameters: the disease-induced mortality, the latent period (the time between infection and infectiousness of a host) and two other non-disease associated parameters, namely non-disease induced mortality and the maximum lifespan of an individual (the farm slaughter time in the case of broilers).

Here we present methods for parameter estimation for MDV isolates, allowing a formal quantification of infected host lifespan, viral shedding rates and viral latent periods. The parameters determined from the analysis allow comparison between isolates of different virulence in both unvaccinated and vaccinated birds. It is anticipated that the parameter estimates will be used for a better understanding of the pathology and aetiology of the disease itself and as a platform for investigation into the causes of virulence evolution.

RESULTS

In the first section, we used survival analysis with a Weibull mortality function to build a statistical model for the lifespan of birds infected with MDV. In the second section, we developed a dynamic model to simulate the shedding of virus by a group of birds and used Bayesian techniques to estimate the four key infectiousness parameters: primary and secondary latent periods and viral shedding rates.

Mortality

An additive covariate Weibull regression model was fitted to the bird survival data, since the interaction terms were not significant. The model estimates are given in Table 1. The Weibull model captures the data significantly better with the covariates used than without ($p < 0.01$). The resulting graphs showing the data and Weibull model fit are displayed in Figure 1. The model showed that host lifespan decreased with virulence score and increased with vaccination (Figure 2 and Table 1), although there was no significant difference between the effect of the two vaccines on lifespan (Table 1). On most days there were either zero or one birds dying from MDV-related illness from the initial 52 or 53 individual birds in each group (Figure 1). These data and further information are published in separate experimental work [30].

Table 1: Survival analysis: Estimated coefficients of the covariates in the survival analysis

Symbol	Coefficient	Value (95% CI via McMC)	*z*	*p*
r	Shape Parameter	4.18 (3.38,4.99)	-11.60	< 0.001
β_0	Intercept	4.54 (4.23,4.87)	26.08	< 0.001
β_1	Virulence Score	-0.53 (-1.01,-0.06)	-2.04	0.040
β_2	HVT vaccine	0.44 (0.24,0.66)	3.95	< 0.001
β_3	Bivalent vaccine	0.35 (0.19,0.52)	4.28	< 0.001

The transformed virulence score is a continuous covariate and the vaccine treatment is a categorical covariate in an additive regression, giving the test statistic, z and the corresponding p value. The numbers in brackets are the 95% credible intervals.

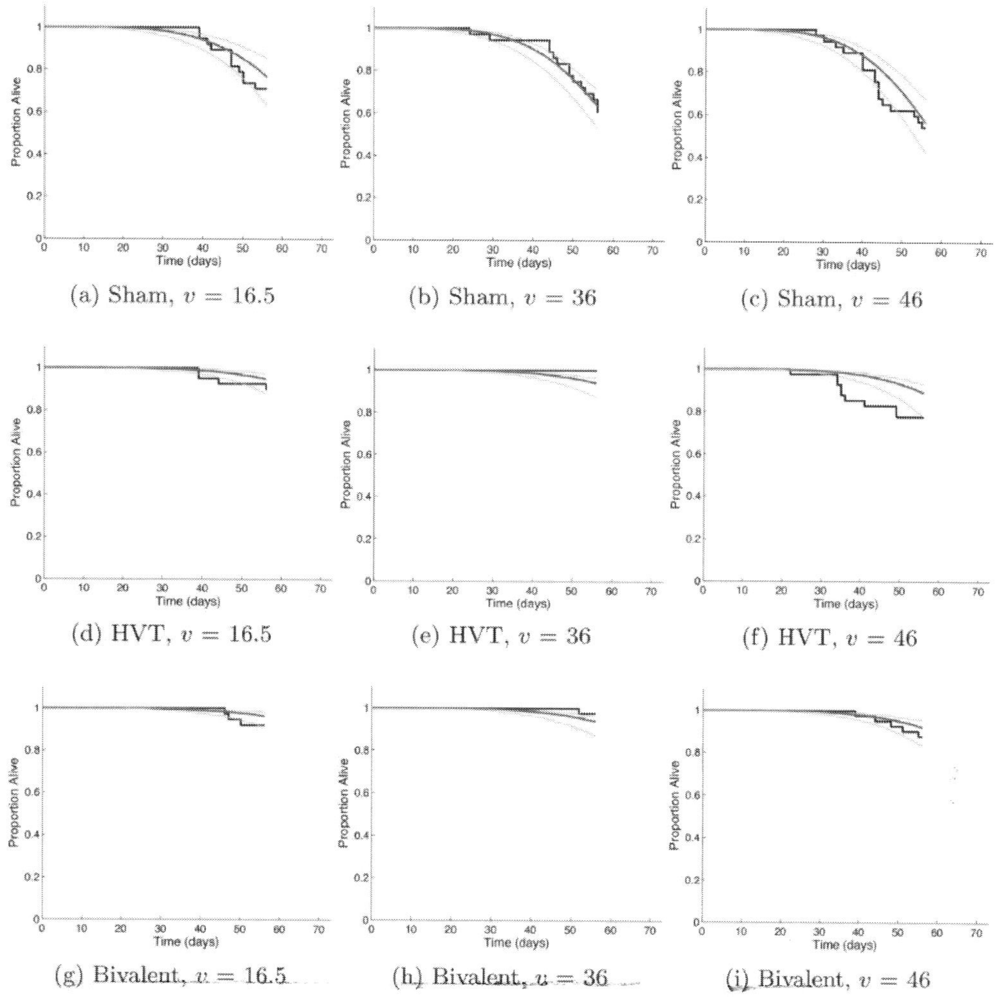

Figure 1: Mortality: Three different treatments crossed with three different virulence scores as stated. Graphs show data (black), maximum likelihood estimate of Weibull distribution (red) and 95% credible interval (green). Transformed virulence score (continuous) and vaccine treatment (categorical) are the (additive) covariates.

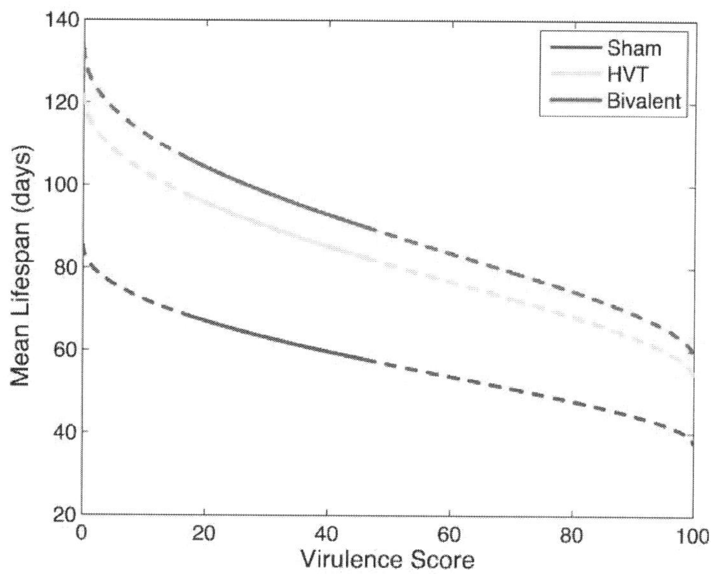

Figure 2: Mortality: Maximum likelihood estimates of λ and r give the mean lifespan ($\lambda\Gamma(1+1r)$, where $\Gamma(z)$ is the Gamma function as it varies with virulence score. The dashed lines represent extrapolation outside virulence scores given in data.

Shedding

The dynamic model output records the density of virus (measured in viral copy number (VCN)/mg dust), which fitted to the data from each isolator independently. The model fits can be compared directly in Figures 3, 4and 5, which give the day of sampling and the sample value. These data and further information are published in a separate experimental paper [30].

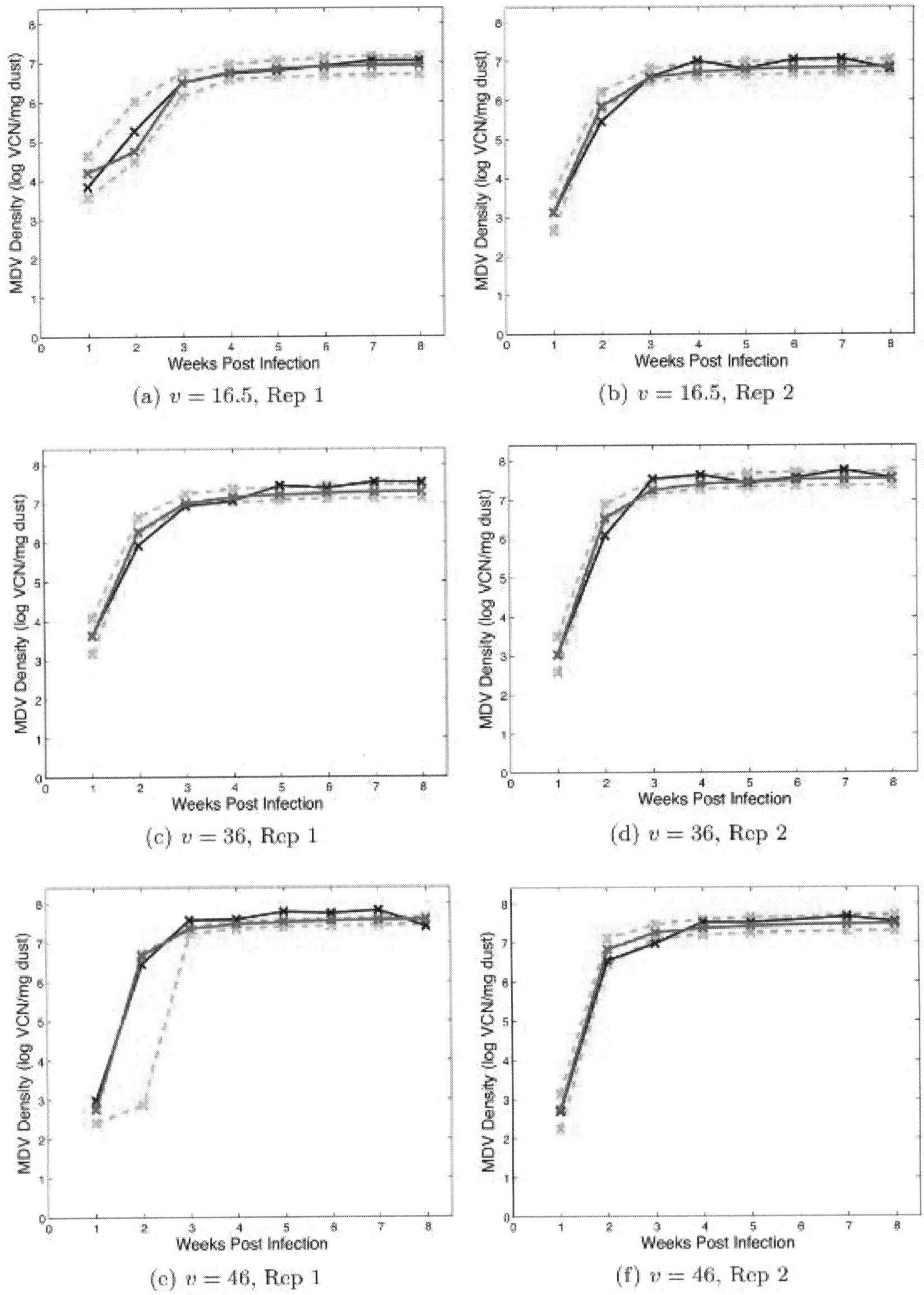

Figure 3: Viral Shedding: Sham vaccinated treatments. Graphs include the data (black), the maximum likelihood realisation (red) and the 95% credible interval (green). The panels show each replicate for the virus strains which have been given the virulence score, v as denoted in the caption.

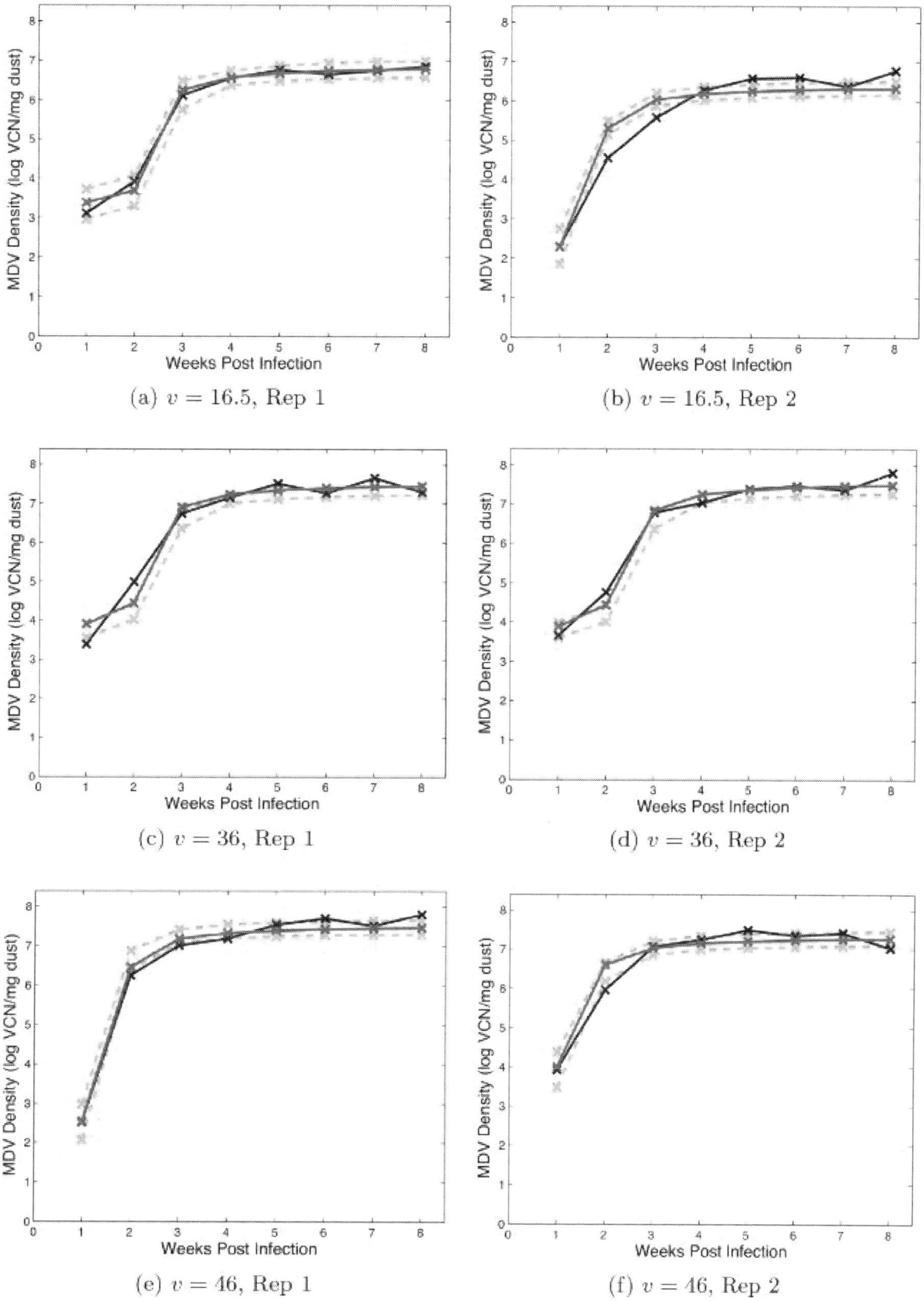

(a) $v = 16.5$, Rep 1
(b) $v = 16.5$, Rep 2
(c) $v = 36$, Rep 1
(d) $v = 36$, Rep 2
(e) $v = 46$, Rep 1
(f) $v = 46$, Rep 2

Figure 4: Viral Shedding: HVT vaccinated treatments. Graphs include the data (black), the maximum likelihood realisation (red) and the 95% credible interval (green). The panels show each replicate for the virus strains which have been given the virulence score, v as denoted in the caption.

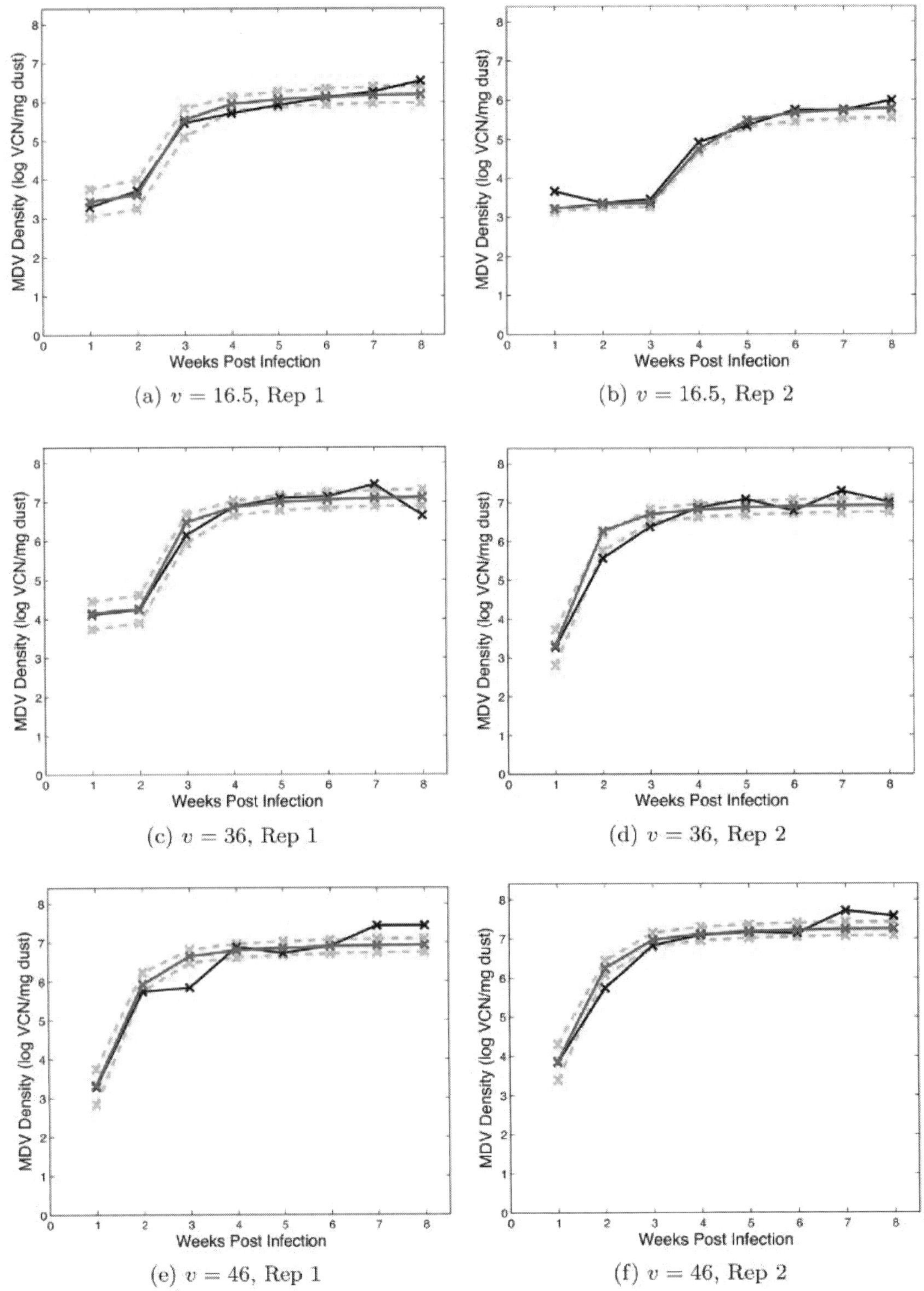

Figure 5: Viral Shedding: Bivalent vaccinated treatments. Graphs include the data (black), the maximum likelihood realisation (red) and the 95% credible interval (green). The panels show each replicate for the virus strains which have been given the virulence score, v as denoted in the caption.

For each isolator, the median of the posterior distribution for the four parameters in the model (two latent periods and two rates of viral shedding) are given in Tables 2, 3 and 4, where the number of air changes per hour, α, is set to 15. Assuming $\alpha \in [5, 25]$, the results obtained in the McMC method did not change, whatever value of α is chosen.

Table 2: Viral shedding unvaccinated: median and 95% credible intervals for the posterior distribution of the four parameters for both replicates of the sham vaccinated factorial study for different virulence scores, v

	VIRULENCE SCORE					
PARAMETER	**16.5**		**36**		**46**	
	Rep 1	**Rep 2**	**Rep 1**	**Rep 2**	**Rep 1**	**Rep 2**
1 Latent Period (days)	6 (2,6)	4 (0,6)	4 (0,6)	4 (0,6)	3 (0,6)	3 (0,6)
2 Latent Period (days)	10 (7,14)	9 (6,13)	9 (6,13)	9 (6,13)	9 (6,13)	9 (5,13)
1 Shedding Rate (logVCN/mg dust)	4.87 (3.79,5.25)	3.37 (2.79,4.05)	3.86 (3.30,4.56)	3.28 (2.71,3.95)	3.20 (2.64,3.90)	2.89 (2.34,3.59)
2 Shedding Rate (logVCN/mg dust)	6.99 (6.72,7.21)	6.89 (6.72,7.07)	7.34 (7.17,7.51)	7.58 (7.41,7.76)	7.69 (7.51,7.85)	7.52 (7.32,7.72)

The numbers in brackets are the 95% credible intervals.

Table 3: Viral shedding HVT vaccination: median and 95% credible intervals for the posterior distribution of the four parameters for both replicates of the HVT vaccinated factorial study for different virulence scores, v

	VIRULENCE SCORE					
PARAMETER	**16.5**		**36**		**46**	
	Rep 1	**Rep 2**	**Rep 1**	**Rep 2**	**Rep 1**	**Rep 2**
1 Latent Period (days)	5 (0,6)	2 (0,6)	6 (4,6)	6 (2,6)	3 (0,6)	3 (0,6)
2 Latent Period (days)	12 (8,17)	10 (7,13)	11 (8,14)	12 (8,16)	10 (6,13)	9 (6,13)
1 Shedding Rate (logVCN/mg dust)	3.79 (3.34,4.20)	2.44 (1.93,3.02)	4.58 (4.19,4.92)	4.56 (4.13,4.91)	2.70 (2.16,3.42)	4.15 (3.58,4.82)
2 Shedding Rate (logVCN/mg dust)	6.83 (6.61,7.04)	6.35 (6.18,6.52)	7.49 (7.28,7.70)	7.53 (7.31,7.75)	7.50 (7.32,7.67)	7.29 (7.11,7.46)

The numbers in brackets are the 95% credible intervals.

Table 4: Viral shedding bivalent vaccination: median and 95% credible intervals for the posterior distribution of the four parameters for both replicates of the Bivalent vaccinated factorial study for different virulence scores, v

	VIRULENCE SCORE					
PARAMETER	**16.5**		**36**		**46**	
	Rep 1	**Rep 2**	**Rep 1**	**Rep 2**	**Rep 1**	**Rep 2**
1 Latent Period (days)	4 (0,6)	2 (0,5)	3 (0,6)	3 (0,6)	3 (0,6)	3 (0,6)
2 Latent Period (days)	14 (9,19)	23 (18,27)	15 (10,19)	9 (6,13)	10 (6,13)	10 (6,13)
1 Shedding Rate (logVCN/mg dust)	4.87 (3.27,5.19)	3.37 (3.27,3.83)	3.86 (2.92,4.70)	3.28 (2.93,4.14)	3.20 (2.94,4.18)	2.89 (2.49,4.69)
2 Shedding Rate (logVCN/mg dust)	6.23 (6.02,6.45)	5.85 (5.61,6.09)	7.16 (6.94,7.37)	6.95 (6.78,7.13)	6.96 (6.78,7.12)	7.28 (7.11,7.45)

The numbers in brackets are the 95% credible intervals.

A linear model was fitted to account for the variation in the primary and secondary latent periods, with the latent period as the response variable and arcsine square-root transformed virulence score and vaccine treatment as multiplicative covariates. For the primary latent period, none of the results were significant ($p = 0.41$), showing that the primary latent period did not differ among virus strains and nor was it altered by vaccine treatment. For the secondary latent period, neither virulence score nor HVT vaccine affected the duration of the latent period, however, Bivalent vaccine increased the duration of the second latent period significantly and to a greater extent for smaller virulence scores (see Table 5).

Table 5: Secondary latent period: linear regression results for effect of transformed virulence score and vaccine treatment on the secondary latent period (days)

	Estimate	Std. Error	t value	Pr(> |*t*|)
Intercept	10.16	4.30	2.36	0.04
Virulence Score	-1.64	6.98	-0.24	0:82
VaccHVT	2.60	6.10	0.43	0:68
VaccBiv	19.29	6.10	3.16	< 0.01
Virulence Score:VaccBiv	-24.82	9.88	-2.51	0.03
Virulence Score:VaccHVT	-1.82	9.88	-0.18	0.86

Adjusted $R^2 = 0.54$, $p = 0.01$.

An association between the virulence score and the estimated viral shedding rates was found. A linear regression was used to estimate the viral shedding rate as a function of arcsine square-root transformed virulence score and vaccine treatment. There was no significant association between the primary viral shedding rate and HVT or Bivalent vaccine treatment ($p = 0.085$, 0.10 respectively) nor virulence score ($p = 0.055$). There was, however, a strong positive association between the virulence score of a strain and the secondary viral shedding rate ($p = 0.00076$). There was no association between the secondary viral shedding rate and either HVT or Bivalent vaccine treatment ($p = 0.66$, 0.39 respectively, see Table 6). Figure 6 shows the estimated secondary viral shedding rate as a function of virulence score for different vaccine treatments.

Table 6: Secondary viral shedding: linear regression results for effect of transformed virulence score and vaccine treatment on the second viral shedding rate (VCN/mg dust)

	Estimate	Std. Error	t value	Pr(> |*t*|)
Intercept	-3.24×10^7	1.35×10^7	-0.24	0.03
Virulence Score	9.79×10^7	2.19×10^7	4.47	< 0.01
VaccHVT	8.51×10^6	1.91×10^7	0.45	0.66
VaccBiv	1.70×10^7	1.91×10^7	0.89	0.39
Virulence Score:VaccBiv	-5.73×10^7	3.19×10^7	-0.19	0.09
Virulence Score:VaccHVT	-2.37×10^7	3.10×10^7	-0.77	0.46

Adjusted $R^2 = 0.74$, $p < 0.01$.

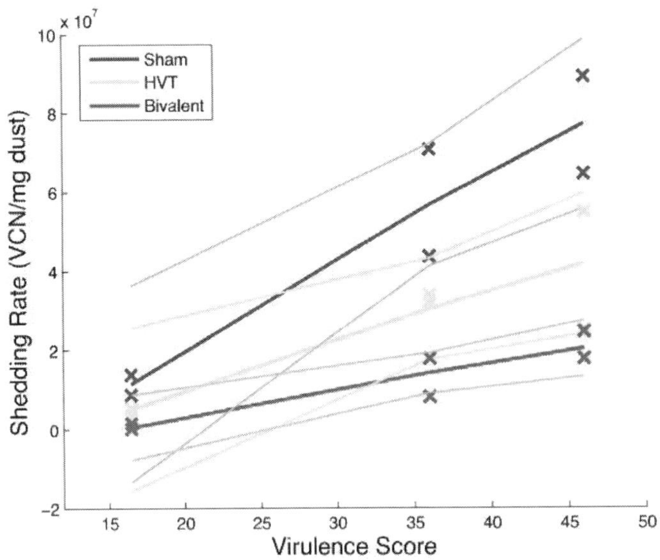

Figure 6: Viral Shedding: Linear regression analysis of the virulence score and the corresponding estimated secondary viral shedding rate for all treatments. The 95% confidence intervals for the mean response are given in fainter lines. The linear regression results are shown in Table 6.

DISCUSSION

The survival model fits the nine survival curves well on the whole, with both replicates combined into one fit. The delay which characterises the mortality profile for Marek's disease is captured well by the Weibull model. In most cases, the viral shedding model captures the data well.

The viral shedding model assumes that all birds become initially infected with MDV. While it is difficult to model individual birds based on the resolution of the data available, we do point out that the assumption that all birds become infected precipitates the estimated per bird viral shedding rates being perhaps an over-estimate of the actual rates. On the other hand, the qPCR methods used estimate total viral copies which is not necessarily equivalent to total infectious material and therefore serves as an upper limit of viable infectious DNA.

Our estimate of the primary latent period of between 2 and 6 days is much shorter than earlier estimates of around 13 days [31] but only slightly shorter than more recent estimates based on new PCR techniques which have detected significant quantities of virus in feather tips [32] and feather dander [27, 28] at 7 days post infection. This discrepancy may be due to a number of factors: detectability of virus at low quantities, the sampling time of the experiments and heterogeneities between the experiments. The estimates here also have confidence intervals between 0-6 days which suggest the data are not able to give a great deal of certainty on the exact time of onset of viral shedding. Further analysis might take into account biological knowledge through prior distributions to get a more exact value, and past experimental studies would suggest shedding begins towards the end of the estimated interval [27, 28].

The pathotyping of MDV isolates, similar to that pioneered by Witter and colleagues [1, 16], is a useful tool for comparing the pathogenicity of strains with respect to others in vaccinated birds. However several of the more recent isolated, highly virulent strains have ranks above 80, in some cases 90 on Witter's ranking system. Should more virulent strains emerge, it will be difficult to categorise them via this method, as the scale is truncated at 100. The method of fitting a Weibull distribution proposed in this paper allows for a greater flexibility of pathotyping, in that an isolate may be categorised according to its mean time to kill the host. The subsequent effect of vaccination of the host on this metric can then be established.

Biologically, it may seem more sustainable to pathotype according to the lifespan of an unvaccinated bird, but experimentally it may also be a solution to the costly procedure of having two or three groups of birds, each vaccinated with a different treatment. Even a simple 'time to death' metric can be performed in relatively few birds, as was the case for the myxoma virus in rabbits in Australia. Myxoma strains were pathotyped effectively when the case fatality rate proved experimentally intensive and ineffective in discriminating between similarly virulent strains [33, 17] noted this restriction of current pathotyping regimes, instead favouring a system based on neuropathology to circumvent these methodological and experimental limitations. The analysis presented here differs from both of these past approaches due to its epidemiological focus on virulence. However, we point out that the methodology adopted here only analyses three distinct strains. Therefore, a wider range of strain virulences and places of origin will be required to sufficiently extrapolate the current results to the whole virulence spectrum.

CONCLUSIONS

This paper uses a mathematical approach to estimate key epidemiological parameters for Marek's Disease Virus (MDV). We used survival analysis with a Weibull mortality function to build a statistical model for the lifespan of birds infected with MDV. We were able to compare the relative impact of virulence score and vaccine covariates by calculating credible intervals via Bayesian methods. We developed a dynamic model to simulate the shedding of virus by a group of birds and used Bayesian techniques to estimate the four key infectiousness parameters: primary and secondary latent periods and viral shedding rates.

We applied these methods to the data for three MDV strains varying moderately in virulence. From the survival analysis, we found that increasing the virulence score of a virus strain decreases the lifespan of the host (Figure 2). Vaccinating the host with either the first or second generation vaccine increases its lifespan, but neither one to any greater degree (Table 1). From the viral shedding analysis, we concluded primary latent period is not affected by vaccination or virulence score, however, secondary latent period increases with Bivalent vaccine treatment (Table 5). Secondary (long term) viral shedding rate (measured in VCN per mg dust) is higher for strains of higher virulence (Table 5 and Figure 6). Vaccination does not reduce long term viral shedding rate significantly. Together, these results imply that more virulent viruses shed more infectious material into the environment and Bivalent vaccines reduce the infectious period, especially for less virulent strain infections.

The results described here are qualitatively in line with current thinking on MDV pathogenesis, but the new methods presented allow a formal comparison of strains in an epidemiological meaningful way. It is anticipated that this analysis will allow a more quantitative understanding of the aetiology of MDV and the mechanisms governing a functional rise in virulence for a strain. With this analysis it is hoped that a more epidemiologically significant classification of MDV strains will emerge which will eventually provide not only insight into the reasons for evolution to highly virulent strains but also a metric to evaluate future risks and the efficacy of possible control strategies.

METHODS

Experimental Data

We look at a single experiment [30] with two response variables measured. Groups of maternal antibody positive IsaBrown layer chicks were inoculated with one of three vaccine treatments: sham (diluent only), HVT (first generation industry vaccine) or Bivalent (second generation industry vaccine) at one day old. At 5 days post vaccination, the birds were challenged with 500 pfu (plaque forming units) of one of three MDV strains: 04CRE, MPF57 and 02LAR. These independently sampled isolates were Australian in origin and have been pathotyped with virulence scores (the percentage of

HVT or Bivalent vaccinated chickens infected with a particular strain which develop gross lesions or die of MDV within eight weeks [34]) of 16.5, 36 and 46 respectively. The viruses were all isolated from field outbreaks and used at passage level 4 to 7 in chick kidney cells. The experiment was therefore a 3 × 3 factorial design study, with vaccine type fully cross-factored against virus strain. Each treatment combination was replicated in two separate isolators resulting in 26 or 27 birds per combination for the pathotyping component of the study. The air in each isolator was changed 8-20 times per hour. Further details can be found in [30]. The experiment was approved by the University of New England Animal Ethics Committee with approval number AEC05/076. Due to Animal Ethics regulations, birds thought to be nearing death were euthanised and classified as mortality in the experimental results. The experiment lasted 56 days post infection (dpi) and all birds were humanely killed and examined for MD lesions on the last day. Therefore for the purposes of this analysis, the data were censored after day 55. All birds were given a large dose of MDV (500 pfu) via intra-abdominal injection and it was therefore assumed that all the birds were infected and would shed the virus. Differences between unvaccinated and vaccinated birds in detection of MD have been reported [35], but these have been in studies where infection has been by injection with 50 pfu. Increasing the MDV dose fourfold (from 50 to 200 pfu) significantly increased the detected incidence of MD in unvaccinated birds from 64% to 81% [27].

The two variables recorded were the time to death of a bird (measured in days) and the density of virus within each isolator every week (measured in logVCN per mg of dust in the environment). The virus titres in the dust were determined via quantitative real-time PCR of dust collected from the isolator exhaust ducts, the methods of which have been discussed in separate studies [25]. The data are shown in Figures 1 (mortality), 3, 4 and 5 (shedding).

Parameter Estimates

We assumed that each infected bird starts shedding virus after an initial delay, then after a further delay, the rate at which it sheds virus rises to a long-term stable value. We estimated five key epidemiological parameters: host lifespan, primary latent period (the time until an infected bird becomes infectious), secondary latent period (the delay between an infectious bird first shedding virus and when it sheds virus at its secondary long term rate) and viral shedding rates (both primary and secondary, measured in VCN/mg dust). A survival analysis was undertaken for the mortality data because multiple covariates were needed to find the host lifespan. These covariates were vaccine treatment and virus virulence score. Latent period and rate of viral shedding could not be estimated directly from the data since the response variable (measured weekly in VCN per mg dust) could vary with the number of birds housed together, which changed as birds died. Therefore a dynamic model capturing this variability was used to estimate the two infectiousness parameters.

For ease of notation, log is written to represent $\log_{10}$ throughout the paper.

Mortality

The biology of Marek's disease suggests that the probability of death by MDV infection changes through time, since the virus undergoes lytic, latent stage (this is a description of an MDV pathogenesis stage and not equivalent to the 'latent periods' in the epidemiological context which this study estimated), late lytic (where infection is extended to other organs of the host), and transformative (tumour-induction) stages. These stages are pathologically distinct and may lead to a different mortality rate. Therefore, a Weibull distribution was chosen as a candidate distribution for modelling survivorship curves, which is often used for time to death data since it is flexible and can mimic other distributions but only has two parameters in its non-location form. We therefore assumed that time to death can be modelled as a random variable, T, such that $T \sim W(r, \lambda)$ wherer, $\lambda > 0$ are the shape and scale parameters respectively. The associated probability density function is

$$f(t) = \begin{cases} \frac{r}{\lambda}\left(\frac{t}{\lambda}\right)^{r-1} e^{-\left(\frac{t}{\lambda}\right)^r} & \text{if } t \geq 0 \\ 0 & \text{if } t < 0 \end{cases}$$

A Weibull model was fitted to the survival data in the study. In the case when r = 1, the distribution collapses to an exponential distribution. When r > 1 or r < 1 there is an increasing or decreasing chance of death over time respectively. Coefficients, $\boldsymbol{\beta}$, are such that $\lambda = \exp(\boldsymbol{\beta} \cdot \mathbf{x})$, where $\boldsymbol{\beta} = [\beta_0, \beta_1, \beta_2, \beta_3]$ are the covariate coefficients and$\mathbf{x} = [1, x_1, x_2, x_3]^T$ are the covariates. In this analysis, there were three covariates (one continuous: the transformation of the virulence score of an isolate (x_1) and two binary: the presence or absence of HVT (x_2) and Bivalent (x_3) vaccine). Therefore there were 9 combinations on the set of possible covariates, thus $\lambda_j = \boldsymbol{\beta} \cdot \boldsymbol{x}_j$ where $j \in [1, 9]$ The likelihood function can therefore be written

$$\mathcal{L}(\boldsymbol{\lambda}, r) = \prod_{j=1}^{9} \prod_{i=i}^{n} \left\{ \frac{r}{\lambda_j} \left(\frac{t_i}{\lambda_j} \right)^{r-1} \right\}^{\delta_i} \exp\left(-\frac{t_i}{\lambda_j} \right)^r$$

where δ_i is zero when the ith observation is censored and unity elsewhere. This function was maximized via the Newton-Raphson algorithm such that $\mathcal{L}(\tilde{\boldsymbol{\lambda}}, \tilde{r}) = \max \mathcal{L}(\boldsymbol{\lambda}, r)$ where $\tilde{\boldsymbol{\lambda}}$ and $\tilde{r}$ are the maximum likelihood estimates. Note that the virulence score, v, of an isolate is a percentage measure and to be used as an explanatory variable in a regression analysis, it should be transformed such that $v_T = \arcsin\sqrt{0.01v}$. The regression fitted the maximum likelihood estimates for $\boldsymbol{\beta}$, however further Bayesian analysis was required to estimate the associated credible intervals when the covariate covariance matrix does not approximate the identity matrix. An McMC framework was set up in WinBUGS [36] to calculate the posterior distributions and the credible intervals of the Weibull survival function.

The prior distributions of each parameter were assumed to be uninformative and were thus taken as uniform, to ensure equivalence to maximum likelihood estimation. The burn-in was set to 22,000 iterations with every 10th sample taken from the subsequent 100,000 iterations [37, 38].

Viral Shedding

The dynamic model tracked the density of MDV over time in each isolator. In this model it is assumed:

1) After MDV infection, there is a delay before virus is first shed [31]. This is termed the primary latent period.

2) After this primary latent period, virus is shed at a constant rate (the primary viral shedding rate, measured in VCN per mg dust) for a set period of time, termed the secondary latent period.

3) Once this secondary latent period is over, virus is shed at a constant rate (the secondary viral shedding rate) until the termination of the experiment.

Additionally, the density of virus (VCN per mg dust) was calculated at the end of each day and any removal of birds was assumed to occur at the start of the day. While all parameters are assumed to be constant across all the birds in a single isolator, they can vary between isolators.

There are therefore four parameters estimated per isolator: the primary/secondary latent periods (in days) and the primary/secondary rates of viral shedding (VCN/mg dust). The primary and secondary latent periods are denoted by T^1 and T^2 respectively. Birds shed virus at a rate, a^1 (VCN per mg dust) from T^1+1 to T^1+T^2 days post infection; and at a rate a^2 (VCN per mg dust) from T^2+1 until the end of the experiment (at 8 weeks post infection). The density of MDV in the dynamic model could be compared to the amount of virus recorded in the data (sampled every 7 days). We therefore estimated $a_j^1(v), a_j^2(v), T_j^1(v)$ and $T_j^2(v)$ for each virulence score $v \in \{16.5, 36, 46\}$ and each vaccine status $j \in \{$Sham, HVT, Bivalent$\}$.

The model accounts for the removal of birds given in the mortality data and the number of complete hourly air changes (varied as a parameter α). The daily quantity of dust shed per bird was calculated by fit ting a cubic spline to dust MDV data provided from the same experiment [30].

To calculate the measurement error of a quantity of virus in a sample of dust we noted that a final MDV density was reached some time before the end of the sampling period. Such data can therefore be assumed to be identically and independently distributed. If Y_1 and Y_2 are samples from the experiment when the virus density and the data are lognormally distributed then:

$$logY_1, logY_2 \sim N(\mu, \sigma^2)$$

$$\Rightarrow logY_1 - logY_2 \sim N(0, 2\sigma^2)$$

To find the time point after which the data is assumed to have plateaued, four sets of points were examined: weeks 5, 6, 7 and 8. The differences between the logged data of weeks 7-8 provided the highest probability of both being drawn from a normal distribution ($p = 0.8$, Anderson-Darling $n = 18$) and that the distribution has a mean of zero ($p = 0.40$). The resulting standard deviation of the difference between log-transformed data was estimated to be 0.33 logVCN/mg dust.

Posterior distributions were found by means of McMC realisations by calculating the likelihood of the data given the parameter values, with lognormal errors of $\sigma^2 = 0.33^2/2$. The burn-in time was set to 30,000 iterations and the posterior distribution was taken as every 10th sample from the following 90,000 iterations [37, 38].

ACKNOWLEDGEMENTS

We thank BBSRC, Pfizer, the Australian Poultry CRC, R. Howey for supplying Bayesian sampling code and M. Chase-Topping for statistical advice.

AUTHORS' CONTRIBUTIONS

KEA performed the statistical analysis and drafted the manuscript. AFR was involved in the drafting and revision of the manuscript for intellectual content. NJS was involved in the Bayesian statistical analysis. SWB and KGR were involved in supplying data and associated technical advice. SWB was involved in the revision of the manuscript for intellectual content. MEJW was involved in design of the statistical analysis and the drafting and revision of the manuscript for intellectual content. All authors approved the final manuscript.

COMPETING INTERESTS

The authors declare that they have no competing interests.

REFERENCES

1. Witter RL: Increased virulence of Marek's disease virus field isolates. Avian Diseases. 1997, 41: 149-163. 10.2307/1592455.
2. Osterrieder N, Kamil JP, Schumacher D, Tischer BK, Trapp S: Marek's disease virus: from miasma to model. Nature Reviews Microbiology. 2006, 4: 753-761.
3. Witter RL: The changing landscape of Marek's disease. Avian Pathology. 1998, 27: S46-S53. 10.1080/03079459808419292.
4. Calnek BW, Adldinger HK, Kahn DE: Feather Follicle Epithelium: a Source of Enveloped and Infectious Cell-Free Herpesvirus from Marek's Disease. Avian Diseases. 1970, 14 (2): 219-233. 10.2307/1588466.
5. Carrozza JH, Fredrickson TN, Prince RP, Luginbuhl RE: Role of desquamated epithelial cells in transmission of Marek's disease. Avian Diseases. 1973, 17: 767-781. 10.2307/1589045.
6. Baigent SJ, Smith LP, Nair VK, Currie RJW: Vaccinal control of Marek's disease: current challenges, and future strategies to maximize protection. Veterinary Immunology and Immunopathology. 2006, 112: 78-86. 10.1016/j.vetimm.2006.03.014.
7. Morrow C, Fehler F: Marek's disease: a worldwide problem. Marek's disease: an Evolving Problem. Elsevier Academic Press 2004.
8. Nair V, Kung HJ: Marek's disease virus oncogenicity: molecular mechanisms. Marek's Disease: an Evolving Problem. Elsevier Academic Press 2004.
9. Churchill A, Payne L, Chubb R: Immunization against Marek's Disease using a live attenuated virus. Nature. 1969, 221 (5182): 744-747. 10.1038/221744a0.
10. Bublot M, Sharma J: Vaccination against Marek's disease. Marek's Disease: an Evolving Problem. Edited by: Nair V, Davison F. Elsevier Academic Press 2004.
11. Dieckmann U: Adaptive dynamics of pathogen-host interactions. Adaptive Dynamics Of Infectious Diseases: In Pursuit of Virulence Management. Cambridge University Press 2002.
12. Read AF: The Evolution of Virulence. Trends in Microbiology. 1994, 2: 73-76. 10.1016/0966-842X(94)90537-1.
13. Levin S, Pimentel D: Selection of Intermediate Rates of

Increase in Parasite-Host Systems. American Naturalist. 1981, 117: 308-315. 10.1086/283708.

14. Anderson RM, May RM: Directly transmitted diseases: Control by vaccination. Science. 1982, 215 (4536): 1053-1060. 10.1126/science.7063839.

15. Tompkins DM, Dobson AP, Arneberg P, Begon ME, Cattadori IM, Greenman JV, Heesterbeeck JAP, Hudson PJ, Newborn D, Pugliese A, Rizzoli AP, Rosa R, Rosso F, Wilson K: Parasites and Host Population Dynamics. The Ecology of Wildlife Diseases. Oxford University Press 2002.

16. Witter RL, Calnek BW, Buscaglia C, Gimeno IM, Schat KA: Classification of Marek's disease viruses according to pathotype: philosophy and methodology. Avian Pathology. 2005, 34 (2): 75-90. 10.1080/03079450500059255.

17. Gimeno IM, Witter RL, Neumann U: Neuropathotyping: A New System to Classify Marek's Disease Virus. Avian Diseases. 2002, 46 (4): 909-918. 10.1637/0005-2086(2002)046[0909:NANSTC]2.0.CO;2. [http://www.jstor.org/stable/1593315]

18. Gimeno IM, Witter RL, Hunt HD, Reddy SM, Lee LF, Silva RF: The pp38 gene of Marek's disease virus (MDV) is necessary for cytolytic infection of B cells and maintenance of the transformed state but not for cytolytic infection of the feather follicle epithelium and horizontal spread of MDV. Journal of Virology. 2005, 79 (7): 4545-4549. 10.1128/JVI.79.7.4545-4549.2005.

19. Calnek BW, Witter RL: Marek's disease - a model for herpesvirus oncology. CRC Critical Reviews of Microbiology. 1986, 12: 293-319.

20. Yunis R, Jarosinski KW, Schat KA: Association between rate of viral genome replication and virulence of Marek's disease herpesvirus strains. Virology. 2004, 328: 142-150. 10.1016/j.virol.2004.07.017.

21. Baigent SJ, Petherbridge LJ, Howes K, Smith LP, Currie RJW, Nair VK: Absolute quantification of Marek's disease virus genome copy number in chicken feather and lymphocyte samples using real-time PCR. Journal of Virological Methods. 2005, 123: 53-64. 10.1016/j.jviromet.2004.08.019.

22. Bumstead N, Sillibourne J, Rennie M, Ross N, Davison F: Quantification of Marek's disease virus in chicken lymphocytes using the polymerase chain reaction with fluorescence detection. Journal of Virological Methods. 1997, 65: 75-81. 10.1016/S0166-0934(96)02172-6.

23. Burgess SC, Davison TF: A quantitative duplex PCR technique for measuring amounts of cell-associated Marek's disease virus: difference s in two populations of lymphomas cells. Journal of Virological Methods. 1999, 82: 27-37. 10.1016/S0166-0934(99)00075-0.

24. Islam A, Harrison B, Cheetham BF, Mahony TJ, Young PL, Walkden-Brown SW: Differential amplification and quantification of Marek's disease viruses using real-time polymerase chain reaction. Journal of Virological Methods. 2004, 119: 103-113. 10.1016/j.jviromet.2004.03.006.

25. Islam A, Cheetham BF, Mahony TJ, Young PL, Walkden-Brown SW: Absolute quantification of Marek's disease virus and herpesvirus of turkeys in chicken lymohocyte, feather tip and dust samples using real-time PCR. Journal of Virological Methods. 2006, 132: 127-134. 10.1016/j.jviromet.2005.10.009.

26. Reddy SM, Witter RL, Gimeno I: Development of quantitative-competitive polymerase chain reaction assay for serotype 1 Marek's disease virus. Avian Diseases. 2000, 44: 770-775. 10.2307/1593048.

27. Islam AFMF, Walkden-Brown SW, Groves PJ, Underwood GJ: Effects of vaccine dose, virus challenge dose and interval from vaccination to challenge on protection of broiler chickens against Marek's disease virus challenge. Australian Veterinary Journal. 2007, 85 (9): 348-355. 10.1111/j.1751-0813.2007.00195.x.

28. Islam AFMF, Walkden-Brown SW, Groves PJ, Underwood GJ: Kinetics of Marek's disease virus (MDV) infection in broiler chickens 1: effect of varying vaccination to challenge interval on vaccinal protection and load of MDV and herpesvirus of turkey in the spleen and feather dander over time. Avian Pathology. 2008, 37 (3): 225-235. 10.1080/03079450701802230.

29. Witter R, Solomon JJ, Nazerian Champion: Long-term studies of Marek's disease infection in individual chickens. Avian Diseases. 1971, 15 (2): 346-65. 10.2307/1588707.

30. Renz KG: In vitro and in vivo characterisation of selected Australian isolates of Marek's disease virus. 2008, Ph.D Thesis, University of New England, Australia, [https://e-publications.une.edu.au/vital/access/services/Download/une:1909/SOURCE04?view=true]

31. Baigent SJ, Davison F: Marek's disease virus: Biology and life cycle. Marek's Disease: an Evolving Problem. Elsevier Academic Press 2004.

32. Baigent SJ, Smith LP, Currie RJW, Nair VK: Replication kinetics of Marek's disease vaccine virus in feathers and lymphoid tissus using PCR and virus isolation. Journal of General Virology. 2005, 86: 2989-2998. 10.1099/vir.0.81299-0.

33. Fenner F, Fantini B: Biological Control of Vertebrate Pests: The History of Myxomatosis - an Experiment in Evolution. CABI Publishing 1999.

34. Walkden-Brown S, Cooke J, Islam A, Renz K, Hussain Z, Islam A, Tannock G, Groves P: Pathotyping of Australian isolates of MDV. Proceedings of the 8th International Marekís Disease Symposium: 6-10th July, Townsville, Australia. 2008

35. Islam AFMF, Walkden-Brown SW, Wong CW, Groves PJ, Burgess SC, Arzey KE, Young PL: Influence of vaccine deposition site on post-vaccinal viraemia and vaccine efficacy in broiler chickens following in ovo vaccination against Marek›s disease. Avian Pathology. 2001, 30: 525-533. 10.1080/03079450120078725.

36. Lunn D, Thomas A, Best N, Spiegelhalter D: WinBUGS Bayesian modelling framework: concepts, structure, and extensibility. Statistics and Computing. 2000, 10: 325-337. 10.1023/A:1008929526011.

37. Raftery A, Lewis S: How many iterations in the Gibbs sampler?. Bayesian Statistics. Oxford University Press, 4: 1992.

38. Carlin B, Lewis T: Bayesian Methods for Data Analysis. Chapman & Hall/CRC 2008.

CHAPTER 7

MAREK'S DISEASE VIRUS INFECTION INDUCES WIDESPREAD DIFFERENTIAL CHROMATIN MARKS IN INBRED CHICKEN LINES

Apratim Mitra, Juan Luo, Huanming Zhang, Kairong Cui, Keji Zhao and Jiuzhou Song

Keywords: Histone modifications, Thymus, Differential marks, Bivalent domain, Chromatin signature, Marek's disease

BACKGROUND

Rapid advances in epigenetics have led to the discovery of complex mechanisms of gene regulation involving phenomena such as DNA methylation and chromatin modifications. Methylation of particular histone residues has been found to correlate with specific and often opposing cellular functions, e.g. trimethylation of histone H3 lysine 4 (H3K4me3) is associated with transcriptional start sites (TSSs) of active genes while trimethylation of histone H3 lysine 27 (H3K27me3) is found to mark broad genomic regions for repression. Recent studies have also suggested that characteristic combinations of histone modifications or 'chromatin states' define functional elements of the genome and determine their contribution to transcriptional regulation[1–3]. Moreover, the epigenetic state of host genes have been observed to be affected by viral infection leading to tumors in humans[4–6]. Thus, epigenetics constitute a dynamic regulatory framework linking genotypes with environmental factors that could play a major role in differential disease responses among individuals having high genetic similarity.

Marek's disease (MD) is a highly contagious disease caused by an oncogenic α-herpesvirus MD virus (MDV) and characterized by T-cell lymphomas in chickens[7]. Major losses to the poultry industry as a result of MD have largely been averted due to the success of various vaccination strategies which, remarkably, is also the first instance of the successful control of a natural cancer-causing agent using vaccines[8–10]. However, the virulence of the virus may have progressively increased as a consequence of vaccine development[11–13]. Several reported instances of vaccine breaks or reduced efficacy of vaccination, therefore, underlines the importance of investigating resistance to the disease as a long-term strategy to control MDV.

Natural resistance to MDV can be divided into two categories: major histocompatibility complex (MHC)-associated resistance, wherein different MHC haplotypes at the B blood group locus confer varying levels of resistance and non-MHC associated resistance in which birds having the same MHC haplotype exhibit vastly different responses to MDV infection. Inbred lines 6_3 and 7_2developed at the Avian Disease and Oncology Laboratory (ADOL, East Lansing, MI) that we used in this study, fall into the latter category. These lines share a high degree of genetic similarity but have divergent responses to MDV infection completely independent of the MHC. Several studies have attempted to pinpoint factors responsible for conferring resistance[14–16], but confounding factors, such as, tissue types, virus strains and ages of birds have made it difficult to find a consensus. Multiple risk elements are possibly at play in this complex disease, and increased resistance or susceptibility is likely to be produced by a combination of such factors. In this study, we take a closer look at epigenetic factors behind different responses to MD

Citation: Apratim Mitra, Juan Luo, Huanming Zhang, Kairong Cui, Keji Zhao, Jiuzhou Song, Marek's disease virus infection induces widespread differential chromatin marks in inbred chicken lines, DOI: 10.1186/1471-2164-13-557.

with a view to a deeper understanding of the broader genomic impact of MDV infection.

We utilized the above population of inbred chickens – line 6_3 is highly resistant to MD, while line 7_2 is highly susceptible – and compared the epigenetic effects of MD. Genome-wide maps of H3K4me3 and H3K27me3 in thymus tissues of birds from these chicken lines at the latent stage of MDV infection were generated. We carried out systematic analyses to find differential chromatin marks induced by MDV infection. We also investigated co-localization patterns of the two chromatin modifications to detect putative bivalent domains and the effect of MDV on such domains. The results of our study confirm that Marek's disease has far-reaching effects on the epigenetic landscape of chicken lines with diverse responses to the virus and, thus, furthers our understanding of this complex disease.

RESULTS

Genome-Wide Distribution of H3k4me3 and H3k27me3

We performed ChIP-Seq experiments on infected and uninfected birds from lines 6_3 and 7_2 to investigate the epigenetic effects of MDV infection. More than 76 million reads from eight samples were mapped to the chicken genome yielding 14418 and 24950 significantly enriched regions (SERs) for H3K4me3 and H3K27me3, respectively (Table1). We further classified these regions as follows: Ubiquitous SERs were found in all samples and were likely due to similarities in the genetic background of the chickens. Line-specific SERs were present in only one line either before or after MDV infection, while condition-specific SERs appeared in both lines but only in individuals with the same infection status.

Table 1: Significantly enriched regions (SERs) and associated genes in each sample

		H3K4me3		H3K27me3	
	Samples	**SERs (%)**	**Genes**	**SERs (%)**	**Genes**
Line-Specific	**63I**	647 (4.5)	78	3477 (13.9)	615
	63N	594 (4.1)	71	2514 (10.1)	896
	63I,63N	924 (6.4)	190	577 (2.3)	150
	72I	105 (0.7)	16	1658 (6.6)	451
	72N	126 (0.9)	11	2506 (10)	346
	72I,72N	73 (0.5)	17	330 (1.3)	89
Condition-specific	**63I,72I**	97 (0.7)	35	2061 (8.3)	579
	63N,72N	47 (0.3)	9	66 (0.3)	22
Ubiquitous	**63I,63N,72I,72N**	10691 (74.2)	9475	5831 (23.4)	2942
	Total	**14418**	**10206**	**24950**	**7904**

63I: line 6_3 infected, 63N: line 6_3 control, 72I: line 7_2 infected, 72N: line 7_2 control.

Ubiquitous SERs formed the largest percentage of all enriched regions, accounting for 74.2% and 23.3% in H3K4me3 and H3K27me3 samples, respectively. In the case of H3K4me3, there were large differences in the number of specific SERs - more than 2000 line-specific SERs were found in line 6_3, compared to about 300 in line 7_2. Similarly, we found 50% more line-specific SERs of H3K27me3 in line 6_3 (6568) compared to line 7_2 (4494). However, upon closer examination, most of the line-specific and condition-specific SERs were revealed to have low read counts (Additional file1: Figure S1) corresponding to regions of low enrichment.

Genes were divided into five regions – promoter, 5' untranslated region (UTR), exons, introns and 3' UTR – and the distribution of SERs across these elements was probed (Figure1A). We found a large number of intergenic regions marked by H3K27me3, consistent with high levels of this mark associated with areas of silent heterochromatin. In the case of H3K4me3, a larger proportion of SERs were found around the promoter, exons and the 5' UTR while similar proportions of H3K4me3 and H3K27me3 SERs were present in introns and 3' UTRs. A comparison of the genomic distributions of SERs in the different samples (Additional file1: Figures S2A, B) showed a similar number of H3K4me3 SERs across the promoter, exons and the 5' and 3' UTRs of genes. Line 6_3 contained a higher number of intronic and intergenic SERs as compared to line 7_2 although this did not appear to change as a result of MDV infection. On the other hand, a greater number of H3K27me3 SERs were found in the infected samples although these levels were similar in the two different lines.

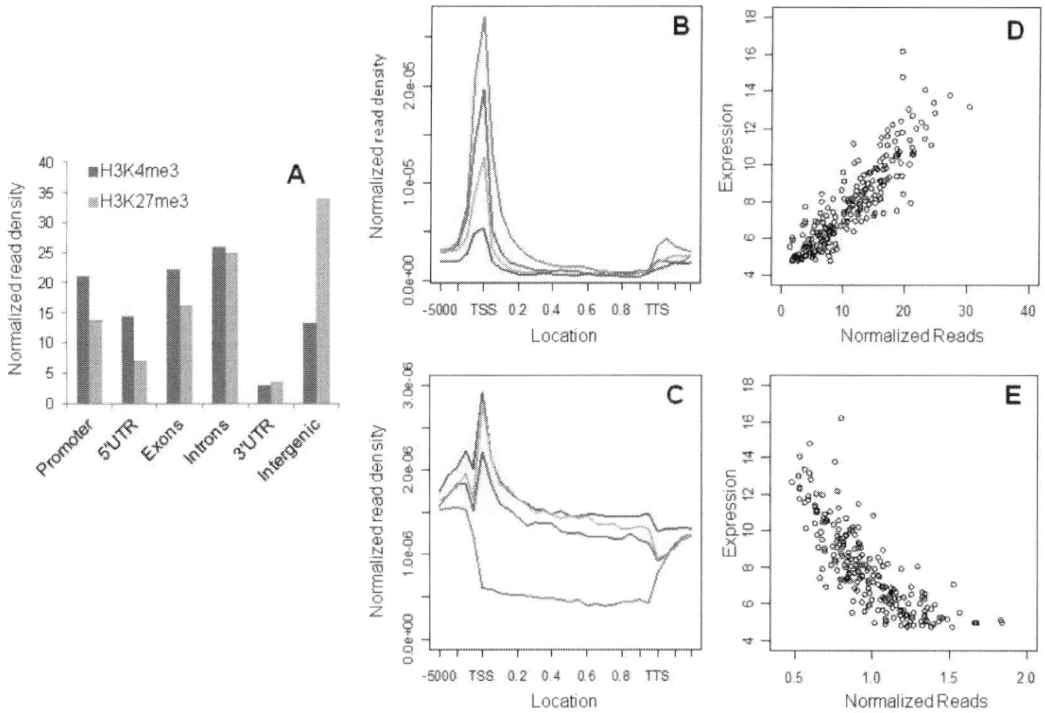

Figure 1: Genomic distribution of SERs and relationship between histone marks and gene expression. (A) Distribution of SERs over different genomic elements shows increased levels of H3K4me3 around the promoter region and exons while there are increased levels of H3K27me3 on intergenic regions. (B-E) Relationship between gene expression and histone marks in infected line 6_3 birds. Plots of histone modifications around the gene body (B, C) in genes having high (blue), medium (red), low (green) and no activity (brown): H3K4me3 shows positive correlation with gene expression levels while H3K27me3 exhibits a negative relationship. A comparison of epigenetic marks and transcriptional levels (D, E) confirms the same. Similar trends were observed in other experimental groups (Additional file1: Figures S3-5).

To analyze the relationship between histone modifications and gene expression, histone modification profiles surrounding the TSS and gene body were plotted for genes ranked by their expression level (Additional file1: Figures1B-E and S3-5). As expected, a strong positive correlation was observed between gene expression and H3K4me3 marks with a distinct peak around the TSS and little to no enrichment within the gene body. On the other hand, H3K27me3 showed negative correlation with gene expression with a peak near the TSS followed by a broad plateau across the gene body. However, the latter relationship was non-linear – genes with lower expression had similar levels of H3K27me3 marks and levels were markedly distinct only at higher expression levels (Figure1C, E).

Table 2: Differential SERs identified in thymus

	H3K4me3		H3K27me3	
Comparison	**Differential SERs***	**Genes**	**Differential SERs***	**Genes**
63I vs 63N	9	4	42	1
72I vs 72N	30	13	5	0
63N vs 72N	148	46	1094	65
Total	**179**	**59†**	**1116**	**66**

*FDR < 0.4. † Some genes are shared between different comparisons. 63I: line 6_3 infected, 63N: line 6_3 control, 72I: line 7_2 infected, 72N: line 7_2 control.

Differential H3K4me3 marks on genes related to MD

We conducted a comprehensive analysis of genome-wide chromatin marks to find significant differences in MDV-induced responses in line 6_3 and 7_2. We used two sets of comparisons: First, to assess the influence of MDV infection within each line, we compared the infected and the non-infected samples from the same line. Secondly, the non-infected samples from the two lines were compared to detect line-specific effects. As a result of this analysis we found 179 differential H3K4me3 SERs and 1116 differential H3K27me3 SERs that mapped to 59 and 66 genes, respectively (Table2). A majority of differential SERs were found in the comparison between non-infected samples of the two lines (Additional files2 and3) with several observed on genes that have been associated with MDV infection.

MX1 is a zinc-finger transcription factor that has antiviral properties against a large number of viruses. *MX1* was upregulated after MDV infection[17] although its contribution to MD progression is unknown. MDV infection induced a highly significant increase in H3K4me3 enrichment in the promoter region of *MX1* in both lines (line 6_3: $p = 1.28 \times 10^{-7}$, line 7_2: $p = 4.26 \times 10^{-9}$; Figure2A). We observed a concurrent increase in transcript levels after MDV infection in line 7_2 ($p = 0.0264$; Figure2B); *MX1* expression in line 6_3 showed a similar trend (fold change = 38.22, p=0.085) although mRNA levels were much lower.

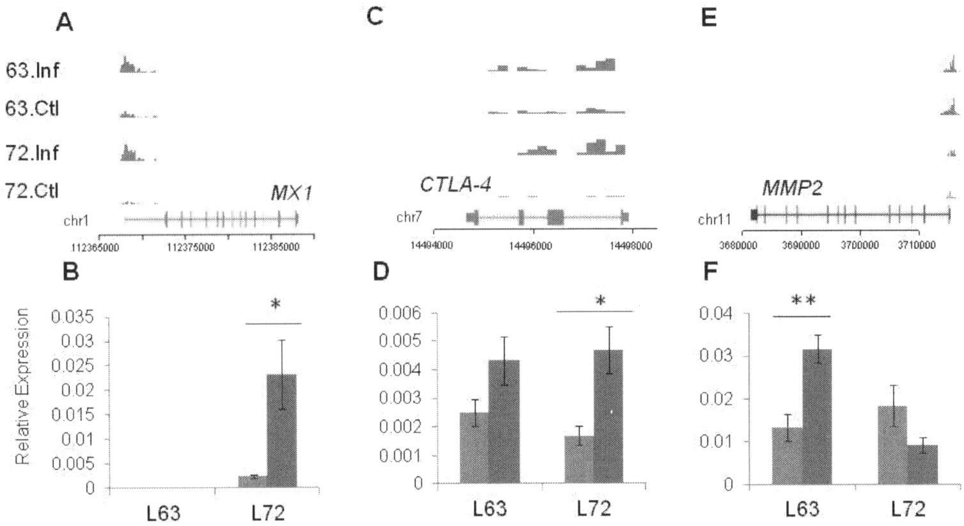

Figure 2: Genes related to MD show differential H3K4me3 marks. *MX1* (A, B) and *CTLA-4*(C, D) show increased H3K4me3 marks and higher expression in infected individuals from both lines; *MMP2* (**E**, **F**) exhibits higher levels of H3K4me3 in susceptible line 7_2. n = 4; * = significant at $p < 0.05$; ** = significant at $p < 0.01$; *** = significant at $p < 0.001$.

CTLA-4 is a cell surface glycoprotein expressed on CD4+ and CD8+ T lymphocytes that is a powerful negative regulator of T-cell activation[18]. The CTLA4 protein is expressed on T lymphocytes soon after activation and regulates T-cell proliferation, production of IL-2 and also supports the function of T_{reg} cells that suppress immune response[19]. Previous studies have reported an increase in *CTLA-4* expression after MDV infection[20]. We found an increase in H3K4me3 enrichment in line 7_2 as a result of MDV infection ($p = 0.0003$) and there was a similar trend in line 6_3 (Figure2C). Consistent with the above, there was a significant increase in transcript levels after MDV infection in line 7_2 ($p = 0.04$) and a small increase in line 6_3(Figure2D).

MMP2 plays a key role in the degradation of the extra-cellular matrix, and an increase in expression has been associated with increasing tumor cell migration and tumor angiogenesis[21, 22]. *MMP2* was upregulated during the neoplastic stage of MD infection in susceptible birds[23] but downregulated in response to MDV infection during early lytic infection in susceptible and resistant chickens[17]. We observed a slight increase in H3K4me3 enrichment after MDV infection in both lines, while line 7_2 exhibited significantly lower levels than line 6_3 ($p = 0.0016$; Figure2E). This was coupled with increased *MMP2* expression in line 6_3 after infection ($p = 0.0068$) while there was no such change in line 7_2 (Figure2F).

Genes Related To Cancers Show Epigenetic Changes in Response to MD

We observed differential histone marks on several genes that have been associated with other cancers but not in the context of MDV infection. Insulin-like growth factor 2 binding protein 1 (*IGF2BP1*) is an RNA-binding factor that regulates the translation of mRNAs produced by certain genes like *IGF2* and *ACTB*. Increased expression of *IGF2BP1* has been implicated in the development and progression of cancers of various organs, e.g. lung, brain, breast and colon[24–27]. There was no change in the H3K4me3 enrichment levels induced by MDV infection although a significantly higher level of enrichment was present in line 7_2 ($p = 4.21 \times 10^{-13}$; Figure3A). Transcript levels in line 7_2 were much higher than in line 6_3, but reduced in response to MDV infection ($p = 0.044$) (Figure3B).

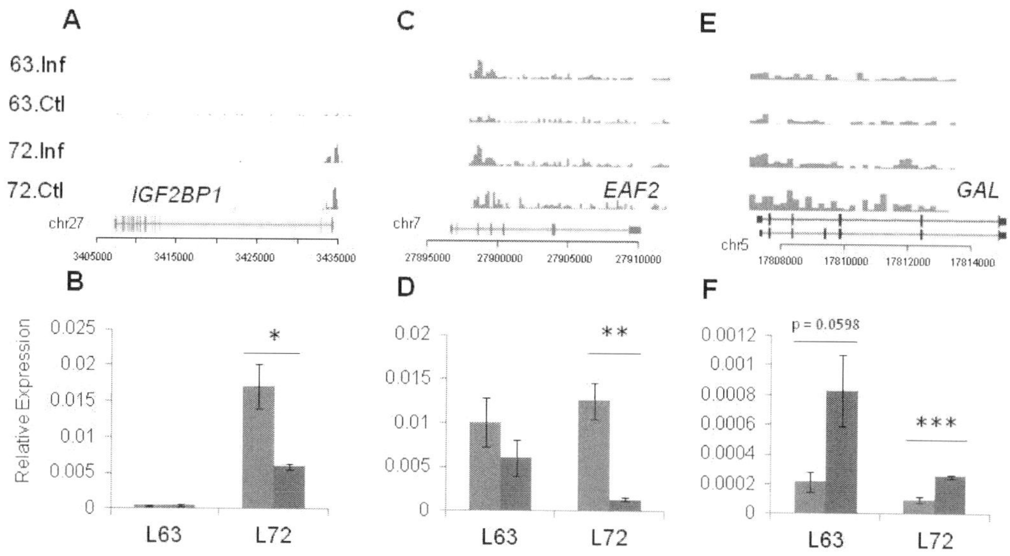

Figure 3: MD induces epigenetic changes in genes related to various cancers.*IGF2BP1*(A, B) shows differential H3K4me3 marks and increased expression in susceptible birds while *EAF2* (**C**, **D**) and *GAL* (**E**, **F**) have differential H3K27me3 levels on the gene body. n = 4; * = significant at $p < 0.05$; ** = significant at $p < 0.01$; *** = significant at $p < 0.001$.

ELL associated factor 2 (*EAF2*) is a testosterone regulated apoptosis inducer and tumor suppressor. Inactivation of *EAF2* has been shown to lead to tumors in multiple organs[28]. There was a significant increase in H3K27me3 levels after MDV infection in line 6_3 (p = 0.0414) while among uninfected chickens these levels were markedly higher in line 7_2 (p = 0.0138; Figure3C). However, *EAF2* expression was drastically reduced after MDV infection in line 7_2(p=0.0016) but showed only a small decrease in line 6_3 (Figure3D).

Galanin (*GAL*) is a neuropeptide that modulates various physiological functions, such as, inhibition of insulin secretion and stimulation of growth hormone secretion. Three galanin receptors are known (*GALR1*, *2* and *3*): the expression of *GALR1* has anti-proliferative effects while *GALR2* can be both anti- or pro-proliferative in function. Therefore, the *GAL* system is considered to be a promising candidate for detection and treatment of various cancers[29, 30]. We observed an increase in H3K27me3 levels on *GAL* after infection in both lines (Figure3E). Also, expression levels were significantly lowered after MDV infection in line 7_2 (p = 0.00087) while there was a similar trend in line 6_3 (p = 0.051; Figure3F). Interestingly, *GALR1* had both H3K4me3 and H3K27me3 enrichments (Figure4) although *GALR2* showed no significant histone marks (Additional file1: Figure S6).

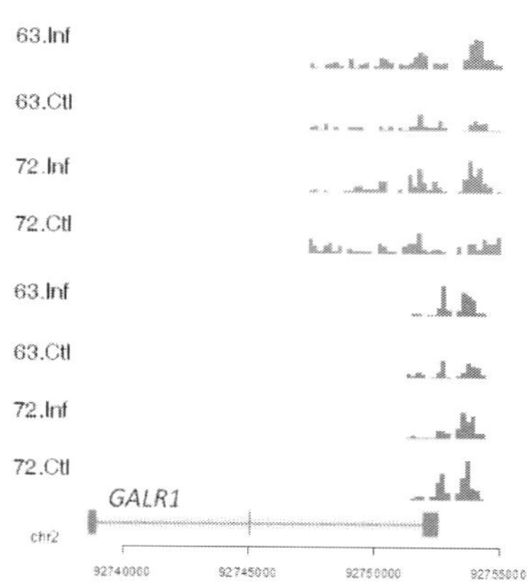

Figure 4: Significant H3K4me3 and H3K27me3 enrichment around*GALR1*. The anti-proliferative *GAL* receptor *GALR1* exhibited both active and repressive marks. There is no change in H3K4me3 levels but a definite increase in H3K27me3 levels after infection in line 7_2.

Chromatin Co-Localization Patterns Reveal Putative Bivalent Genes

Regions of chromatin having both active and repressive marks are said to be bivalent and have been shown to play important roles in development and genetic imprinting[31, 32]. For example, bivalent domains have been shown to mark promoters of genes that are subsequently silenced in tumors by DNA hypermethylation indicating their importance in cancer[33]. A mono-allelic bivalent chromatin domain that controls tissue-specific genomic imprinting at a specific locus was recently found in mice[32]. To investigate the presence of such bivalent chromatin states and the possible effect of MDV infection, we defined bivalent genes as those having H3K4me3 reads (TSS ± 500 bp) greater than 30 reads per million mapped reads (RPM) and H3K27me3 reads (gene body) greater than 2 RPM, respectively (~85th percentile). This filtering process yielded a list of 99 putative bivalent genes (Additional file4).

Functional annotation clustering of the above genes using DAVID[34, 35] revealed significant enrichment of several immune-related functions. These included transcription factor *EGR1*which is reported to have tumor suppressor properties, genes involved in lymphocyte activation and differentiation such as *BCL6*, *CD4* and *SMAD3* and genes *TLR3* and *TIRAP* that are part of the toll-like receptor signaling pathway. Bivalent domains were also present on a variety of transcription factors with immune-related functions such as *CITED2*, a transactivator that regulates NF-κB, *MYC* a transcription factor associated with hematopoetic tumors and *RHOB* a Ras family homolog that mediates apoptosis in tumor cells after DNA damage. Moreover, all genes involved in the top five functional annotation clusters showed higher chromatin levels in line 7_2 primarily after MDV infection (Additional file5).

Bivalent Domains are Altered in Response to MD

We further investigated the effect of MD on bivalent chromatin domains observed on *BCL6*,*CITED2*, *EGR1*, *CD4* and *TLR3* (Additional file1: Figure5 and S7). Interestingly, three of these genes, *CITED2, BCL6* and *EGR1*, showed identical epigenetic and transcriptional signatures.

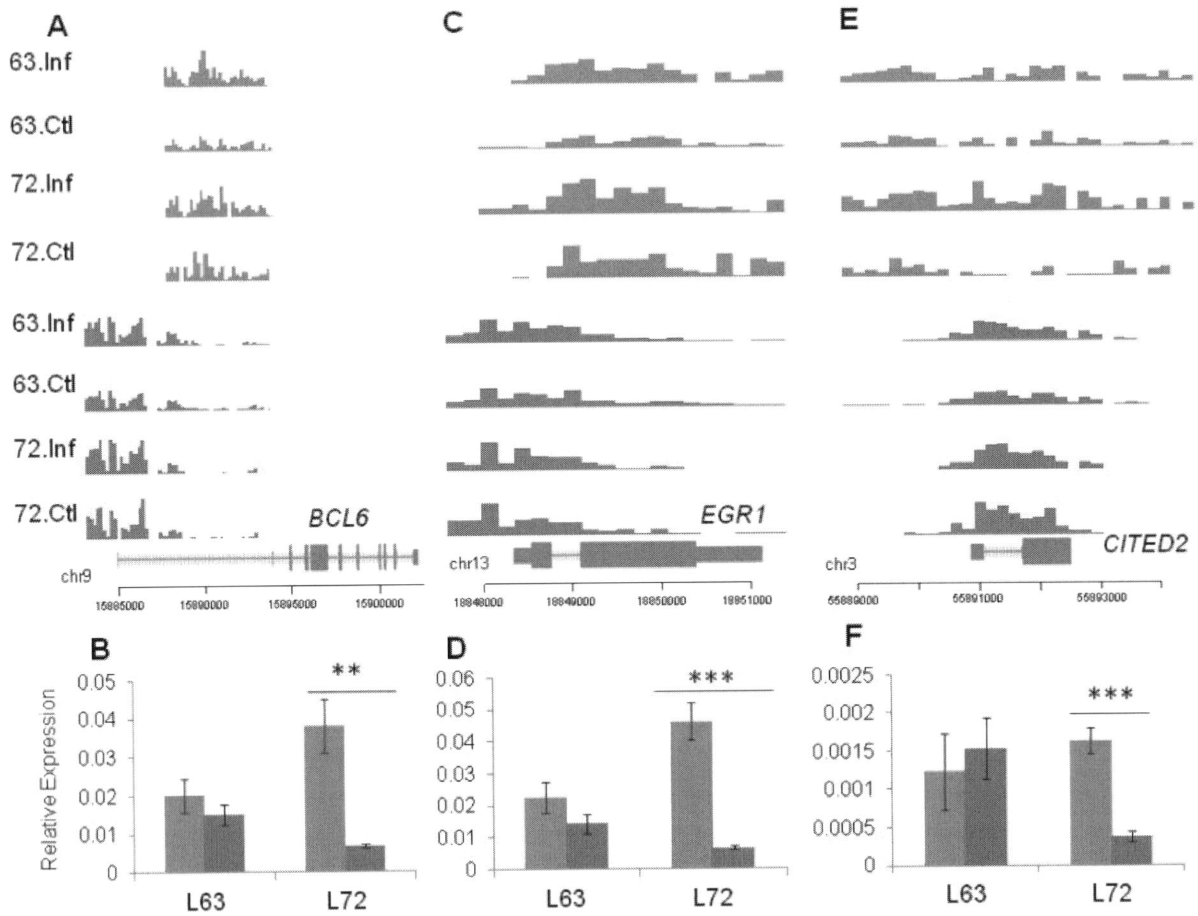

Figure 5: Bivalent domains on transcriptional regulators are altered by MD. H3K4me3 (orange) and H3K27me3 (green) profiles and associated transcript levels of *BCL6* (A, B),*EGR1* (C, D) and *CITED2* (E, F). In all three cases we observe a slight increase in H3K27me3 induced by MDV infection in line 7_2 and a concurrent significant decrease in transcript levels while increase in active and repressive marks appear to cancel each other out in line 6_3.

CITED2 is a member of the p300/CBP co-activator family that has intrinsic histone acetyltransferase activity and plays a major role in regulating and coordinating multiple complex cellular signals to determine the expression level of a gene[36]. B-cell CLL/lymphoma 6 (*BCL6*) is a zinc finger protein that functions as a transcriptional repressor which was downregulated at 15 dpi in spleen tissues from F1 progeny ($15I_5$ X 7_1) of MD-susceptible chickens[20]. *EGR1* belongs to a group of early response genes induced by a variety of signaling molecules such as growth factors, hormones and neurotransmitters that is believed to play a major role in cell proliferation and apoptosis[37]. Overexpression of *EGR1* promotes tumor growth in kidney cells[38] but suppresses growth and transformation in other cell types, e.g. fibroblasts and glioblastoma cells[39].

In each of the above genes, both active and repressive chromatin marks were increased in response to infection in line 6_3 chickens. However, in line 7_2, there was a definite increase in H3K27me3 marks but no change in H3K4me3 (Figures5A, C, E). Transcript levels were in agreement with this observation: infected line 7_2 chickens showed a significant downregulation (*CITED2*: p=0.0004; *BCL6*: p=0.0048; *EGR1*: p=0.0005; Figures5B, D, F), while there were no such changes in line 6_3.

On the other hand, *TLR3* and *CD4* showed a slight increase in H3K4me3 marks after MDV infection while there were no appreciable changes in H3K27me3 levels. In keeping with the epigenetic changes, there was a small increase in gene expression in infected birds from both lines (Additional file1: Figure S7).

DISCUSSION

Immune parameters that are known to play a major role in genetic resistance to MDV are correlated with innate immune responses, such as NK cell activity, production of nitric oxide and cytokines, such as, interferons. Recent studies have identified host cytokines such as IL-18 and IFN-γ that contribute to the initiation and continuation of latency[40]. However, cytokine levels can undergo rapid flux in response to infection, and consistent with this, we did not observe any epigenetic changes associated with these genes in the MHC-congenic lines used in our study (Additional file1: Figure S8). This suggests the existence of other extrinsic factors responsible for transcriptional variations between resistant and susceptible chickens at this stage of the disease.

A global comparison of histone modification levels in the two inbred chicken lines yielded some interesting results. As expected, a majority of SERs were found in all the experimental groups, indicating a high level of epigenetic similarity between the lines in addition to inherent genetic similarity. In the case of H3K27me3, the percentage of ubiquitous SERs was relatively low (23.4%), although this was likely due to lower precision of the peak caller for broad chromatin marks. Besides, most of the SERs detected in a subset of samples corresponded to regions of low enrichment, which may also be the reason behind the relatively low number of differential SERs detected in our study.

Genes that have been previously associated with MD and various other cancers showed differential marks that are either MD-induced (*MX1, CTLA-4, EAF2* and *GAL*) or line-specific (*IGF2BP1* and *MMP2*). The increase in H3K4me3 marks around the TSS of *MX1,* a gene with known antiviral properties, appeared to be correlated with upregulated expression in both lines in response to MDV infection. However, lowered overall mRNA levels in the resistant line suggest additional factors could be involved in the regulation of this gene. Similarly, increased mRNA levels of the lymphocyte surface marker *CTLA4* is possibly due to the presence of larger numbers of T cells in MDV infected birds as a result of higher levels of infection. *EAF2* functions as an apoptosis inducer in addition to being a tumor suppressor, and therefore, its downregulation could contribute to higher tumor incidence rates in line 7_2. However, it is not clear why a significant increase in H3K27me3 levels did not have any effect on transcript levels in the resistant line.

IGF2BP1 is believed to act by stabilizing the mRNA of the *c-myc* oncogene and therefore, the higher expression of this gene and a more stable c-myc gene product might play a role in increasing MD susceptibility in line 7_2 birds via increased cell proliferation and transformation. The matrix metalloprotease *MMP2* is upregulated after infection in the resistant line 6_3, similar to the previously observed increase at the neoplastic stage of MD. However, mRNA levels were similar in the two lines before infection contrary to the difference in H3K4me3 levels suggesting that additional factors are responsible for regulating this gene.

The correlation between observed differential histone marks and transcript levels was moderate at best. Indeed, differential H3K4me3 marks were strongly predictive of gene expression levels but the correlation between H3K27me3 and mRNA levels was relatively poor. There could be various reasons for this – indeed, H3K27me3 levels had a non-linear relationship with gene expression with higher marks showing little difference in the effect on expression. Therefore, in this tissue, the levels of H3K27me3 may not be a very good indicator of gene expression levels. Also, the transcription of these genes might be controlled by other factors with the change in H3K27me3 levels only incidental.

Bivalent domains were detected on transcriptional regulators *BCL6, CITED2* and *EGR1* and the galanin receptor *GALR1*. The epigenetic and transcriptional signatures observed on these genes indicated that they were poised at the latent stage of the disease, but with crucial differences in the two lines. Increased repressive marks in the susceptible line correlated with significant downregulation of the genes, while in line 6_3, the increase in both marks appeared to compensate for each other with no eventual effect on gene transcription. This suggests that some 'poised' bivalent genes can become highly repressed even with a relatively small increase in H3K27me3 marks. The change in the chromatin levels could also be correlated with an increase in cell populations having the repressive mark. Taken together, the above findings point towards the existence of finely balanced epigenetic control of transcription, which may be necessary to mount a rapid response by the immune system. However, this machinery could potentially be hijacked by a pathogen and result in an aberrant phenotype. The effect of MDV infection on the bivalent domain on *GALR1*, in particular, suggests the repression and potential loss of its anti-proliferative effects. Thus, the galanin system possibly plays an important and hitherto unknown role in MD progression and could be a novel target for long-term control of the disease.

One of the major players in MDV-induced malignant transformation is Meq, a virus-encoded oncoprotein that has diverse functions, e.g. transactivation, chromatin remodeling and regulation of transcription. Meq interacts with and sequesters the tumor suppressor protein p53, resulting in the dysregulation of cell-cycle control[7] and inhibition of the transcriptional and apoptotic activities of the protein[41]. Several genes that show epigenetic changes in response to MDV infection have been associated with p53. Downregulation of *CITED2* stabilizes the p53 protein leading to its accumulation[42]. The *BCL6* gene product is believed to contribute to lymphomagenesis by inactivation of p53[43]. Besides, *EAF2* has also been shown to interact with and suppress the function of p53[28]. The downregulation of all of the above genes in susceptible birds after MDV infection points towards the increased production of p53 and a robust anti-tumor response. That we still observe higher tumor incidence rates in this line, suggests the presence of large amounts of inactivating viral Meq protein which, in turn, indicates that increased numbers of MD-infected cells are present in the susceptible line at this stage of the disease. A majority of tumors are believed to result from the viral transformation of CD4+ T cells some of which are infected at the latent stage of MD[44]. The larger number of virus-infected cells produced in the susceptible line is possibly due to lowered immunocompetence as a result of the early stages of infection. Thus, a more detailed investigation of the early cytolytic stage of MD is necessary to shed further light on the causes behind the divergent response to MD observed in these birds.

Whole tissues represent a mixture of various cell populations, and observed epigenetic changes might be due to a change in chromatin marks in a particular cell type or a variation in the relative number of cells carrying active or repressive histone marks. However, such *in vivo*studies are representative of the host response at a systems level wherein different cell types might interact in a cooperative manner to fight infection. Thus, while the study of pure cell populations is likely to yield greater discriminative power, the investigation of tissue macroenvironments is, perhaps, closer to reality.

This study focused on the thymus tissue as it is a major immune organ and contains a significant population of T lymphocytes in various stages of differentiation. Our earlier study of the MDV-induced transcriptome in these birds indicated the presence of line-specific differences at the latent stage of MD[45]. In addition, birds susceptible to MD suffer thymic atrophy during the early stages of infection[46], indicating the importance of understanding changes in this organ to elucidate the mechanisms involved in disease progression. Ongoing studies in our lab include other tissues, e.g. spleen[47], and a time-course through the various stages of infection, to further investigate the systemic effects of MD and the epigenetic basis of MD resistance.

CONCLUSIONS

We studied the effect of latent MDV infection on two chromatin modifications in inbred chicken lines exhibiting different degrees of resistance to MD. Several genes showed changes in histone enrichment and this response was often significantly different between the two chicken lines. A detailed analysis of co-localization patterns of the chromatin marks revealed the presence of bivalent domains on a number of immune-related transcriptional regulators. More importantly, these domains showed marked changes in response to MDV infection and provide further evidence of the far-reaching epigenetic effects of MD. Our results suggest putative roles for the*GAL* system in MD progression. A majority of the differential chromatin marks are also suggestive of increased levels of viral infection in the susceptible line symptomatic of lowered immunocompetence in these birds at early stages of the disease. In summary, our study provides further insight into the mechanisms of MD progression while revealing a complex landscape of epigenetic regulatory mechanisms. Further work is necessary to fully elucidate the underlying mechanisms of MD, but our results suggest that this is a promising step towards a deeper understanding of this complex disease.

METHODS

Animals and Viruses

Two specific-pathogen-free inbred lines of White Leghorn either resistant (6_3) or susceptible (7_2) to MD were hatched, reared and maintained in Avian Disease and Oncology Laboratory (ADOL, Michigan, USDA). Four chickens from each line were injected intra-abdominally with a partially attenuated very virulent plus strain of MDV (648A passage 40) at 5 days after hatch with a viral dosage of 500 plaque-forming units (PFU). Infected and control chickens from both lines (n = 4) were terminated at 10dpi to collect thymus tissues. All procedures followed the standard animal ethics and use guidelines of ADOL.

Quantification of MDV Loads In Thymus

The MDV gene *ICP4* was used for quantification of viral genomic DNA in thymus as previously described[48]. Quantitative PCR was performed by using 100 ng/μl of genomic DNA on the iCycler iQ PCR system (Bio-Rad, USA) and QuantiTect SYBR Green PCR Kit (Qiagen, USA) (Additional file1: Figure S9). The relative MDV loads were determined by normalizing to a single-copy gene *Vim* (vimentin)[49]. The primers for *Vim* are as follows: Forward: 5'-CAGCCACAGAGTAGGGTAGTC-3'; Reverse: 5'-GAATAGGGAAGAACAGGAAAT-3'.

Chromatin Immunoprecipitation and Illumina Sequencing

Chromatin immunoprecipitation (ChIP) was carried out using thymus samples from MDV infected and controls birds[50]. About 30 mg thymus samples from three individuals were cut into small pieces (1 mm^3) and digested with MNase to obtain mononucleosomes. PNK and Klenow enzymes (NBE, Ipswich, MA, USA) were used to repair the ChIP DNA ends pulled down by the specific antibody. A 3′ adenine was then added using Taq polymerase and a pair of Solexa adaptors (Illumina, USA) ligated to the repaired ends. Seventeen cycles of PCR was performed on ChIP DNA using the adaptor primers and fragments with a length of about 190 bp (mononucleosome + adaptors) were isolated from agarose gel. Subsequently, cluster generation and ChIP-sequencing (ChIP-Seq) using the purified DNA was performed on the Solexa 1G Genome

Analyzer (Illumina, USA) following manufacturer protocols. The antibodies used and the total number of reads obtained for each sample is listed in Additional file6.

Read Mapping and Summary Counts

Sequence reads obtained from the Illumina 1G Genome Analyzer were aligned to the May 2006 version of the chicken genome (galGal3) using Maq version 0.7.1[51]. Default alignment policies of Maq were enforced: a valid alignment could have a maximum of two mismatches and if a read aligned equally well to multiple places in the genome, one was chosen at random. If multiple reads mapped to the same genomic location only one was kept to avoid amplification bias. Summary read counts were calculated using non-overlapping windows of 200 bp for visualization and normalized to per million mapped reads in each sample for the purpose of comparisons.

Identification of Enriched Regions

Summarized read counts were subjected to peak calling with SICER[52]. The source code was modified to include support for the chicken genome. Fragment length was specified to be 190. A window size of 200 bp and gap size of 400 bp was used for the analysis. The E-value for estimating significant peaks was set to 100. For the purposes of comparing different samples, SERs found in similar genomic regions of different samples were merged to obtain a consolidated list as follows: SERs from one sample were used to initialize the list. For each such region *M,* we searched for overlapping SERs in the next sample. In the case of an overlap between *M* and a significant region, *S*, the merged region was updated to include both *M* and *S*. This procedure was iterated over all samples to obtain a consolidated list of merged SERs.

Gene Annotation and Genomic Distribution of SERS

RefSeq and Ensembl gene annotations were downloaded from UCSC genome browser[53]. As there were only 4306 RefSeq genes in the database, we included Ensembl genes in our analysis to improve genome-wide coverage. There were 17858 annotated genes in the Ensembl database, which include validated and predicted genes. Redundancies between the databases were listed and accounted for, yielding a non-redundant list of 18198 genes with 4306 RefSeq genes and 13892 Ensembl genes. We then searched for overlaps between merged SERs and the non-redundant list of annotated genes. For H3K4me3, an SER was annotated with a gene if it overlapped the TSS region of the gene whereas in the case of H3K27me3, a valid overlap constituted an SER overlapping the gene body. To calculate the genomic distribution we counted all SERs having an overlap with one of the following regions: promoter (TSS ± 1 kb), exons, introns, 5' UTR and 3' UTR.

Histone Modification Profiles and Differential Chromatin Marks

Genes were divided into 10 sets based on their absolute expression and representative sets corresponding to high, medium, low and no expression were chosen for visualization. We defined the gene body as the region between the transcription start site (TSS) and the transcription termination site (TTS). Histone modification profiles surrounding the gene body were calculated in 3 distinct regions: 5000 bp upstream of the 5' end, gene body and 5000 bp downstream of the 3' end of the gene. For reads falling within the gene body, read counts were obtained in bins 5% of the gene length while 1000 bp windows were used for the 5' and 3' flanking regions. The read counts in all cases were normalized to the total number of genes in the categories and total number of reads in the sample. We also compared gene expression to histone modification levels by plotting normalized microarray data (Zhang, H. unpublished data) against reads mapping to (i) TSS ± 500 bp and (ii) the gene body for each gene.

Reads mapping to merged SERs were tested for epigenetic changes induced by MDV infection in lines 6_3 and 7_2 using DESeq[54]. We used the method 'blind' for variance estimation and p-values were corrected for multiple testing using the Benjamini-Hochberg FDR procedure[55]. Statistical significance was defined using a false discovery rate (FDR) threshold of 0.4.

Validation of ChIP, ChIP-Seq and Gene Transcription by Q-PCR

Quantitative real-time RT-PCR was used to validate the quality of the ChIP and gene transcript levels on the iCycler iQ PCR system (Bio-Rad, Hercules, CA, USA). The real-time RT-PCR reactions were performed with a QuantiTect SYBR Green PCR Kit (Qiagen, Valencia, CA, USA) according to the manufacturer's instructions. An online primer system (http://frodo.wi.mit.edu/primer3/) was used to design the primers and four biological and four technical replicates were performed. The primer sequences are shown in Additional file7.

AUTHOR CONTRIBUTIONS

AM performed the data analysis and wrote the manuscript. JL performed the ChIP experiments, sequence library preparation, validation of ChIP-Seq results and revised the manuscript. HMZ collected samples and revised the manuscript. JZS designed the experiments and revised the manuscript. All authors read and approved the final manuscript.

ACKNOWLEDGEMENTS

This project was supported by National Research Initiative Competitive Grant no. USDA-NRI/NIFA 2008-35204-04660 and USDA-NRI/NIFA 2010-65205-20588 from the USDA National Institute of Food and Agriculture.

REFERENCES

1. Ernst J, Kellis M: Discovery and characterization of chromatin states for systematic annotation of the human genome. Nat Biotechnol. 2010, 28 (8): 817-825. 10.1038/nbt.1662.
2. Ernst J, Kheradpour P, Mikkelsen TS, Shoresh N, Ward LD, Epstein CB, Zhang X, Wang L, Issner R, Coyne M, et al: Mapping and analysis of chromatin state dynamics in nine human cell types. Nature. 2011, 473 (7345): 43-49. 10.1038/nature09906.
3. Weishaupt H, Sigvardsson M, Attema JL: Epigenetic chromatin states uniquely define the developmental

plasticity of murine hematopoietic stem cells. Blood. 2010, 115 (2): 247-256. 10.1182/blood-2009-07-235176.

4. Flanagan JM: Host epigenetic modifications by oncogenic viruses. Br J Cancer. 2007, 96 (2): 183-188. 10.1038/sj.bjc.6603516.
5. Knight JS, Lan K, Subramanian C, Robertson ES: Epstein-Barr virus nuclear antigen 3C recruits histone deacetylase activity and associates with the corepressors mSin3A and NCoR in human B-cell lines. J Virol. 2003, 77 (7): 4261-4272. 10.1128/JVI.77.7.4261-4272.2003.
6. Tsai CN, Tsai CL, Tse KP, Chang HY, Chang YS: The Epstein-Barr virus oncogene product, latent membrane protein 1, induces the downregulation of E-cadherin gene expression via activation of DNA methyltransferases. Proc Natl Acad Sci U S A. 2002, 99 (15): 10084-10089. 10.1073/pnas.152059399.
7. Osterrieder N, Kamil JP, Schumacher D, Tischer BK, Trapp S: Marek's disease virus: from miasma to model. Nat Rev Microbiol. 2006, 4 (4): 283-294. 10.1038/nrmicro1382.
8. Epstein MA, Achong BG, Churchill AE, Biggs PM: Structure and development of the herpes-types virus of Marek's disease. J Natl Cancer Inst. 1968, 41 (3): 805-820.
9. Churchill AE, Payne LN, Chubb RC: Immunization against Marek's disease using a live attenuated virus. Nature. 1969, 221 (5182): 744-747. 10.1038/221744a0.
10. Okazaki W, Purchase HG, Burmester BR: Protection against Marek's disease by vaccination with a herpesvirus of turkeys. Avian Dis. 1970, 14 (2): 413-429. 10.2307/1588488.
11. Davison F, Nair V: Use of Marek's disease vaccines: could they be driving the virus to increasing virulence?. Expert Rev Vaccines. 2005, 4 (1): 77-88. 10.1586/14760584.4.1.77.
12. Gimeno IM: Marek's disease vaccines: a solution for today but a worry for tomorrow?. Vaccine. 2008, 26 (Suppl 3): C31-C41.
13. Witter RL: Increased virulence of Marek's disease virus field isolates. Avian Dis. 1997, 41 (1): 149-163. 10.2307/1592455.
14. Liu HC, Cheng HH, Tirunagaru V, Sofer L, Burnside J: A strategy to identify positional candidate genes conferring Marek's disease resistance by integrating DNA microarrays and genetic mapping. Anim Genet. 2001, 32 (6): 351-359. 10.1046/j.1365-2052.2001.00798.x.
15. Liu HC, Kung HJ, Fulton JE, Morgan RW, Cheng HH: Growth hormone interacts with the Marek's disease virus SORF2 protein and is associated with disease resistance in chicken. Proc Natl Acad Sci U S A. 2001, 98 (16): 9203-9208. 10.1073/pnas.161466898.
16. Wain HM, Toye AA, Hughes S, Bumstead N: Targeting of marker loci to chicken chromosome 16 by representational difference analysis. Anim Genet. 1998, 29 (6): 446-452. 10.1046/j.1365-2052.1998.296374.x.
17. Smith J, Sadeyen JR, Paton IR, Hocking PM, Salmon N, Fife M, Nair V, Burt DW, Kaiser P: Systems Analysis of Immune Responses in Marek's Disease Virus Infected Chickens Identifies a Gene Involved in Susceptibility and Highlights a Possible Novel Pathogenicity Mechanism. J Virol. 2011, 85 (21): 11146-11158. 10.1128/JVI.05499-11.
18. McCoy KD, Le Gros G: The role of CTLA-4 in the regulation of T cell immune responses. Immunol Cell Biol. 1999, 77 (1): 1-10. 10.1046/j.1440-1711.1999.00795.x.
19. Curran MA, Montalvo W, Yagita H, Allison JP: PD-1 and CTLA-4 combination blockade expands infiltrating T cells and reduces regulatory T and myeloid cells within B16 melanoma tumors. Proc Natl Acad Sci U S A. 2010, 107 (9): 4275-4280. 10.1073/pnas.0915174107.
20. Heidari M, Sarson AJ, Huebner M, Sharif S, Kireev D, Zhou H: Marek's disease virus-induced immunosuppression: array analysis of chicken immune response gene expression profiling. Viral Immunol. 2010, 23 (3): 309-319. 10.1089/vim.2009.0079.
21. Fingleton B: Matrix metalloproteinases: roles in cancer and metastasis. Front Biosci. 2006, 11: 479-491. 10.2741/1811.
22. Rath NC, Parcells MS, Xie H, Santin E: Characterization of a spontaneously transformed chicken mononuclear cell line. Vet Immunol Immunopathol. 2003, 96 (1–2): 93-104.
23. Chen C, Li H, Xie Q, Shang H, Ji J, Bai S, Cao Y, Ma Y, Bi Y: Transcriptional profiling of host gene expression in chicken liver tissues infected with oncogenic Marek's disease virus. J Gen Virol. 2011, 92: 2724-2733. 10.1099/vir.0.034066-0.
24. Ioannidis P, Kottaridi C, Dimitriadis E, Courtis N, Mahaira L, Talieri M, Giannopoulos A, Iliadis K, Papaioannou D, Nasioulas G, et al: Expression of the RNA-binding protein CRD-BP in brain and non-small cell lung tumors. Cancer Lett. 2004, 209 (2): 245-250. 10.1016/j.canlet.2003.12.015.
25. Ioannidis P, Mahaira L, Papadopoulou A, Teixeira MR, Heim S, Andersen JA, Evangelou E, Dafni U, Pandis N, Trangas T: 8q24 Copy number gains and expression of the c-myc mRNA stabilizing protein CRD-BP in primary breast carcinomas. Int J Cancer. 2003, 104 (1): 54-59. 10.1002/ijc.10794.
26. Kato T, Hayama S, Yamabuki T, Ishikawa N, Miyamoto M, Ito T, Tsuchiya E, Kondo S, Nakamura Y, Daigo Y: Increased expression of insulin-like growth factor-II messenger RNA-binding protein 1 is associated with tumor progression in patients with lung cancer. Clin Cancer Res. 2007, 13 (2 Pt 1): 434-442.
27. Ross J, Lemm I, Berberet B: Overexpression of an mRNA-binding protein in human colorectal cancer. Oncogene. 2001, 20 (45): 6544-6550. 10.1038/sj.onc.1204838.
28. Su F, Pascal LE, Xiao W, Wang Z: Tumor suppressor U19/EAF2 regulates thrombospondin-1 expression via p53. Oncogene. 2010, 29 (3): 421-431. 10.1038/onc.2009.326.
29. Berger A, Santic R, Hauser-Kronberger C, Schilling FH, Kogner P, Ratschek M, Gamper A, Jones N, Sperl W, Kofler B: Galanin and galanin receptors in human cancers. Neuropeptides. 2005, 39 (3): 353-359.

10.1016/j.npep.2004.12.016.

30. Rauch I, Kofler B: The galanin system in cancer. EXS. 2010, 102: 223-241.

31. Bernstein BE, Mikkelsen TS, Xie X, Kamal M, Huebert DJ, Cuff J, Fry B, Meissner A, Wernig M, Plath K, et al: A bivalent chromatin structure marks key developmental genes in embryonic stem cells. Cell. 2006, 125 (2): 315-326. 10.1016/j.cell.2006.02.041.

32. Sanz LA, Chamberlain S, Sabourin JC, Henckel A, Magnuson T, Hugnot JP, Feil R, Arnaud P: A mono-allelic bivalent chromatin domain controls tissue-specific imprinting at Grb10. EMBO J. 2008, 27 (19): 2523-2532. 10.1038/emboj.2008.142.

33. Rodriguez J, Munoz M, Vives L, Frangou CG, Groudine M, Peinado MA: Bivalent domains enforce transcriptional memory of DNA methylated genes in cancer cells. Proc Natl Acad Sci U S A. 2008, 105 (50): 19809-19814. 10.1073/pnas.0810133105.

34. da Huang W, Sherman BT, Lempicki RA: Systematic and integrative analysis of large gene lists using DAVID bioinformatics resources. Nat Protoc. 2009, 4 (1): 44-57.

35. da Huang W, Sherman BT, Lempicki RA: Bioinformatics enrichment tools: paths toward the comprehensive functional analysis of large gene lists. Nucleic Acids Res. 2009, 37 (1): 1-13. 10.1093/nar/gkn923.

36. Chan HM, La Thangue NB: p300/CBP proteins: HATs for transcriptional bridges and scaffolds. J Cell Sci. 2001, 114 (Pt 13): 2363-2373.

37. Thiel G, Cibelli G: Regulation of life and death by the zinc finger transcription factor Egr-1. J Cell Physiol. 2002, 193 (3): 287-292. 10.1002/jcp.10178.

38. Scharnhorst V, Menke AL, Attema J, Haneveld JK, Riteco N, van Steenbrugge GJ, van der Eb AJ, Jochemsen AG: EGR-1 enhances tumor growth and modulates the effect of the Wilms' tumor 1 gene products on tumorigenicity. Oncogene. 2000, 19 (6): 791-800. 10.1038/sj.onc.1203390.

39. Huang RP, Liu C, Fan Y, Mercola D, Adamson ED: Egr-1 negatively regulates human tumor cell growth via the DNA-binding domain. Cancer Res. 1995, 55 (21): 5054-5062.

40. Buscaglia C, Calnek BW: Maintenance of Marek's disease herpesvirus latency in vitro by a factor found in conditioned medium. J Gen Virol. 1988, 69 (Pt 11): 2809-2818.

41. Deng X, Li X, Shen Y, Qiu Y, Shi Z, Shao D, Jin Y, Chen H, Ding C, Li L, et al: The Meq oncoprotein of Marek's disease virus interacts with p53 and inhibits its transcriptional and apoptotic activities. Virol J. 2010, 7: 348-10.1186/1743-422X-7-348.

42. Wu Z-Z, Sun N-K, Chao CCK: Knockdown of CITED2 using short-hairpin RNA sensitizes cancer cells to cisplatin through stabilization of p53 and enhancement of p53-dependent apoptosis. J Cell Physiol. 2011, 226 (9): 2415-2428. 10.1002/jcp.22589.

43. Phan RT, Dalla-Favera R: The BCL6 proto-oncogene suppresses p53 expression in germinal-centre B cells. Nature. 2004, 432 (7017): 635-639. 10.1038/nature03147.

44. Schat KA, Chen CL, Calnek BW, Char D: Transformation of T-lymphocyte subsets by Marek's disease herpesvirus. J Virol. 1991, 65 (3): 1408-1413.

45. Yu Y, Luo J, Mitra A, Chang S, Tian F, Zhang H, Yuan P, Zhou H, Song J: Temporal Transcriptome Changes Induced by MDV in Marek's Disease-Resistant and -Susceptible Inbred Chickens. BMC Genomics. 2011, 12 (1): 501-10.1186/1471-2164-12-501.

46. Morimura T, Ohashi K, Kon Y, Hattori M, Sugimoto C, Onuma M: Apoptosis and CD8-down-regulation in the thymus of chickens infected with Marek's disease virus. Arch Virol. 1996, 141 (11): 2243-2249. 10.1007/BF01718230.

47. Luo J, Mitra A, Tian F, Chang S, Zhang H, Cui K, Yu Y, Zhao K, Song J: Histone Methylation Analysis and Pathway Predictions in Chickens after MDV Infection. PLoS One. 2012, 7 (7): e41849-10.1371/journal.pone.0041849.

48. Yunis R, Jarosinski KW, Schat KA: Association between rate of viral genome replication and virulence of Marek's disease herpesvirus strains. Virology. 2004, 328 (1): 142-150. 10.1016/j.virol.2004.07.017.

49. Zehner ZE, Paterson BM: Characterization of the chicken vimentin gene: single copy gene producing multiple mRNAs. Proc Natl Acad Sci U S A. 1983, 80 (4): 911-915. 10.1073/pnas.80.4.911.

50. Barski A, Cuddapah S, Cui K, Roh TY, Schones DE, Wang Z, Wei G, Chepelev I, Zhao K: High-resolution profiling of histone methylations in the human genome. Cell. 2007, 129 (4): 823-837. 10.1016/j.cell.2007.05.009.

51. Li X, Chiang HI, Zhu J, Dowd SE, Zhou H: Characterization of a newly developed chicken 44K Agilent microarray. BMC Genomics. 2008, 9: 60-10.1186/1471-2164-9-60.

52. Zang C, Schones DE, Zeng C, Cui K, Zhao K, Peng W: A clustering approach for identification of enriched domains from histone modification ChIP-Seq data. Bioinformatics. 2009, 25 (15): 1952-1958. 10.1093/bioinformatics/btp340.

53. Fujita PA, Rhead B, Zweig AS, Hinrichs AS, Karolchik D, Cline MS, Goldman M, Barber GP, Clawson H, Coelho A, et al: he UCSC Genome Browser database: update 2011. Nucleic Acids Res. 2010, 39 (Database issue): D876-D882.

54. Anders S, Huber W: Differential expression analysis for sequence count data. Genome Biol. 2010, 11 (10): R106-10.1186/gb-2010-11-10-r106.

55. Benjamini Y, Hochberg Y: Controlling the False Discovery Rate: A Practical and Powerful Approach to Multiple Testing. J Royal Stat Soc Series B (Methodological). 1995, 57 (1): 289-300.

CHAPTER 8

DIRECT DETECTION OF MAREK'S DISEASE VIRUS IN POULTRY DUST BY LOOP-MEDIATED ISOTHERMAL AMPLIFICATION

Grzegorz Woźniakowski, Elzbieta Samorek-Salamonowicz

Keywords: Marek's disease virus, Poultry dust, Direct detection, Loop-mediated isothermal amplification.

Marek's disease virus (MDV) is a serious concern for poultry production and represents a unique herpesvirus model. MDV can be shed by doubly infected chickens despite vaccination. The fully infectious MDV particles are produced in the feather follicle epithelium (FFE), and MDV remains infectious for many months in fine skin particles and feather debris. Molecular biology methods including PCR and real-time PCR have been shown to be valuable for the detection of MDV DNA in farm dust. Recently, loop-mediated isothermal amplification (LAMP) was found to be useful in the detection of MDV in feathers and internal organs of infected chickens. LAMP is also less affected by the inhibitors present in DNA samples. Taking into account the advantages of LAMP, direct detection of MDV DNA in poultry dust has been conducted in this research. The detection of MDV DNA was possible in 11 out of the 12 examined dust samples without DNA extraction. The DNA was retrieved from dust samples by dilution and incubation at 95 C for 5 min. The direct detection of MDV DNA in the dust was possible within 30 min using a water bath and UV light. The results were confirmed by electrophoresis and melting curve analysis of the LAMP products. Our results show that LAMP may be used to test for the presence of virulent MDV in poultry farm dust without DNA extraction.

Citation: Grzegorz Wozniakowski, Elzbieta Samorek-Salamonowicz, Direct detection of Marek's disease virus in poultry dust by loop-mediated isothermal amplification, DOI 10.1007/s00705-014-2157-5.

Marek's disease virus (MDV; species Gallid herpesvirus 2, genus Mardivirus, family Herpesviridae [5, 21]) is one of the most contagious agents in poultry production. Consequently, Marek's disease (MD) has a huge economic impact [15, 17, 22, 25]. Recently, increasing incidence of MD caused by virulent MDV strains with increased pathogenicity has been observed [2, 8, 22, 25, 27]. Initially, MDV that is present in poultry farm dust infects the host via the respiratory tract [5, 13, 21]. The virus is then transferred by the bloodstream in alveolar macrophages to B and T lymphocytes. After the primary infection stage, between 7 and 14 days postinfection (dpi), MDV may become latent in infected lymphoid cells [5], which proliferate in different parts of host, especially in the liver, spleen, kidney, proventriculus and ovaries. This leads to tumour formation after reactivation of the virus to the transformation stage [17, 21]. Most MDV transmission occurs in the fully productive stage of infection and takes place in the feather follicle epithelium (FFE) [5, 7, 12, 19]. Subsequently, the virus is then transferred to the environment as fine particles of skin and feather debris [3, 4, 7, 11]. The DNA of MDV can be detected in poultry dust as early as 7 dpi [3, 4, 11]. Infectious MDV can persist in dust particles for many months and therefore be a potential source of infection for the next flock of chickens. Vaccination against MD with live attenuated vaccines safeguards against its clinical form as well as against tumours [21, 25]. However, this does not exclude the possibility of superinfection with very virulent MDV (vvMDV) and shedding of the virus into the environment [5]. Poultry farm dust may be also a good source of MDV DNA in order to monitor any possible future outbreak of infection in the flock following decontamination of an affected farm [7, 11, 13, 19, 22]. Conventional detection methods of viral DNA in dust such as polymerase chain reaction (PCR) and real-time PCR require the extraction of nucleic acids [3, 4, 6, 9–12, 19, 22–24]. PCR-based techniques are dependent on laboratory equipment, including thermocyclers or complex real-time PCR systems. Recently, loop-mediated isothermal amplification (LAMP) has been described as a powerful and rapid tool for MDV detection in infected chickens [1, 26]. LAMP partially fulfils expectations as an ideal diagnosis method because it can be used without access to advanced laboratory equipment [16, 18]. The bottleneck in the LAMP procedure is the extraction of nucleic acid. The objective of this study was to apply the LAMP method for direct detection of MDV in poultry farm dust without DNA purification from crude samples. This is the first report of a LAMP application used for the monitoring of MDV in poultry farm dust. The standard 31/07 vv?MDV strain (GenBank accession number HQ204806.1) with a titer of 104.1 TCID50 (8812 PFU) was used as positive control for the MDV LAMP. The strain was propagated in SPF chicken embryo fibroblasts (CEFs). The stock of the 31/07 strain was stored in liquid nitrogen at (-196 C). The DNA of the 31/07 strain was extracted from 200 lL of the virus stock using a QIAamp Mini Kit (QIAGEN, Hilden, Germany) as described previously [27].

Dust samples (about 20 g each) were collected from twelve different farms where MD had previously been diagnosed in chickens vaccinated with FC126 HVT and CVI988/Rispens. The dust from the farms was collected into plastic sealed bags, mainly from air fans, window sills and floors. The dust from each farm was then pooled, and the dust samples were then stored at -20 C for further examination. About 20 mg of each pooled dust sample was resuspended in 480 ll of phosphate-buffered saline (PBS) by vigorous vortexing in a 1.5-ml Eppendorf tube. The samples were incubated for 5 min at 95 C and then centrifuged at 60009g for 3 min (Mikro 22R, Hettich Zetrifugen, Tuttlingen, Germany). The supernatants were then transferred to fresh microtubes and diluted 10-fold in PBS. The DNA yield and purity were measured by determining the A_{260}/A_{280} ratio using a Nanophotometer™, P-Class (Implen, Westlake Village, CA, USA). Three pairs of primers complementary to the meq gene sequence of the RB1B very virulent plus MDV strain (GenBank accession number AY571783) were designed using Primer Explorer version 4 (NetLaboratory, Tokyo, Japan) (Table 1) [26]. The reactions were set up on ice in 0.2-ml OptiAmp optical tubes (Applied Biosystems, Foster City, CA, USA). The LAMP volume was 15 lL and contained 7.5 lL of Isothermal Mastermix (OptiGene, Horsham, West Sussex, United Kingdom), 50 pmol each of inner primers FIP and BIP, 10 pmol each of outer primers F3 and B3, 25 pmol each of loop primers LF and LB, 1 ll of 1:10,000-diluted ROX passive reference dye (EurX, Gdansk, Poland), 1 of PCR-grade water (EurX, Gdansk, Poland) and 2 ll of DNA template. After incubation of the reaction mixtures in a water bath (Memmert, Schwabach, Germany) at 62.3 C for 30 min, 0.5 ll of a 1:100 stock dilution of 10,000-fold concentrated SYBR Green I dye in DMSO (Invitrogen) was added to each sample. The samples were observed under UV illumination to detect green fluorescence in positive samples. At the same time, the reaction was run in an ABI 7500 system (Applied Biosystems, Foster City, CA, USA) at 62.3 C for 40 cycles (1 min each). Next, melting curve analysis was conducted at temperatures ranging from 95 C to 60.0 C. The positive control was DNA from the 31/07 strain, and the negative control DNA was extracted from uninfected CEFs. Pictures were taken using a G5 camera (Canon, O¯ ta, Tokyo, Japan). The cycle threshold values (CT) were recorded and melting curve analysis was done using the software from the ABI 7500 system (version 2.0.1). LAMP amplicons were separated in 2 % agarose gels stained with GelRedTM dye (Biotum, Hayward, CA, USA).

Table 1: Sequences of LAMP primers used for detection of MDV-1 in dust samples

Primer	Sequence (5′-3′)
F3	TTCCCTCTTCTGCCCTCC
B3	TCCTGTTCGGGATCCTCG
FIP	GTAAACCGTCCCCGGCGATG *TTTT*GGGCATCTTCCCTGCATTG
BIP	CTTTGTCCTGTTGGCCAGGCTC *TTTT*GACGAGCATAAAGCCTCTCC
LF	TACACGGCTCGGTAACAGGA
LB	CCACATCCGGCTCCGGAGCC

TTTT is a thymidine linker in the FIP and BIP primers

The primers were designed based on the *meq* gene sequence of the RB1B very virulent plus MDV strain (GenBank accession number AY571783), using Primer Explorer version 4 (NetLaboratory, Tokyo, Japan)

In this study, we describe the first direct detection of MDV in crude dust samples using LAMP. The previously described application of LAMP for MDV DNA detection was conducted using samples extracted from feather tips using commercial kits or phenol-chloroform based procedures [26]. Over the past few years, LAMP has been found to be a powerful method for detecting a number of viral pathogens of poultry and free-ranging birds [1, 18, 26]. However, the application of an inexpensive method for infarm detection of viral genetic material has been limited by the necessity of obtaining purified nucleic acids [16, 18]. Therefore, the development of a ‘portable’ technique that can be used even when access to laboratory equipment is limited has not been fully accomplished. The DNA polymerases used for LAMP include Bsm, Bst, GspSSD polymerase [1, 16, 26]. The last of these has been known to be less prone to inhibition than the other. Therefore, we decided to examine whether the simple process of dilution and heat treatment of dust samples from MDV-contaminated farms would be sufficient to perform LAMP using GspSSD polymerase.

Table 2: Detection of MDV DNA in dust samples by LAMP

Sample	DNA purity A_{260}/A_{280} ratio/ concentration (ng/μL)	Ct value LAMP MDV-1	Melting temperature
Negative control	1.83/125.0	40.0	-
Farm 1	0.47/24.3	30.4	87.4
Farm 2	0.38/15.9	18.3	86.9
Farm 3	0.41/19.0	21.4	86.8
Farm 4	0.42/20.5	15.4	87.4
Farm 5	0.45/27.5	40.0	-
Farm 6	0.41/19.9	16.7	86.9
Farm 7	0.45/20.1	20.5	87.4
Farm 8	0.41/19.4	20.7	87.3
Farm 9	0.39/7.9	14.0	86.5
Farm 10	1.0/12.7	20.9	86.8
Farm 11	0.46/18.4	17.0	87.1
Farm 12	0.40/10.5	15.9	86.3
Positive control DNA of 31/07 MDV strain	1.85/200.5	11.8	87.4

The measured DNA purity ratio (A_{260}/A_{280}), concentration (ng/μL), cycle threshold value and melting temperature of LAMP products are given. Negative control, DNA extracted from uninfected SPF chicken embryo fibroblasts (CEFs)

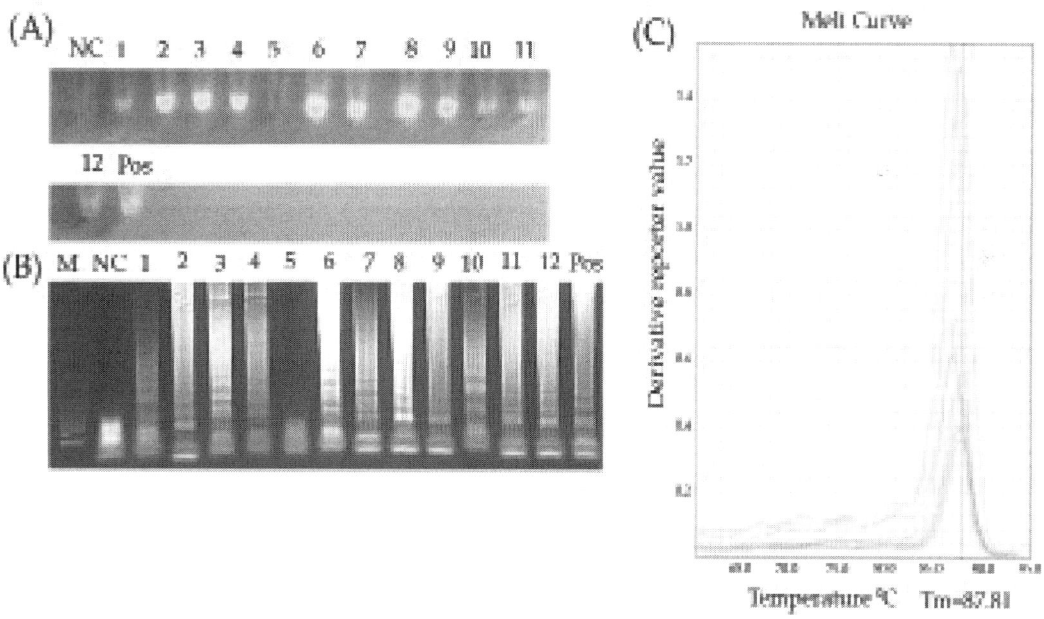

Figure. 1: Direct detection of MDV DNA in poultry farm dust by LAMP. (A) Detection of MDV under UV light illumination after the addition of SYBR Green I dye, (B) electrophoresis of LAMP products in a 2 % agarose gel stained with GelRed™ dye (Biotum, Hayward, CA, USA), (C) melting curve analysis of LAMP products conducted in a real-time PCR 7500 system (Applied Biosystems, Foster City, CA, USA). NC, negative control DNA extracted from uninfected SPF chicken embryo fibroblasts (CEF); Pos, positive control DNA of the standard 31/07 vv?MDV strain (accession number: HQ204806.1); 1-12, poultry dust samples collected from potentially MDV-contaminated farms; M, molecular size marker (GeneRuler™ 100bp DNA Ladder Plus, Thermo-scientific, Waltham, Massachusetts, USA). The common melting temperature point for all LAMP products was 87.8 C. The derivative reporter value is plotted on the y-axis, whilst the temperature is plotted on the x-axis.

Indeed, by dilution and incubation at 95 C of dust samples collected from 12 different chicken farms, it was possible to retrieve from 7.9 to 27.5 ng of DNA per μL. However, the crude samples appeared to contain protein contamination, since the A_{260}/A_{280} ratio ranged from 0.38 to 1.0 (Table 2). In spite of this, the LAMP was found to be resistant to contamination, as we obtained a positive signal for MDV, visible as green fluorescence in 11 out of 12 dust samples (Fig. 1). The one negative dust sample that was obtained was also extracted using a commercial DNA extraction kit. However in spite of this, the sample remained MDV negative, possibly due to the presence of inhibitors. Indeed, attempts to amplify MDV DNA from dust samples using conventional PCR also failed (data not shown). The reliability of the results was confirmed by the CT values in the ABI 7500 system (Table 2) as well as gel electrophoresis and melting curve analysis (Fig.1B and C). The common melting temperature for all of the MDV dust samples examined was about 87.8 C. Interestingly, the CT values that were obtained did not seem to be dependent on the purity of the DNA used, suggesting that the LAMP assay is resistant to inhibitors present in crude dust samples. Recently, Yoshikawa et al. [28] applied a dry-reagents LAMP assay for the direct detection of human herpesvirus 6B in human serum samples. They found the direct detection system to be a potentially perfect solution for major tropical diseases.

Likewise, Segawa et al. [20] developed a direct LAMPbased detection technique for canine distemper virus (CDV), pseudorabies virus (PRV), feline calicivirus (FCV) and feline parvovirus (FPV) [20]. The authors used a RNA GEM Tissue Kit to extract mammalian DNA and RNA using heat treatment. A similar simplified DNA release method was used to detect Aspergillus nomius and A. flavus from samples of shelled Brazil nuts [14]. In the case of MDV, a good source of viral DNA is poultry dust. PCR and real-time PCR have been widely used for detection, monitoring and modelling of shedding patterns of MDV [2–4, 6, 9–12, 19, 23, 24]. The recently published study by Walkden-Brown et al. [24] showed that all three MDV serotypes could be detected in farm dust using quantitative real-time PCR. They found that 23.1 % of field dust samples contained DNA of MDV-1. The concentration of the extracted DNA was low and reached approximately 5 ng/ lL, while the median value of the DNA purity reached 0.44, which supports the results obtained in our study (Table 2). Our results demonstrate the robustness of LAMP for MDV detection, even without DNA purification using crude dust samples. This procedure simplifies monitoring for the presence of MDV under farm conditions when access to laboratory equipment is limited. In summary, this study shows that the use of simple diagnostic methods including simplified DNA extraction combined with LAMP may aid in the monitoring of chicken farm contamination with MDV. This potentially has an economic aspect, taking into account the high rate of transmission of MDV through poultry dust and the losses caused by MD. In the future, it would be interesting to extend our study to LAMP detection of the remaining two MDV serotypes and attenuated CVI988/Rispens strains in crude dust samples.

CONFLICT OF INTEREST

The authors declare that they have no competing interests.

REFERENCES

1. Angamuthu R, Baskaran S, Gopal DR, Devarajan J, Kathaperumal K (2012) Rapid detection of the Marek's disease viral genome in chicken feathers by loop-mediated isothermal amplification. J Clin Microbiol 50:961–965
2. Atkins KE, Read AF, Walkden-Brown SW, Savill NJ, Woolhouse ME (2013) The effectiveness of mass vaccination on Marek'sdisease virus (MDV) outbreaks and detection within a broilerbarn: a modeling study. Epidemics 5:208–217. doi:10.1016/j.epidem.2013.10.001
3. Baigent SJ, Petherbridge LJ, Howes K, Smith LP, Currie RJ, Nair VK (2005) Absolute quantitation of Marek's disease virus genome copy number in chicken feather and lymphocyte samplesusing real-time PCR. J Virol Methods 123:53–64
4. Baigent SJ, Smith LP, Currie RJ, Nair VK (2005) Replication kinetics of Marek's disease vaccine virus in feathers and lymphoid tissues using PCR and virus isolation. J Gen Virol 86:2989–2998
5. Baigent SJ, Davison F (2004) Marek's disease virus: biology and life cycle. In: Davidson F, Nair V (eds) Marek's disease: an evolving problem, 1st edn. Elsevier Academic Press, London, pp 62–76
6. Becker Y, Asher Y, Tabor E, Davidson I, Malkinson M, Weisman Y (1992) Polymerase chain reaction for differentiation between pathogenic and non-pathogenic serotype 1 Marek's disease viruses (MDV) and vaccine viruses of MDV-serotypes 2 and 3. J Virol Methods 40:307–322
7. Calnek BW, Adldinger HK, Kahn DE (1970) Feather follicle epithelium: a source of enveloped and infectious cell-free herpesvirus from Marek's disease. Avian Dis 14:219–233
8. Gong Z, Zhang L, Wang J, Chen L, Shan H, Wang Z, Ma H (2013) Isolation and analysis of a very virulent Marek's disease virus strainin China. Virol J 10:155. doi:10.1186/1743-422X-10-155
9. Handberg KJ, Nielsen OL, Jørgenson PH (2001) Use of serotype1-and serotype 3-specific polymerase chain reaction for the detection of Marek's disease virus in chickens. Avian Pathol 30:243–249
10. Islam A, Harrision B, Cheetham BF, Mahony TJ, Young PL,Walkden-Brown SW (2004) Differential amplification and quantitation of Marek's Disease viruses using real-time polymerase chain reaction. J Virol Methods 119:103–113
11. Islam A, Walkden-Brown SW (2007) Quantitative profiling of the shedding rate of the three Marek's disease virus (MDV) serotypesreveals that challenge with virulent MDV markedly increasesshedding of vaccinal viruses. J Gen Virol 88:2121–2128
12. Islam T, Renz KG, Walkden-Brown SW, Ralapanawe S (2013) Viral kinetics, shedding profile and transmission of serotype 1 Marek's disease vaccine Rispens/CVI988 in maternal antibodyfree chickens. Avian Dis 57(Suppl. 1):454–463
13. Jurajda V, Klimes B (1970) Presence and survival of Marek's disease agent in dust. Avian Dis 14:188–190
14. Luo J, Taniwaki MH, Iamanaka BT, Vogel RF, Niessen L (2014) Application of loop-mediated isothermal amplification assays for direct identification of pure cultures of Aspergillus flavus, A. nomius, and A. caelatus and for their rapid detection in shelled Brazil nuts. Int J Food Microbiol 172:5–12. doi:10.1016/j.ijfoodmicro.2013.12.001
15. Morrow C, Fehler F (2004) Marek's disease: a worldwide problem.In: Davidson F, Nair V (eds) Marek's disease:

an evolving problem, 1st edn. Elsevier Academic Press, London, pp 49–61

16. Notomi T, Okayama H, Masubuchi H, Yonekawa T, Watanabe K, Amino N, Hase T (2000) Loop-mediated isothermal amplificationof DNA. Nucleic Acids Res 28:E63

17. Osterrieder N, Kamil JP, Schumacher D, Tischer BK, Trapp S (2006) Marek's disease virus: from miasma to model. Nat RevMicrobiol 4:283–294

18. Parida M, Posadas G, Inoue S, Hasebe F, Morita K (2004) Realtimereverse transcription loop-mediated isothermal amplification for rapid detection of West Nile Virus. J Clin Microbiol 42:257–263

19. Renz KG, Islam A, Cheetham BF, Walkden-Brown SW (2006) Absolute quantification of Marek's disease virus serotype 2 using real-time polymerase chain reaction in field dust samples, spleens, and feather tips. J Virol Methods 135:186–191

20. Segawa T, Kobayashi Y, Sase Y, Itou T, Suzuki M, Endoh T, Nakanishi T, Sakai T (2014) Easy-to-use rapid gene amplification method for direct detection of RNA and DNA viruses in sera and feces from various animals. J Virol Methods. doi:10.1016/j.jvir omet.2014.01.019

21. Venugopal K (2000) Marek's disease: an update on oncogenic mechanisms and control. Res Vet Sci 69:17–23

22. Wajid SJ, Katz ME, Renz KG, Walkden-Brown SW (2013) Prevalence of Marek's disease virus in different chicken populations in Iraq and indicative virulence based on sequence variation in the EcoRI-Q (meq) gene. Avian Dis 57:562–568

23. Walkden-Brown SW, Hunt PW, McNally J, Cressman MD, Burgess SK, Islam AFMF (2013) Development of methods for recovery and quantitation of viral nucleic acids from broiler litter. Proc Aust Poult Sci Symp 24:171–174

24. Walkden-Brown S, Islam A, Groves P, Rubite A, Sharpe S, Burgess S (2013) Development, application and results of routine monitoring of Marek's disease virus in broiler house dust using real-time quantitative PCR. Avian Dis 57(Suppl. 2):544–554

25. Witter RL (1997) Increased virulence of Marek's disease virus field isolates. Avian Dis 41:149–163

26. Woz′niakowski GJ, Samorek-Salamonowicz E, Kozdrun′ W (2011) Rapid detection of Marek's disease virus in feather follicles by loop-mediated amplification (LAMP). Avian Dis 55:462–467

27. Woz′niakowski G, Samorek-Salamonowicz E, Kozdrun′ W (2011) Molecular characteristics of Polish field strains of Marek's disease herpesvirus isolated from vaccinated chickens. Acta Vet Scand 53:10. doi:10.1186/1751-0147-53-10

28. Yoshikawa T, Matsuo T, Kawamura Y, Ohashi M, Yonekawa T, Kanda H, Notomi T, Ihira M (2014) Direct detection of human herpesvirus 6B by the LAMP method using newly developed dryreagents. J Virol Methods. doi:10.1016/j.jviromet.2014.02.017

CHAPTER 9

DETECTION AND CHARACTERIZATION OF CHICKEN ANEMIA VIRUS FROM COMMERCIAL BROILER BREEDER CHICKENS

Zerihun Hailemariam, Abdul Rahman Omar, Mohd Hair-Bejo, and Tan Ching Giap

BACKGROUND

Chicken anemia virus (CAV) is a small DNA virus with a circular, covalently linked, single negative-strand genome. It is the causative agent of chicken infectious anemia (CIA) and classified in the family Circoviridae, genus Gyrovirus [1].

CAV is an economically important pathogen with a worldwide distribution. CAV infections are manifested by either clinical or subclinical signs [2]. The clinical disease is mainly noticed in young chicks of 10–14 days of age, which usually acquire the infection vertically. Chickens older than 2–3 weeks of age are also susceptible to infection, but will only develop a subclinical disease evidenced by poor vaccine response, increased severity of other infections [2, 3], and decreased cell mediated immune responses [4, 5]. Outbreaks of the disease are characterized by anemia, thymus atrophy, bone marrow aplasia and immunosuppression [3, 6].

In general, no significant antigenic or pathogenic difference was reported among the CAV isolates in the past. Thus, until lately, CAV was known as a much conserved virus of one serotype [7] with several genetic groups [8]. However, an antigenically different isolate (CAV-7) has been reported from USA [9, 10], which could be a prototype virus of serotype 2. In Malaysia, previous studies undertaken indicated high prevalence of the virus in commercial broiler and layer farms [11]. Subsequently, CAV isolates were isolated from broilers farms and some of these isolates have been characterized based on pathogenicity and molecular analysis [12, 13]. However, there is no study conducted in the broiler breeder farms regarding the extent of occurrence of the virus, type of isolates present and their genetic diversity. In the present study we report detection of CAV and characterization of isolates based on sequence and phylogenetic analysis of partial VP1 gene from commercial broiler breeder chickens in Malaysia. Level of transmission to the progeny and level of protection afforded in the commercial broiler breeder chickens were also analyzed and discussed.

RESULTS

Distribution of CAV DNA in Various Organs in Commercial Broiler Breeder Hens

A total of 420 organ samples collected from 60 commercial broiler breeder hens were tested by nested PCR assay for the presence of CAV DNA. The data are summarized in Additional file 1. The highest percentage of positive samples was detected in spleen where 45 samples out of 60 (75%) were positive for CAV DNA. Duodenum was found to be an organ with the least distribution of CAV DNA in which 28 organs out of 60 (46.7%) were positive for CAV DNA. Even though, there is difference in the percentages of CAV DNA between spleen,

Citation: Zerihun Hailemariam, Abdul Rahman Omar, Mohd Hair-Bejo and Tan Ching Giap, Detection and characterization of chicken anemia virus from commercial broiler breeder chickens, DOI: 10.1186/1743-422X-5-128.

bone marrow, thymus and ovary, the differences were not statistically significant ($P < 0.05$). However, the distributions of viral DNA in liver, duodenum and oviduct were significantly less ($P < 0.05$) from the rest of the organs.

CAV DNA in Embryos and Egg Shell Membranes (ESM)

The nested PCR assay result indicated the presence of positive embryos ranging from 40–100% in different farms from three states of Malaysia (Fig. 1).

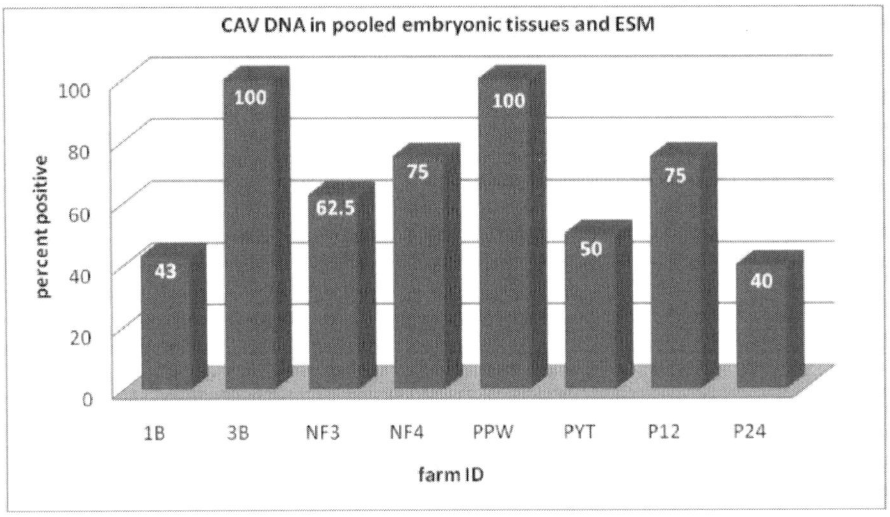

Figure 1: Detection of CAV DNA in pooled embryonic tissues and ESM from eggs collected from commercial broiler breeder farms. As it is shown on the graph, CAV DNA was detected in 40 to 100% of pooled embryonic tissue and ESM samples tested for different commercial broiler breeder farms.

Nucleotide Sequence Analysis

Nucleotide sequence analysis of a 498 bp region of CAV genome from position 892 to 1389; numbering according to Meehan *et al.* [14], encompassing the hypervariable region of VP1 protein revealed total nucleotide variation among the isolates ranging from 0.3 to 6.1% whilst the overall maximum nucleotide variation is 6.5%. The nucleotide sequence alignment with the published isolates considered for comparison revealed 9 to 30 nucleotide substitutions in the isolates from commercial broiler chickens (Table 2). Based on comparisons of percentage homologies, six isolates (MF1A, MF3C, M3B5, NF4A, P12B and P24A) were found to have maximum homology (97.1 to 99.1%) with SMSC-1 isolate, four isolates (M1B1, NF3A, PYT4 and PPW4) were found to have maximum homology (98.1 to 98.9%) with BL-5 isolate and the remaining two (NF1D and NF2C) have similar maximum homology (98.1%) both with isolates 3-1 and BL-5 (Table 1).

Table 1: Nucleotide percentage homologies of studied isolates from the commercial broiler breeder hens in relation to previously characterized Malaysian CAV field isolates

	Previously identified Malaysian CAV isolates		
	SMSC-1	**BL-5**	**3-1**
MF1A	**98.9**	94.9	94.5
MF3C	**97.9**	94.5	94.3
M3B5	**97.1**	95.1	94.9
NF4A	**98.3**	95.1	94.7
P12B	**99.1**	95.1	95.1
P24A	**98.7**	95.1	94.7
M1B1	94.9	**98.9**	98.1
NF3A	94.1	**98.5**	97.7
PYT4	94.1	**98.5**	97.7
PPW4	94.7	**98.1**	97.1
NF1D	94.7	**98.1**	**98.1**
NF2C	94.3	**98.1**	**98.1**

Accordingly, isolates MF1A, MF3C, M3B5, NF4A, P12B and P24A were found to have maximum homology with SMSC-1 isolate, where as four isolates (M1B1, NF3A, PYT4 and PPW4) were found to have maximum homology with BL-5 isolate. However, isolates NF1D and NF2C have similar maximum homology (98.1%) both with isolates 3-1 and BL-5.

Table 2: Number of nucleotide and amino acid substitutions and Ka/Ks ratio of Malaysian CAV isolates

CAV isolates	*No. of Nucleotide Substitutions	*No. of amino acid Substitutions	non-synonymous substitution rate (K_A)	Synonymous substitution rate (K_S)	Omega value (K_A/K_S)	References
MF1A	28	6	0.017	0.199	0.08	This study
MF3C	30	6	0.017	0.225	0.07	This study
M3B5	27	6	0.017	0.191	0.09	This study
NF4A	27	7	0.020	0.179	0.11	This study
P12B	27	6	0.017	0.187	0.09	This study
P24A	18	6	0.017	0.191	0.09	This study
PPW4	12	7	0.020	0.039	0.50	This study
NF1D	13	6	0.017	0.053	0.33	This study
NF2C	13	6	0.017	0.055	0.31	This study
NF3A	11	4	0.012	0.047	0.25	This study
PYT4	11	5	0.015	0.042	0.35	This study
M1B1	9	4	0.012	0.036	0.32	This study
SMSC-1	26	5	0.017	0.187	0.09	[13]
3-1	8	4	0.011	0.060	0.19	[13]
BL-5	4	3	0.009	0.006	1.50	[12]

* compare to the consensus nucleotide and amino acid sequences. Pair wise comparison with the consensus nucleotide and amino acid sequence indicated Ka/Ks ratio less than one for all the studied isolates which indicates the occurrence of negative selection.

Compared to isolates from other geographical places around the world, six out of 12 isolates (MF1A, MF3C, M3B5, NF4A, P12B, and P24A) were found to have maximum homology with Australian isolate 704. The remaining isolates showed maximum homology with isolates from China (NF3A and PYT4), isolate A2 from Japan (NF1D and NF2C) and CAV-B isolate from India (PPW4). Only two of the isolates (M1B1 and NF3A) were found to have maximum homology with isolates 26P4 and Del-Ros from USA.

Amino Acid Sequence Analysis

The amino acid sequence alignment with the published isolates considered for comparison also showed 4 to 7 amino acid substitutions in the isolates from commercial broiler breeder chickens (Table 2). The calculation of synonymous and non-synonymous substitution rate demonstrated omega values (K_a/K_s) of less than one suggesting the occurrence of purifying (negative) selection in all the 12 isolates (Table 2). All the isolates have the omega value ranging from 0.07 to 0.35 except for PPW4 with omega value of 0.50. Eight variable amino acid positions were detected in more than one isolate at amino acid positions 22, 75, 83, 97, 125, 139, 141, 144. Maximum variation among the CAV isolates was observed at amino acid position 144. Proline (P) at position 22 for isolate NF4A and glycine (G) at position 48 for isolate PYT4 were unique amino acid substitutions found only in the studied isolates (Table 3).

Table 3: Amino acid substitutions in VP1 sequence of CAV isolates

Isolate	Amino acid positions									
	22	48	75	83	97	125	139	141	144	157
Consensus	H	A	V	I	M	I	K	Q	Q	V
M1B1	.	.	.	.	.	.	.	.	E	M
NF1D	.	.	.	L	.	L	.	E	E	.
NF2C	.	.	.	L	.	L	.	E	E	.
NF3A	.	.	.	.	.	L	.	.	E	.
PYT4	.	G*	.	.	.	L	.	.	E	.
NF4A	P*	.	I	.	L	.	Q	.	.	.
PPW4	N	.	I	.	L	.	Q	.	.	.
P24A	.	.	I	.	L	.	Q	.	.	.
P12B	.	.	I	.	L	.	Q	.	.	.
M3B5	.	.	I	.	L	.	Q	.	.	.
MF3C	.	.	I	.	L	.	Q	.	.	.
MF1A	.	.	I	.	L	.	Q	.	.	.
NIE/19.04/118/Nigeria	.	.	I	.	L	.	Q	.	.	.
BD-3/Bangladesh	.	.	I	.	L	.	Q	.	.	.
ISOLATE704/Australia	.	.	I	.	L	.	Q	.	.	.

Isolate	Amino acid positions									
	22	48	75	83	97	125	139	141	144	157
Consensus	H	A	V	I	M	I	K	Q	Q	V
130/Slovenia	.	.	I		L	.	Q	.	.	.
CAV-B/India	.	.	I	.	L	.	Q	.	.	.
SMSC-1/Malaysia	.	.	I	.	L	.	Q	.	.	.
BL5/Malaysia	.	.	.	.	.	.	.	.	E	.
BL5/P90/Malaysia	.	.	.	.	.	.	.	.	K	.
SMSC-1/P60/Malaysia	.	.	.	L	.	.	.	E	E	M
3-1/Malaysia	.	.	.	.	.	.	.	.	E	.
3-1/P60/Malaysia	.	.	.	.	.	.	.	E	E	M
CUX-1(M)/Germany	.	.	.	.	.	.	.	.	D	.
CUX-1(N)Germany	.	.	.	.	.	.	.	.	D	.
CIA-1/USA	N	.	I	.	L	.	Q	.	.	.
ConnB/USA	.	.	.	.	.	.	.	.	N	.
Del-Ros/USA	.	.	.	.	.	.	.	.	E	.
26P4/USA	.	.	.	.	.	.	.	.	E	M
CAV-A/India	.	.	.	.	.	L	.	.	D	.
A2/Japan	.	.	.	.	.	.	.	E	E	M
AF448446/China	.	.	.	.	.	.	.	.	E	.

Amino acids marked with an asterisk (*) are unique substitutions for isolates from commercial broiler breeder chickens. The isolates can be grouped into two distinct groups based on their amino acid profile at positions 75, 97, 139 and 144. Five isolates (M1B1, NF1D, NF2C, NF3A and PYT4) had amino acid profile of 75-V, 97-M, 139-K and 144-E.

Phylogenetic Analysis

Phylogenetic analysis based on 165 deduced amino acid sequences of VP1 protein of the 12 isolates in comparison to 20 previously identified isolates revealed the formation of two clusters. All the 12 isolates were found both in cluster I and II (Fig. 2). Seven of the isolates with amino acid profile of 75-I, 97-L, 139-Q and 144-Q were clustered together in cluster I while the remaining five isolates with amino acid profile of 75-V, 97-M, 139-K and 144-E were grouped under cluster II.

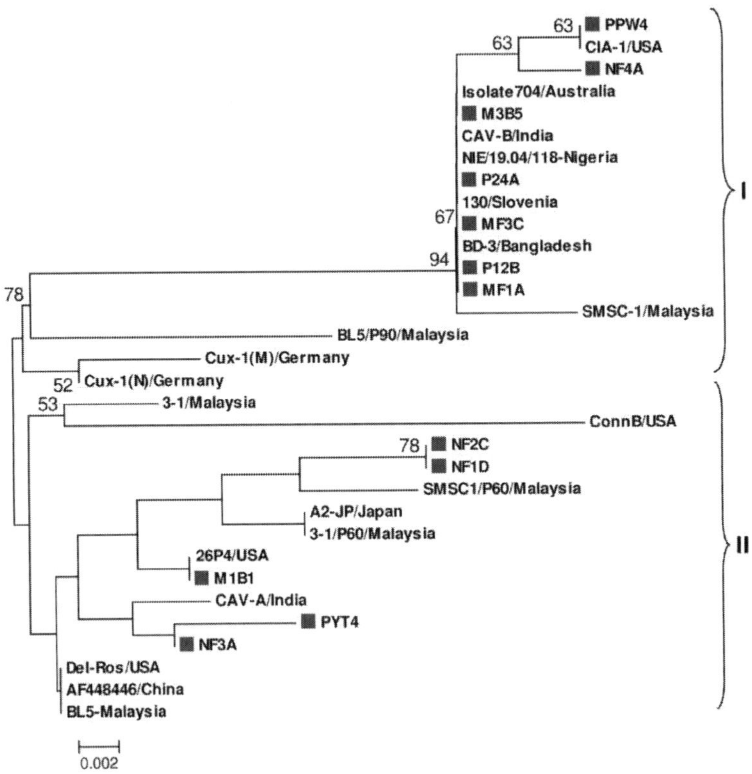

Figure 2: Phylogenetic relationship among 32 different CAV isolates based on partial VP1 amino acid sequences. Note: The boxes (■) indicate isolates identified in this study. The isolates were found both cluster I and II. Seven of the isolates with amino acid profiles of 75-I, 97-L, 139-Q and 144-Q clustered together in cluster I. The remaining five studied isolates with amino acid profiles of 75-V, 97-M, 139-K and 144-E were grouped under cluster II.

ELISA

The ELISA result shows that CAV infection is widespread in these unvaccinated commercial broiler chicken farms in Malaysia. Out of 52 chickens from which serum sample was collected, 50 (96.15%) were positive for antibodies against CAV (Fig. 3).

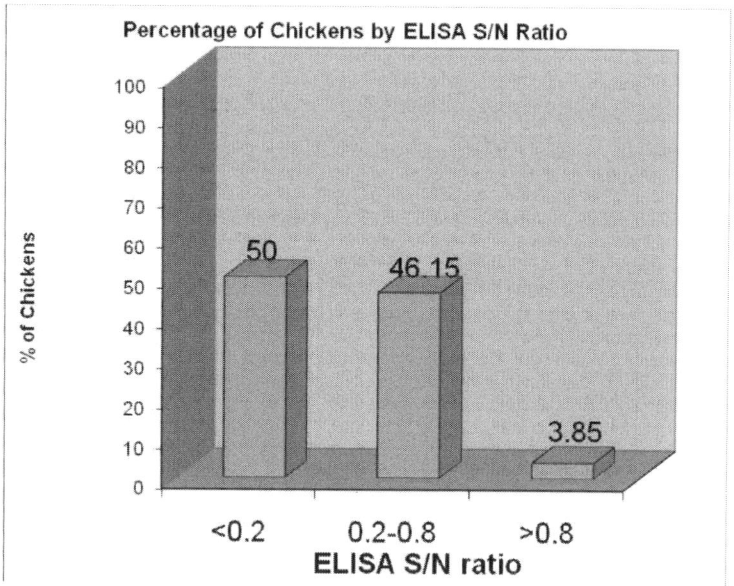

Figure 3: ELISA results of serum collected from commercial broiler breeder chickens. Fifty percent of the chickens have ELISA S/N < 0.2 which indicates the presence of high protective titers and able to afford high level protection to the progeny, meanwhile 46.15% of the chickens have ELISA S/N in the range of 0.2 to 0.8 affording low levels of protection to the progeny. Only 3.85% of the chickens have ELISA S/N > 0.8 indicating negative result for antibody against CAV.

Out of the total number of hens sampled 26 hens (50%) have anti-CAV antibody titers above 8600 (high protective titers), 24 hens (46.15%) have the titer below 8600 but above 1000 (moderate protective titers) and 2 hens (3.85%) have anti-CAV antibody titers below 1000 (negative for antibodies against CAV). The ELISA-based analysis indicated that all farms had neutralizing antibodies. Based on the correlation of ELISA titer to virus neutralizing antibody titer as recommended by the manufacturer of the ELISA kit, 50% of hens found out to have high neutralizing antibody titers that able to confer high level protection to the progeny whilst 46.15% of hens have low neutralizing antibody titers affording low levels of protection to the progeny (Fig. 3).

Indirect Immunoperoxidase Staining

All the experimentally infected chickens from hyperimmune serum production produced the highest readable antibody titer (>8661) which remained constant starting from two weeks after the second inoculation. Sera from the inoculated and control chickens were used as primary antibodies for the indirect immunoperoxidase staining. Specific positive staining was demonstrated in thymus and bone marrow tissue sections from commercial broiler breeder chickens that were detected positive for CAV DNA. Positive staining was observed in lymphoblasts in the cortex of the thymus (Fig. 4) and in hemocytoblasts in the sinuses of the bone marrow (Fig. 5). Specific staining was not demonstrated in tissue sections from spleen, liver, duodenum, ovary and oviduct.

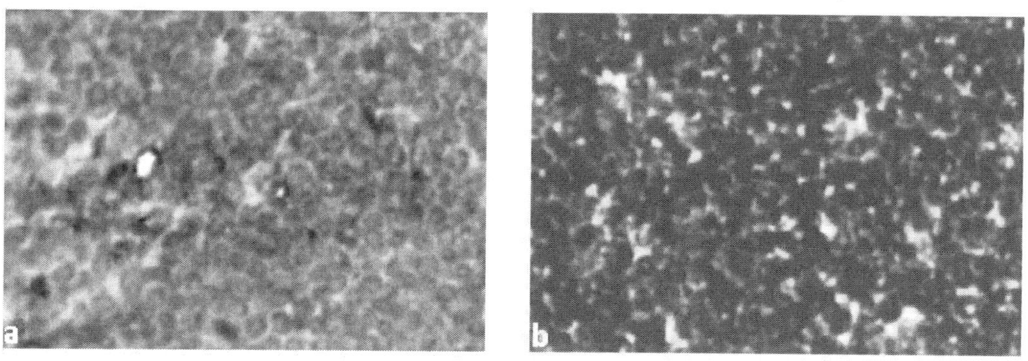

Figure 4: IPS performed on formalin fixed paraffin-embedded thymic tissues. Thymic tissue slide from commercial broiler breeder hen: a) infected lymphoblasts in the cortex demonstrated by brown staining (400×) b) IPS using CAV negative serum as primary antibody and devoid of any specific brown staining (400×).

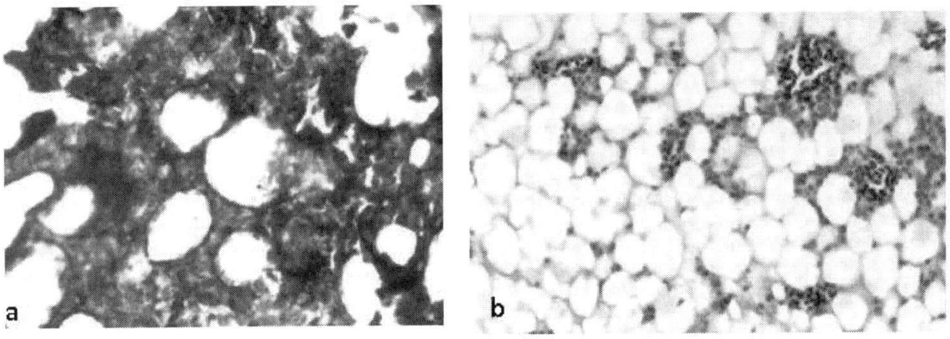

Figure 5: IPS performed on formalin fixed paraffin-embedded tissues from bone marrow. Bone marrow tissue slide from commercial broiler breeder hen: a) infected hemocytoblasts in the sinuses of the bone marrow demonstrated by brown staining (400×) b) IPS using CAV negative serum as primary antibody and devoid of any specific brown staining (200×).

DISCUSSION

In the present study, out of 420 organ samples tested, 75% of spleen, 68.3% of bone marrow, 70% of thymus, 53.3% of liver, 46.7% of duodenum, 66.7% of ovary and 48.3% of oviduct tested were positive for CAV DNA. CAV replicates mostly in lymphoid tissues of susceptible chickens [15, 16, 17]. Cardona et al. [18] found out a few CAV positive cells in spleen by *in situ* PCR and managed to detect CAV in ovaries by *in situ* PCR even in the absence of splenic virus. They also indicated that ovaries and to a lesser degree infundibulum of the oviduct are sites for persistence of CAV in hens. The finding of significantly higher positive tissues in the spleen, bone marrow, thymus and ovary as compared to duodenum, liver and oviduct in the present study is similar to the aforementioned findings. Therefore, we can suggest that spleen, thymus, and bone marrow could serve as an excellent choice of organs for the diagnosis of CAV infection while ovary representing more favourable tissue for the persistence of CAV in the reproductive organs in the broiler breeder hens.

Results from ELISA reading showed that 96.15% of blood samples collected from the same farms were positive for antibody against CAV indicating the widespread occurrence of CAV infection in these unvaccinated commercial broiler breeder chicken farms. Testing pooled embryonic tissue samples (thymus, bursa of Fabricius and spleen) together with ESM showed positive embryos for CAV DNA in the range of 40% to 100% for different commercial broiler breeder chickens despite the presence of neutralizing antibodies in majority of the hens (96.15%) tested for CAV antibodies suggesting high level of occurrence of vertical transmission of viral DNA to the progeny. The detection of CAV DNA in the ovary and oviduct of commercial broiler breeder hens with virus neutralizing antibodies and in their embryos supports the previous evidence that CAV may remain in the gonads of antibody positive chickens and can be vertically transmitted to their progeny [18, 19,20, 21]. The finding by Schat et al. [2] indicated low level of viral transcripts can be detected in the developing embryo during specific developmental periods supporting the current hypothesis that, it may be possible that a limited viral replication occurs in the embryos, but if the embryos have VN antibodies, the VN antibodies prevent the development of viremia in the embryos [19].

Viral antigens were identified only in individual lymphocytes in the cortex of the thymus and infected hemocytoblasts in the bone marrow of tissues collected from commercial broiler breeder chickens by the IPS. However, consistent and observable differences on the intensity of staining were not observed on those positive slides from individual chickens with moderate and high protective antibody titers at the same level of dilution of primary antibodies. Specific staining could not be detected in spleen, liver, duodenum, ovary and oviduct. Most of the tissue sections obtained from commercial broiler breeder chickens and tested positive by the nested PCR assay turned out to be negative in IPS. This could be due to virus replication in those tissues might be limited and below the detection limit of the assay. It might also be related with the age of commercial broiler breeder chickens or the poor sensitivity of the technique compared to the other two detection methods used in the present study namely the nested PCR assay for detecting CAV DNA and ELISA for detection of antibodies against CAV. Smyth *et al.* [17] demonstrated viral antigens in the lymphoid tissues of other organs including proventriculus, the ascending part of duodenum, kidney and lung. However, they also confirmed that, infected cells in these tissues usually cannot be detected for more than 22 days after infection at one day of age.

Comparisons of percentage homologies of the studied CAV isolates with previously characterized local CAV isolates showed diverse similarity among the local isolates. The phylogenetic analysis of 165 deduced amino acid sequences of the VP1 protein also revealed grouping of the Malaysian CAV isolates into two major clusters (Fig. 2). An overall similarity with CAV isolates circulating in south and south-east Asia and Australia was also observed while still having limited variation with isolates from different geographical parts of the world. Natesan et al. [22] also found similar result in that Indian isolate (CAV-E) having maximum similarity with Australian isolate isolate-704, Japanese isolate TR-20 and Malaysian SMSC-1 isolate. Unique amino acid residues observed in isolates from commercial broiler breeder chickens include proline (P) at amino acid position 22 and glutamine (G) at amino acid position 48 in isolates NF4A and PYT4, respectively.

Islam et al. [8] identified amino acid residues at positions 75-I/T, 97-L, 139-Q and 144-Q can be used to group CAV isolates into different groups. In the present study, two distinct groups were observed in the current isolates based on their amino acid profile at these positions. Seven of the isolates from the commercial broiler breeder chickens including previously characterized Malaysian isolate SMSC-1, had 75-

I, 97-L, 139-Q and 144-Q and clustered together in cluster I of the deduced amino acid phylogenetic tree. Together in this group also included other isolates from different geographical places. This includes, CIA-1 from USA, CAV-B from India, BD-3 from Bangladesh, isolate 704 from Australia, isolate 130 from Slovenia and NIE/19.04/118 isolate from Nigeria. The remaining five current isolates including previously characterized Malaysian isolates BL-5 [12] and 3-1 [13] had amino acid profile of 75-V, 97-M, 139-K and 144-E. These isolates were found in cluster II of the deduced amino acid phylogenetic tree with other isolates from around the world. In addition, there was no evidence of recombination effect observed in the analysis of Malaysian CAV isolates as reported by Van Santen et al. [23] on CAV isolates from USA. In that study, they indicated that different CAV isolates from Alabama can be divided into two groups with one isolate showing an exceptionally different amino acid profile of I-75, L-97, K-139 and E-144 suggesting of a possible evidence of recombination event.

The analysis of the ratio of synonymous and non-synonymous substitution rate (omega value) indicated the presence of only purifying (negative) selection in the studied isolates. This is similar to the result by Ducatez *et al.* [24] where they indicated a very slow CAV virus evolution at amino acid level corresponding to a strong negative selection (0.04 to 0.20) of VP1 gene in China and worldwide. However, in this study, we found that one of the isolates has omega value of 0.50 meanwhile five out of 12 isolates have omega value between 0.25 to 0.35 and the rest with omega value of 0.07 to 0.11 (Table 2). In our previous study, we suggested that the BL-5 isolate was distantly related to other Malaysian CAV isolates, SMSC-1 and 3-1 [12], has omega value of 1.50 suggesting of a positive selection of VP1 protein in this isolate (Table 2).

The overall phylogenetic pattern and clustering of different CAV isolates based on the partial VP1 gene in this study was similar to previous one based on complete sequence of VP1 gene [24] or the entire CAV genome [25] for the common CAV isolates considered in all the three studies. This suggests that relationships of different CAV isolates can be determined on the basis of partial sequence of VP1 gene due to the fact that most of the amino acid substitutions in comparisons between isolates lies in VP1 gene and more specifically on the N-terminal half of VP1 gene.

CONCLUSION

Generally, from the present study we can conclude the widespread occurrence of CAV infection in commercial broiler breeder farms at least in the three states of Malaysia. Detection of significantly higher percentage of positive DNA from spleen, thymus and bone marrow make these organs an excellent choice of organs in the screening and diagnosis of flocks for CAV infection. The finding of CAV DNA in embryos from broiler breeder chickens with neutralizing antibodies supports the previous finding that CAV may remain in the gonads of antibody positive chickens and can be vertically transmitted to their progeny. However, the importance of transmission of viral DNA detected by nested PCR assay still needs further study and explanation. The result also indicated the occurrence of genetic variability in local CAV isolates that can be divided at least into two groups based on characteristic amino acid substitutions at positions 75, 97, 139 and 144 of the VP1 protein. However, the CAV isolates showed only negative selection based on the calculated omega value of the partial sequences of the VP1 gene. The characterized CAV isolates show overall similarity with CAV isolates circulating in South East Asia and Australia while still having limited variations with isolates from different geographical parts of the world.

METHODS

Broiler Breeder Farms

Tissue and blood samples were collected from 12 commercial broiler breeder chicken farms located at three states of Peninsular Malaysia. The farms were not vaccinated against CAV and the samples were collected from a total of 60 broiler breeder hens that range in age from 25–35 weeks.

Sample Collection

A total of 420 organ samples were collected. Spleen, thymus, liver, bone marrow, duodenum, ovaries and oviduct were organs collected from each hen. Blood samples were collected from 52 broiler breeder hens by veno-puncture of the wing vein. Sera were separated and stored at -20°C until used.

Ten eggs were also collected from each farm and incubated for 18–20 days. Prior to hatching pooled embryonic organ samples consisting of thymus, bursa of Fabricius and spleen together with egg shell membrane (ESM) were collected from individual embryos. Tissue samples from the hens and embryos were stored at -20°C until DNA extraction.

DNA Extraction from Samples

DNA was extracted from a total of 420 tissue sample from hens and 52 pooled embryo samples. Briefly, tissue samples (1–5 mg) were homogenized in Phosphate Buffered Saline (PBS) solution by grounding with a mortar and pestle. Then the homogenate (~700 μl) was transferred into 1.5 ml eppendorf tube and centrifuged at 13000 rpm for 1 minute. The supernatant was then transferred into a new microcentrifuge tube. DNA extraction was carried out using MasterPure complete DNA and RNA purification kit (Epicentre, Madison, WI), following the instructions of the manufacturer with some modifications. The concentration and purity of the extracted DNAs were determined by a spectrophotometer (Beckman, USA) according to the method described by Sambrook *et al.* [26].

Detection of CAV by Nested PCR Assay

The extracted DNA was first screened for CAV DNA using a highly sensitive nested detection PCR as previously described by Cardona *et al.* [18] with slight modifications. The first-step PCR reaction was carried out using 20 pmol each of the primers O3F and O3R amplifying a 386 bp fragment of the VP3 gene [18]. The PCR reaction was carried out in a total volume of 25 μl using the following cycling parameters: initial denaturation of 94°C for 2 min followed by 35 cycles of denaturation, annealing and extension at 94°C for 2 min, 50°C for 1 min and 72°C for 1 min, respectively, and the final extension was carried out at 72°C for 3 min. An aliquot of the first PCR reaction (1 μl) was then added to 24 µl of a new mastermix (total volume 25 µl) containing 20 pmol of the nested primers N3 and primer N4 for amplification of a 209 bp nested fragment of the VP3 gene as reported by Cardona *et al.* [18]. The nested PCR assay was carried out in MyCycler® Thermal Cycler (Bio-Rad, Hercules, CA, USA). The PCR products were analyzed by 1.8% agarose gel electrophoresis and the photographs were taken using Bio Imaging System in GeneSnap program (SynGene, Cambridge, UK).

Amplification of Partial VP1 Gene for Sequencing

Spleen samples from each farm that were detected CAV positive by VP3 nested PCR assay were used for amplification of partial VP1 gene using primers VP1F and VP1R for the first round amplification as described by Natesan *et al.* [22]. Nested fragment of first round amplification were amplified using primers O1F and PshA1R [18]. The first round PCR condition was carried out using the following cycling parameters: initial denaturation of 94°C for 4 min followed by 35 cycles of denaturation, annealing and extension at 94°C for 1 min, 57°C for 1 min and 72°C for 2 min, respectively, and the final extension was carried out at 72°C for 8 min. The second synthesis was carried out in a 50 µl reaction mixture with 1 µl of the first PCR reaction product and cycling parameters similar to that described for nested detection PCR. The PCR products were run on 1.6% agarose gel electrophoresis and purified from the gel by using GeneAll® kit (General Biosystem Inc., Korea) following the supplied instructions.

Sequence and Phylogenetic Analysis

Using the gel purified PCR products, the partial nucleotide sequences of VP1 gene were determined by direct sequencing in both direction using nested primers O1F and PshA1R. Sequencing reactions were performed in MJ Research PTC-225 Peltier Thermal Cycler using ABI PRISM® BigDyeTM Terminator Cycle Sequencing Kits with AmpliTaq DNA polymerase (FS enzyme) (Applied Biosystems, CA, USA). Each sample was sequenced three times to confirm consistency of the sequencing results.

DNA sequences of the 12 CAV isolates were aligned and compared with 20 local and foreign CAV isolates retrieved from the GenBank database. Sequences of VP1 gene for the studied Malaysian isolates were submitted to GenBank under the following accession numbers: MF1A [FJ167513]; MF3C [FJ167514]; M1B1 [FJ167515]; M3B5 [FJ167516]; NF1D [FJ167517]; NF2C [FJ167518]; NF3A [FJ167519]; NF4A [FJ167520]; P12B [FJ167521]; P24A [FJ167522]; PYT4 [FJ167523]; PPW4 [FJ167524]. The retrieved CAV isolates sequence name, GenBank accession numbers (in square brackets) and country are as follows: Cux-1 [M-M55918], Germany; Cux-1N [NC001427], Germany; SMSC-1 [AF285882], Malaysia; SMSC-1P60 [AF390102], Malaysia; 3-1 [AF390038], Malaysia; 3-1P60 [AY040632], Malaysia; BL-5 [AF527037], Malaysia; BL-5/P90 [AY150576], Malaysia; Isolate 704 [U65414], Australia; CIA-1 [L14767], USA; ConnB [U69548], USA; Del-Ros [AF313470], USA; 26P4 [D10068], USA; China [AF448446]; A2-[AB031296], Japan; BD-3 [AF395114], Bangladesh; CAV-A [AY583755], India; CAV-B [AY583756], India; NIE/19.04/118 [AJ888524], Nigeria; 130 [DQ016138], Slovenia. Percentage of homology, sequence identity matrix and translation from nucleotides to amino acids were determined using BioEdit software package version 7.01 [27]. Multiple sequence alignment of nucleotide and translated amino acids were performed using ClustalX software [28]. The phylogenic analysis of 165 deduced amino acids of the VP1 gene was performed with the software MEGA4 for phylogenetic and molecular evolutionary analyses using the Neighbor Joining Phylogeny reconstruction method with Poisson correction analysis and bootstrap consensus tree inferred from 1000 replicates [29]. Omega values [ratio of non-synonymous (K_A) to Synonymous (K_S) substitution rates] was calculated in comparison to consensus nucleotide sequences using PAL2NAL program [30].

ELISA

The sera were tested using a commercial ELISA kit (Idexx Lab, USA) at a 1:100 dilution and the results were expressed as S/N ratios (sample to negative ratio) according to manufacturer's instructions. Optical density value was read at 650 nm wave length on an ELX 800™ microplate reader (BIO-TEK Instruments, USA). The ELISA antibody titer has 78% correlation to virus neutralization titers [19].

Experimental Infection of Chicks with CAV

Eight 5 days old specific-pathogen-free (SPF) chicks were obtained from Veterinary Research Institute (VRI), Ipoh, Perak, Malaysia. The chicks were divided into 2 groups. Group 1 (n = 5) was inoculated intramuscularly with 1 ml of SMSC-1 CAV isolate cell culture inoculum containing $10^{5.5}$ $TCID_{50}$/ml [13]. Group II (n = 3) was left uninoculated as negative control chicks. Each group was reared separately in different room and the chicks were observed daily, and feed and water were provided *ad libitum*. The chicks were sacrificed at 14 days p.i. (post inoculation) for collection of organs. Tissue samples collected from infected and uninoculated chicks were processed and considered as positive and negative control slides for immunoperoxidase staining (IPS), respectively. All experimental research carried out on animals in this paper (including chicken hyperimmune serum production) followed internationally recognized guidelines and approved by animal care and use committee at the Faculty of Veterinary Medicine in University Putra Malaysia (Ref: UPM/FPV/PS/3.2.1.1551/AUP-R4).

Specimen Preparation for Immunohistochemical Staining

Tissue samples were fixed in 10% (v/v) neutral phosphate-buffered formalin for about 24 hrs and were then trimmed to the thickness of 0.5 cm. The bone marrow samples were decalcified by 5% nitric acid solution following the procedure of Luna [31]. Following tissue processing, tissue blocks were sectioned at a thickness of 4 µm and collected on clean silanized microscope slides [32, 33].

Immunohistochemical Staining

Prior to staining, the tissue slides were deparaffinized to remove embedding media and rehydrated following the supplied instructions. Antigen retrieval was achieved using the microwave-based antigen retrieval technique [33, 34]. The staining procedure of the detection system was carried out following the manufacturer's instruction manual [DakoCytomation Envision®+ Dual Link System-HRP (DAB+), Denmark]. A known negative and positive antigen control and a negative serum control were included in every procedure.

Chicken hyperimmune Serum Production

The hyperimmune serum production was carried out in four 60 weeks old SPF broiler breeder roosters obtained from VRI, Ipoh, Perak, Malaysia, with the following immunization protocol: Briefly, at day 0, prior to inoculation, blood samples were collected from all chickens to test their freedom from CAV antibody. Then the roosters were immunized orally with 2 ml of live CAV vaccine AviPro®THYMOVAC (Lohmann Animal Health, Cuxhaven, Germany). The immunization regimen was repeated at 14, 28, and 42 days after the first immunization in combination with Freund adjuvant (Sigma, USA). At day 56, blood was withdrawn from all the chickens from the wing vein to separate the hyperimmune serum produced. Commercial Idexx ELISA kit (Idexx Lab, Westbrook, Maine, USA) was used to evaluate the antibody titer of the chicken hyperimmune serum at different levels of immunization.

IGY Purification from Chicken Hyperimmune Serum

IgY purification was carried out using Pierce® Thiophilic Adsorption Kit (Pierce, USA). The T-gel was initially used according to the manufacturer›s instruction for mammalian immunoglobulin purification. For the purification of IgY from chicken serum, the manufacturer›s instruction was followed together with optimized protocol of T-gel chromatography for the purification of IgY from chicken sera as described by Constantinoiu*et al.* [35].

Statistical Methods

Data from the distribution of CAV DNA in organ samples from commercial broiler breeder chickens were analyzed by Kruskal-Wallis one-way ANOVA with significance defined at $P < 0.05$. Groups with significance difference in means from the rest were determined using Q-statistics [36].

COMPETING INTERESTS

The isolated Malaysian CAV isolates are currently been modified for development of live attenuated vaccines for poultry.

AUTHORS' CONTRIBUTIONS

Design and conception of the study (ARO, ZH, MHB, TCG); samples detection by PCR assays (ZH, ARO, TCG), Immunohistochemstry (ZH, MHB), sequence alignment and phylogenetic tree analysis (ZH, ARO), manuscript preparation (ZH, ARO, MHB); All the authors read and approved the final manuscript.

REFERENCES

1. Pringle CR: Virus taxonomy at the XIth International Congress of Virology, Sydney, Australia. *Arch Virol* 1999, 144:2065-2069. 10.1007/s007050050728
2. Schat KA, Circovirus infections: Chicken infectious anemia. In *Diseases of Poultry*. Edited by: Saif YM, Barnes HJ, Fadly AM, Glisson JR, McDougald LR, Swayne DE. Iowa State University Press, Ames, IA; 2003:182-202.
3. Adair BM: Immunopathogenesis of chicken anemia virus infection. *Dev Comp Immunol* 2000, 24: 247-255. 10.1016/S0145-305X(99)00076-2
4. Markowski-GrimsrudCJ,SchatKA:Infectionwithchicken anaemia virus impairs the generation of pathogen-specific cytotoxic T lymphocytes. *Immunology* 2003, 109: 283-294. 10.1046/j.1365-2567.2003.01643.x
5. McConnell CD, Adair BM, McNulty MS: Effects of chicken anemia virus on cell-mediated immune function in chickens exposed to the virus by a natural route. *Avian Dis* 1993, 37: 366-374. 10.2307/1591660
6. Yuasa N, Imai K, Watanabe K, Saito F, Abe M, Komi K: Aetiological examination of an outbreak of haemorrhagic syndrome in a broiler flock in Japan. *Avian Pathol* 1987, 16: 521-526. 10.1080/03079458708436401
7. McNulty MS, Connor TJ, McNeilly F, McLoughlin MF, Kirkpatrick KS: Preliminary characterisation of isolates of chicken anaemia agent from the United Kingdom. *Avian Pathol* 1990, 19: 67-73. 10.1080/03079459008418657
8. Islam MR, Johne R, Raue R, Todd D, Muller H: Sequence analysis of the full-length cloned DNA of a chicken anaemia virus (CAV) strain from Bangladesh: evidence for genetic grouping of CAV strains based on the deduced VP1 amino acid sequences. *J Vet Med B Infect Dis Vet Public Health* 2002, 49: 332-337.
9. Spackman E, Cloud SS, Pope CR, Rosenberger JK: Comparison of a putative second serotype of chicken infectious anaemia virus with a prototypical isolate I. Pathogenesis. *Avian Dis* 2002, 46: 945-955. 10.1637/0005-2086(2002)046[0945:COAPSS]2.0.CO;2
10. Spackman E, Cloud SS, Pope CR, Rosenberger JK: Comparison of a putative second serotype of chicken infectious anaemia virus with a prototypical isolate II. Antigenicity. *Avian Dis* 2002, 46: 955-960.
11. Rozanah AS, Aini I, Al-Ajeeli KS, Jalila A, Salim NB: Detection of chicken anemia virus in flocks of commercial chicken in Malaysia. *Journal of Veterinary Malaysia* 1995, 7: 77-79.
12. Hasmah MS, Omar AR, Wan KF, Hair-Bejo M, Aini I: Genetic diversity of chicken anemia virus following cell culture passaging in MSB-1 cells. *Acta Virologica* 2004, 48: 85-89.
13. Chowdhury S, Omar AR, Aini I, Hair-Bejo M, Jamaluddin AA, Md-Zain BM, Kono Y: Pathogenicity, sequence and phylogenetic analysis of Malaysian chicken anemia virus obtained after low and high passages in MSB-1 cells. *Archives of Virology* 2003, 148: 2437-2448. 10.1007/s00705-003-0189-3
14. Meehan BM, Todd D, Creelan JL, Earle JAP, Hoey EM, McNulty MS: Characterization of viral DNAs from cells infected with chicken anaemia agent: Sequence analysis

of the cloned replicative form and transfection capabilities of cloned genome fragments. *Arch Virol* 1992, 124: 301-319. 10.1007/BF01309811

15. Adair BM, McNeilly F, McConnell CD, McNulty MS: Characterization of surface markers present on cells infected by chicken anemia virus in experimentally infected chickens. *Avian Dis* 1993, 37: 943-950. 10.2307/1591898
16. Hoop RK, Reece RL: The use of immunofluorescence and immunoperoxidase staining in studying the pathogenesis of chicken anemia agent in experimentally infected checkens. *Avian Pathol* 1991, 20: 349-355. 10.1080/03079459108418770
17. Smyth JA, Moffett DA, McNulty MS, Todd D, Mackie DP: A sequential histopathologic and immunocytochemical study of chicken anemia virus infection at one day of age. *Avian Dis* 1993, 37: 324-338. 10.2307/1591656
18. Cardona CJ, Oswald WB, Schat KA: Distribution of chicken infectious anemia virus in the reproductive tissues of specific pathogen-free chickens. *J Gen Virol* 2000, 81: 2067-2075.
19. Brentano L, Lazzarin S, Bassi SS, Klein TA, Schat KA: Detection of chicken anemia virus in the gonads and in the progeny of broiler breeder hens with high neutralizing antibody titers. *Veterinary Microbiology* 2005, 105: 65-72. 10.1016/j.vetmic.2004.09.019
20. Cardona CJ, Lucio B, O›Connell P, Jagne J, Schat KA: Humoral immune responses to chicken infectious anemia virus in three strains of chickens in a closed flock. *Avian Dis* 2000, 44: 661-667. 10.2307/1593108
21. Miller MM, Katie AE, Oswald WB, Schat KA: Detection of chicken anemia virus DNA in embryonal tissues and eggshell membranes. *Avian Dis* 2003, 47: 662-671. 10.1637/7007
22. Natesan S, Kataria JM, Dhama K, Rahul S, Baradhwaj N: Biological and molecular characterization of chicken anemia virus isolates of Indian origin. *Virus Res* 2006, 118: 78-86. 10.1016/j.virusres.2005.11.017
23. van Santen VL, Li L, Hoerr FJ, Lauerman LH: Genetic characterization of chicken anemia virus from commercial broiler chickens in Alabama. *Avian Dis* 2001, 45: 73-388. 10.2307/1592977
24. Ducatez MF, Chen H, Guan Y, Muller CP: Molecular epidemiology of chicken anemia virus (CAV) in Southeastern Chinese live birds markets. *Avian Diseases* 2008, 52: 68-73. 10.1637/8049-070407-Reg
25. He C, Ding N, Fan W, Wu Y, Li J, Li Y: Identification of chicken anemia virus putative intergenotype recombinants.*Virology* 2007, 366: 1-7. 10.1016/j.virol.2007.06.007
26. Sambrook J, Fitsch EF, Maniatis T: *Molecular Cloning: A Laboratory Manual*. Cold Spring Harbor, Cold Spring Harbor Press; 1989.
27. Hall TA: BioEdit: a user-friendly biological sequence alignment editor and analysis program for Windows 95/98/NT.*Nucl Acids Symp Ser* 1999, 41: 95-98.
28. Thompson JD, Gibson TJ, Plenwniak F, Jeanmoungin F, Higgins DG: The Clustal_X windows interface: flexible strategies for multiple sequence alignment aided by quality analysis tools. *Nucleic Acids Res* 1997, 25: 4876-4882. 10.1093/nar/25.24.4876
29. Tamura K, Dudley J, Nei M, Kumar S: MEGA4: Molecular Evolutionary Genetics Analysis (MEGA) software version 4.0.*Molecular Biology and Evolution* 2007, 24: 1596-1599. 10.1093/molbev/msm092
30. Suyama M, Torrents D, Bork P: PAL2NAL: robust conversion of protein sequence alignments into the corresponding codon alignments. *Nucleic Acids Res* 2006, 34: W609-W612. 10.1093/nar/gkl315
31. Luna LG: *Manual of Histologic Staining Methods of the Armed Forces Institute of Pathology*. The Blakiston Division, Mc Graw-Hill Book Company, New York; 1968.
32. Bancroft JD, Stevens A, Stevens A: *Theory and practice of histological technique*. New York: Churchill Livingstone; 1996.
33. Key M, Boenisch T: Antigen retrieval. In *Immunohistochemical staining methods*. Edited by: Key M. Dako corporation, Carpinteria; 2006:41-45.
34. Gown AM, de Wever N, Battifora H: Microwave-based antigenic unmasking. A revolutionary new technique for routine immunohistochemistry. *Appl Immunohistochem* 1993, 1: 256-266.
35. Constantinoiu CC, Molley JB, Jorgensen WK, Coleman GT: Purification of immunoglobulins from chicken Sera by thiophilic gel chromatography. *Poultry Science* 2007, 86: 1910-1914.
36. Petrie A, Watson P: *Statistics for veterinary and animal science*. Blackwell Science Ltd. UK; 1999.

CHAPTER 10

GENOMIC ANALYSIS OF THE CHICKEN INFECTIOUS ANEMIA VIRUS IN A SPECIFIC PATHOGEN-FREE CHICKEN POPULATION

Yang Li, Yixin Wang, Lichun Fang, Jiayuan Fu, Shuai Cui, Yingjie Zhao, Zhizhong Cui, Shuang Chang, and Peng Zhao

INTRODUCTION

Chicken infectious anemia virus (CIAV) can cause the atrophy of bone marrow hematopoietic and lymphoid tissues (e.g., thymus) in young chickens, leading to aplastic anemia and immune suppression. The infection has been very common in chicken populations worldwide [1–4] since first reported CIAV in 1979 [5].

CIAV belongs to the Gyrovirus genus of the Circoviridae family, and the genome has three open reading frames (ORFs) which encode VP1, VP2, and VP3 [2]. VP1 and VP2 are protective antigen proteins which are encoded to produce neutralizing antibodies [6]. Except for VP1, the amino acid (AA) composition of CIAV was relatively conserved. Previous studies have indicated that AAs 139–151 are a highly variable region of the VP1 gene, and AA 139 or 144 could especially affect virus replication and cell infection [7]. Generally speaking, if both AA 139 and 144 in the loci are glutamines, replication and infective ability of the virus are relatively weaker [7].

It was reported that CIAV often contaminated attenuated vaccines to infect chicken populations [4, 8], and the use of CIAV-contaminated specific pathogen-free (SPF) chicken embryos is one of the important reasons for CIAV infection of avian attenuated vaccine. In the current study, a CIAV strain was isolated from a Chinese SPF chicken population, and the sequence of its full genome was analyzed.

MATERIALS AND METHODS

Background of the SPF Chicken Population

According to guidelines, the detection of antibodies in SPF chicken flocks must be carried out regularly to evaluate the infection status of different pathogenic bacteria and viruses. CIAV antibody test kit (IDEXX, Beijing, China) was used to investigate the positive rate of CIAV antibody in one of SPF flocks in China. The positive rate of CIAV antibody in this flock was 20%. Blood samples were collected from those chickens, and blood DNA was isolated using a DNA extraction kit (Omega, USA). Subsequently, PCR experiments were carried out according to the methods in previous studies to detect viral genome DNA of CIAV [8].

Isolation and Detection of Virus

Blood samples positive for CIAV antibody were collected from ten individual chickens and then inoculated into MDCC-MSB1 cells. This cell line was purchased from ATCC and passaged in our lab for fifteen generations. The blind passage was carried out every three days, until the emergence of cytopathogenic effects (CPE). After CPE, the media of MSB1 cells were collected and repeatedly frozen and thawed three times. The media were then centrifuged, and DNA was extracted using

Citation: Yang Li, Yixin Wang, Lichun Fang, et al., "Genomic Analysis of the Chicken Infectious Anemia Virus in a Specific Pathogen-Free Chicken Population," BioMed Research International, vol. 2016, Article ID 4275718, 5 pages, 2016. doi:10.1155/2016/4275718.

a DNA extraction kit (Tiangen, Beijing, China) from the supernatant. PCR experiments for identification were carried out according to previous studies [8].

Whole Genome Sequencing

According to published CIAV genome sequences (Table 1), three pairs of primers were designed (Primer Premier 5.0) and three overlapping fragments were amplified, respectively (Table 2). Those fragments were subcloned into PMD-18T vectors (Takara, Dalian, China) and sequenced by Sanger method (Sangon, Shanghai, China). Then, the whole genome sequences of the isolated strains were obtained by assembling those overlapping fragments using Lasergene 7.1 software.

Table 1: Information of CIAV reference strains and amino acids in highly variable regions of VP1 protein

Strains	Country and time	Accession number	Whole length	Amino acid position of VP1	
				139#	144#
M81223	1993, Germany	M81223	2298 bp	Lysine	Aspartate
CAU65414	1996, Australia	CAU65414	2298 bp	Glutamine	Glutamine
TR20	1999, Japan	AB027470	2298 bp	Glutamine	Glutamine
L14767	1999, USA	L14767	2298 bp	Glutamine	Glutamine
A2	2000, Japan	AB031296	2298 bp	Lysine	Glutamine
AF227982	2001, Australia	AF227982	2286 bp	Lysine	Glutamine
AB046590	2001, Japan	AB046590	2298 bp	Lysine	Glutamine
AF475908	2002, China	AF475908	2298 bp	Lysine	Glutamine
clone 33	2002, Germany	AJ297684	2298 bp	Lysine	Glutamine
SMSC-1P60	2003, Malaysia	AF390102	2298 bp	Lysine	Glutamine
BD-3	2004, Bangladesh	AF395114	2298 bp	Glutamine	Glutamine
TJBD40	2004, China	AY846844	2298 bp	Lysine	Glutamine
AH6	2005, China	DQ124935	2298 bp	Glutamine	Glutamine
TJBD33	2005, China	AY843527	2298 bp	Lysine	Glutamine
SMSC-1P123WT	2005, Malaysia	DQ217401	2298 bp	Lysine	Glutamine
CAE26P4	2007, Netherlands	D10068	2298 bp	Lysine	Glutamine
3711	2007, Australia	EF683159	2279 bp	Lysine	Glutamine
C14	2007, China	EF176599	2298 bp	Lysine	Glutamine
01-4201	2007, USA	DQ991394	2298 bp	Lysine	Glutamine
SDLY08	2008, China	FJ172347	2298 bp	Lysine	Glutamine
CAECA123	2008, Japan	D31965	2319 bp	Lysine	Glutamine
AB119448	2009, Japna	AB119448	2298 bp	Glutamine	Glutamine
98D02152	2010, USA	AF311892	2298 bp	Lysine	Glutamine
GD-1-12	2012, China	JX260426	2298 bp	Lysine	Glutamine
GD-K-12	2013, China	KF224935	2298 bp	Glutamine	Glutamine
GD-L-12	2013, China	KF224936	2298 bp	Lysine	Glutamine
CIAV89-69	2013, Korea	JF507715	2298 bp	Lysine	Glutamine
CAV-10	2014, Argentina	KJ872513	2298 bp	Glutamine	Glutamine
SC-MS	2014, China	KM496305	2298 bp	Lysine	Glutamine
NC001427	2015, USA	NC001427	2319 bp	Lysine	Aspartate
SD1403	2016, China	KU221054	2298 bp	Glutamine	Glutamine

Table 2: Primers used for genome amplification

Primers	Sequence	Location	Temperature	Product length
F1	5′-GCATTCCGAGTGGTTACTATTCC-3′	# 1–23	55°C	842 bp
R1	5′-CGTCTTGCCATCTTACAGTCTTAT-3′	# 819–842		
F2	5′-CGAGTACAGGGTAAGCGAGCTAAA-3′	# 743–767	55°C	990 bp
R2	5′-TGCTATTCATGCAGCGGACTT-3′	# 1712–1732		
F3	5′-ACGAGCAACAGTACCCTGCTAT-3′	# 1643–1664	55°C	802 bp
R3	5′-CTGTACATGCTCCACTCGTT-3′	# 87–151		

Sequence Analysis of Isolated Strain

In order to compare the homology of CIAV with different reference strains, the homology of the whole genome and its coding regions (VP1, VP2, and VP3) were analyzed by DNAstar software. Phylogenetic trees were reconstructed on the basis of sequences from the whole genome, VP1, VP2, and VP3 respectively, using MEGA5.1 (NJ) software. The mutation sites of AA from the 139 to 151 region of the VP1 loci in different strains were analyzed statistically.

Nucleotide Sequence Accession Numbers. The GenBank accession number for the sequences of SD1403 was KU221054.

RESULTS

The Isolation and Identification of the Virus

10^6 MDCC-MSB1 cells cultured in six-well plates were inoculated with 100 μL blood from ten individual SPF chickens which might be infected with CIAV, and, after four blind passages, the cells infected with one blood sample displayed CPEs. Some cells became round and enlarged, and lysis or disintegrating of cells could be observed, with the number of viable cells decreasing, while the morphology of uninoculated MSB1 cells remained normal (Figure 1). A specific PCR band of about 600 bp was amplified from the DNA of MSB1 cells showing CPE (Figure 2). The result indicated that a CIAV strain was isolated successfully and named as SD1403.

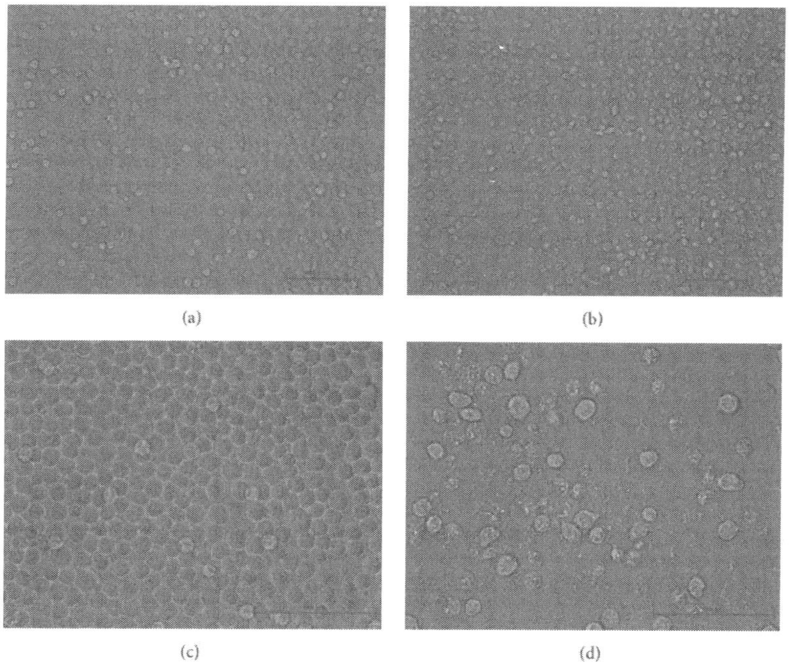

Figure 1: Cytopathic effect of CIAV isolate on MSB1 cells. (a) Normal MSB1 cells (200x). (b) CPE of MSB1 cells infected with CIAV (200x). (c) Normal MSB1 cells (400x). (d) CPE of MSB1 cells infected with CIAV (400x).

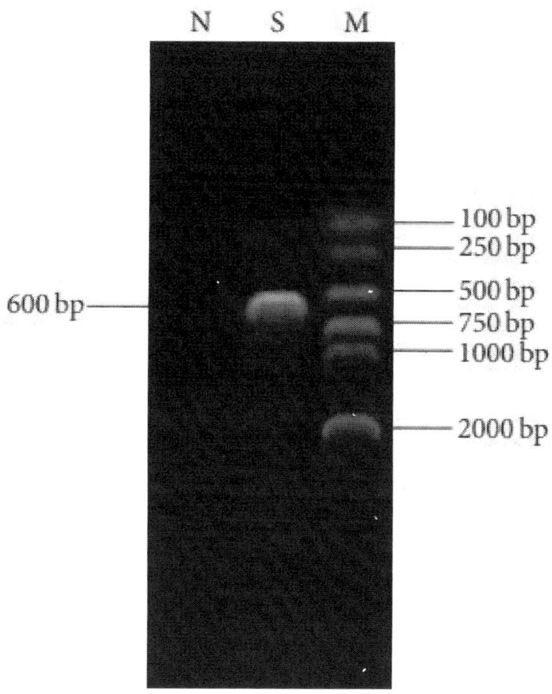

Figure 2: Identification by PCR amplification of the CIAV isolates. N: MSB1 cell without CIAV infection as negative control; S: sample detected in the current study; M: marker.

Genome Sequencing of the Isolate

The full genome of SD1403 strain was 2298 bp, including three ORFs (VP1, VP2, and VP3). VP1, VP2, and VP3 were 1350 bp, 651 bp, and 366 bp in length, respectively. Comparison between the whole genome sequences of the SD1403 strain and 30 other strains was carried out, and the results demonstrated that the homology was 94.8%–98.5%. A phylogenic tree was constructed using whole genome sequence of SD1403 showing that SD1403 was closely related to the GD-K-12 strain, which has low pathogenicity and was isolated in China in 2013 (Figure 3(a)).

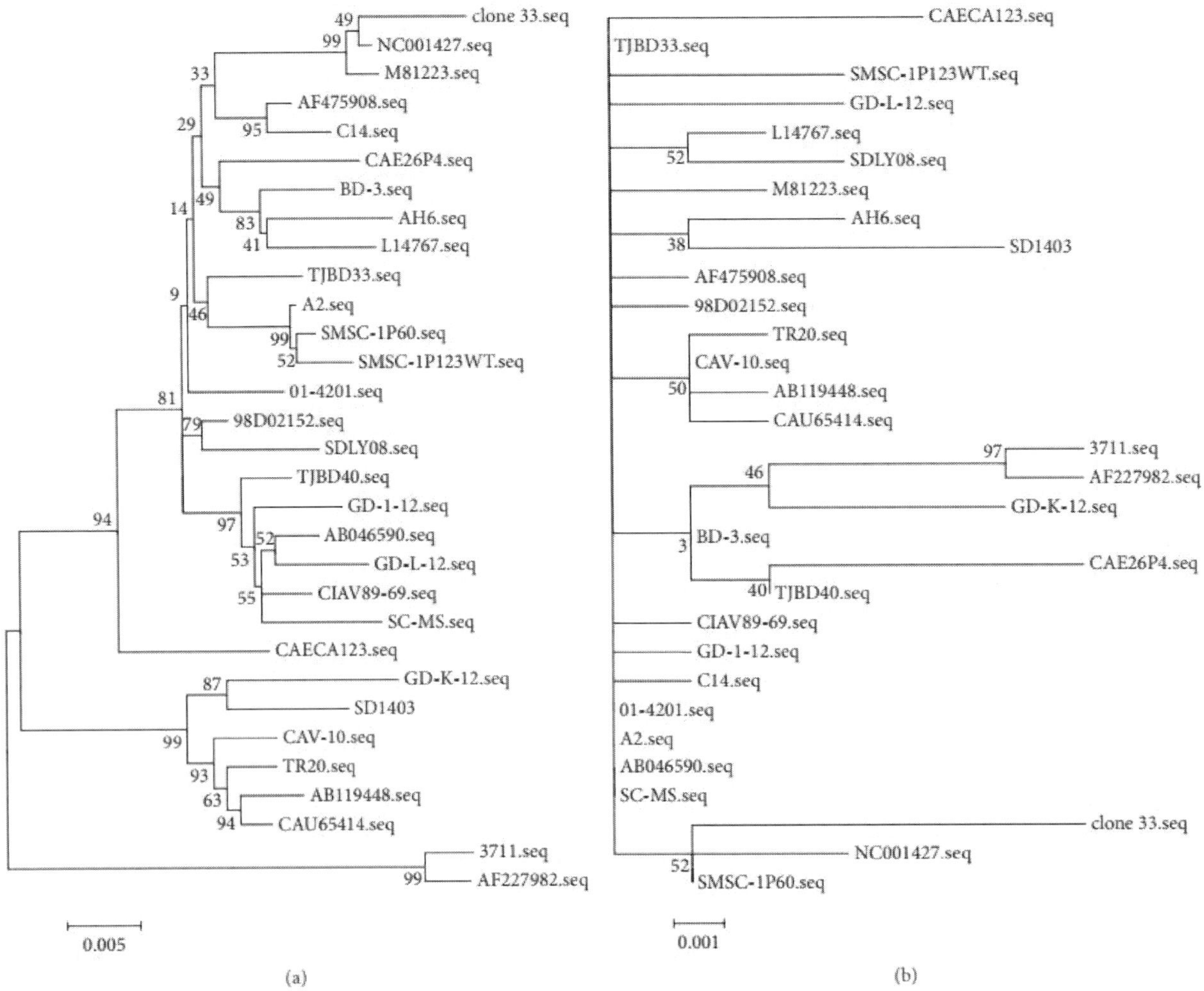

Figure 3: Phylogenetic analysis of different CIAV isolates based on whole genome sequence and VP3 protein. (a) Phylogenetic analysis of CIAV isolates based on whole genome. (b) Phylogenetic analysis of VP3 protein from different CIAV isolates.

Compared with other reference strains, the homology of VP1, VP2, and VP3 from the SD1403 strain was 97.7%–98.3%, 98.3%–99.2%, and 97.8%–99.5%, respectively. It was shown that the sequences of the three genes were highly conserved. VP2 was more conserved than VP1, with the most conserved gene being VP3. The phylogenetic trees were constructed on the basis of VP1, VP2, and VP3 proteins. Phylogenetic tree which was constructed on the basis of VP1 protein was similar to that constructed on the basis of whole genome. It showed that SD1403 and GD-K-12 were in the same clade, indicating that VP1 was more representative between the strains. The SD1403 and GD-K-12 strains were isolated in China and had common characteristics, with AA 139 and 144 of VP1 being glutamine (Table 1). Phylogenetic tree constructed on the basis of VP2 protein was similar to that constructed on the basis of VP3 protein, indicating that both of them were relatively conserved (Figure 3(b)).

DISCUSSION

A lot of attenuated vaccines for poultry have been confirmed to contain some exogenous viruses. The most reported viruses are Avian leukosis virus (ALV) [9, 10] and Reticuloendotheliosis virus (REV) [11–13], with CIAV also being reported [8]. The main reason for this phenomenon is that vaccines made from SPF chicken embryos contaminated by viruses have been used. These viruses (ALV, CIAV, REV, etc.) are mainly transmitted vertically. In China, it has been previously reported that SPF chicken populations were infected with ALV [14]. This is the first report of a CIAV strain isolated and identified in a SPF chicken population in China.

To analyze the genetic variation of this strain and track its possible origin, we sequenced its whole genome. Similar to other CIAV strains, the genomic sequence of SD1403 was highly conserved, and the homology of nucleotide sequence with other reference strains was 94.8%–98.5%. From phylogenetic trees constructed on the basis of genomic sequences, the SD1403 and GD-K-1 were closely related, and results were very similar to the phylogenetic trees constructed on the basis of VP1, showing that VP1 is more representative between the strains.

It was reported that pathogenicity and infectivity of GD-K-12 were weak [3], so we speculated that SD1403 might also be an attenuated strain. Previous studies indicated that if AA 139 and/or AA 144 of VP1 were glutamines, the infectivity and replication were weak. Both AA 139 and 144 of VP1 from SD1403 were glutamines, and these traits were in accordance with GD-K-1 and AH6 strains which were isolated in China. We observed that the SPF chicken population might have been infected with the SD1403 strain for a long time, and the results found that it did not show obvious pathogenicity. It also revealed that SPF chickens raised at the same time did not have pathogenicity, indicating that the pathogenicity and transmissibility of SD1403 strain were weak. According to reports, vaccines contaminated with ALV and REV might cause more serious damage, maybe because ALV and REV have relatively strong pathogenicity [15]. The transmissibility and pathogenicity of SD1403 were weak without showing clinical signs and symptoms. Perhaps this was an important reason why chicken populations were infected with SD1403 latently. Maybe because ALV, REV, and CIAV have different pathogenicity, the avian vaccines contaminated by REV and ALV were more than those by CIAV in the past years.

Although the CIAV infection in SPF chicken was relatively weak, we could not ignore its potential damage. On the one hand, when vaccine contaminated with CIAV is used in a younger chicken population, the pathogenicity of CIAV might be enhanced; on the other hand, if some chicken populations are infected with both ALV and REV, the pathogenicity of CIAV could also be enhanced [16]. Therefore, we must keep detecting CIAV in SPF chicken populations and attenuated vaccines.

COMPETING INTERESTS

The authors declare that they have no competing interests.

AUTHORS' CONTRIBUTIONS

Yang Li and Yixin Wang have contributed equally to this work.

REFERENCES

1. J. K. Rosenberger and S. S. Cloud, "The isolation and characterization of chicken anemia agent (CAA) from broilers in the United States," Avian Diseases, vol. 33, no. 4, pp. 707–713, 1989.
2. S. Natesan, J. M. Kataria, K. Dhama, S. Rahul, and N. Baradhwaj, "Biological and molecular characterization of chicken anaemia virus isolates of Indian origin," Virus Research, vol. 118, no. 1-2, pp. 78–86, 2006.
3. Y. M. Eltahir, K. Qian, W. Jin, P. Wang, and A. Qin, "Molecular epidemiology of chicken anemia virus in commercial farms in China," Virology Journal, vol. 8, article 145, 2011.
4. P. Bhatt, S. K. Shukla, M. Mahendran, K. Dhama, M. M. Chawak, and J. M. Kataria, "Prevalence of Chicken Infectious Anaemia Virus (CIAV) in commercial poultry flocks of Northern India: a serological survey," Transboundary and Emerging Diseases, vol. 58, no. 5, pp. 458–460, 2011.
5. N. Yuasa, I. Yoshida, and T. Taniguchi, "Isolation of a reticuloendotheliosis virus from chickens inoculated with Marek›s disease vaccine," National Institute of Animal Health Quarterly, vol. 16, no. 4, pp. 141–151, 1976.
6. G. Koch, D. J. van Roozelaar, C. A. J. Verschueren, A. J. van der Eb, and M. H. M. Noteborn, "Immunogenic and protective properties of chicken anaemia virus proteins expressed by baculovirus,"Vaccine, vol. 13, no. 8, pp. 763–770, 1995.
7. R. W. Renshaw, C. Soiné, T. Weinkle et al., "A hypervariable region in VP1 of chicken infectious anemia virus mediates rate of spread and cell tropism in tissue culture," Journal of Virology, vol. 70, no. 12, pp. 8872–8878, 1996.
8. H. Shao, P. Wang, M. Wu, W. Jin, K. Qian, and A. Qin, "Rapid PCR approach for detecting the contamination of chicken anemia virus in avian live vaccines," China Poultry, vol. 21, article 10, 2012.
9. M. Fadly, R. L. Witter, E. J. Smith et al., "An outbreak of lymphomas in commercial broiler breeder chickens vaccinated with a fowlpox vaccine contaminated with reticuloendotheliosis virus," Avian Pathology, vol. 25, no. 1, pp. 35–47, 1996.
10. P. Zhao, X. Dong, and Z. Cui, "Isolation, identification, and gp85 characterization of a subgroup A avian leukosis virus from a contaminated live Newcastle Disease virus vaccine, first report in China," Poultry Science, vol. 93, no. 9, pp. 2168–2174, 2014.
11. M. Awad, H. S. A. El-Hamid, A. A. A. Rawash, and H. H. Ibrahim, "Detection of reticuloendotheliosis virus as a contaminant of fowl pox vaccines," Poultry Science, vol. 89, no. 11, pp. 2389–2395, 2010.
12. S. K. Biswas, C. Jana, K. Chand, W. Rehman, and B. Mondal, "Detection of fowl poxvirus integrated with reticuloendotheliosis virus sequences from an outbreak in backyard chickens in India," Veterinaria Italiana, vol. 47, no. 2, pp. 147–153, 2011.
13. FadlyandM.C.Garcia,"Detectionofreticuloendotheliosis virus in live virus vaccines of poultry,"Developments in Biologicals, vol. 126, pp. 301–305, 2006.
14. T. Barbosa, G. Zavala, and S. Cheng, "Molecular characterization of three recombinant isolates of avian leukosis virus obtained from contaminated Marek›s disease vaccines," Avian Diseases, vol. 52, no. 2, pp. 245–252, 2008.
15. K. Wei, Z. Sun, S. Zhu et al., "Probable congenital transmission of reticuloendotheliosis virus caused by vaccination with contaminated vaccines," PLoS ONE, vol. 7, no. 8, Article ID e43422, 2012.
16. Y. Li, Z. Z. Cui, S. Jiang, and H. Guo, "Synergic inhibitory effect of co-infection of CAV and REV on immune responses to vaccines in SPF chickens," Chinese Journal of Veterinary Science, vol. 28, no. 11, pp. 1243–1246, 2008.

CHAPTER 11

PATHOGENICITY AND COMPLETE GENOME CHARACTERIZATION OF FOWL ADENOVIRUSES ISOLATED FROM CHICKENS ASSOCIATED WITH INCLUSION BODY HEPATITIS AND HYDROPERICARDIUM SYNDROME

Jing Zhao, Qi Zhong, Ye Zhao, Yan-xin Hu, Guo-zhong Zhang

INTRODUCTION

Adenoviruses (AdVs) are non-enveloped double stranded DNA-viruses, which belong to the family Adenoviridae. The family Adenoviridae is currently divided into five genera: Mastadenovirus, Aviadenovirus, Atadenovirus, Siadenovirus, and Ichtadenovirus [1, 2]. Chickens can be infected by fowl adenoviruses (FAdVs), belonging to the genus Aviadenovirus, egg drop syndrome (EDS) virus (duck adenovirus A, genus Atadenovirus) and turkey hemorrhagic enteritis (HE) virus (turkey adenovirus A, genus Siadenovirus).

FAdVs have a worldwide distribution and have been grouped into 5 species (FAdV-A to FAdV-E) with 12 serotypes (FAdV-1 to 8a and -8b to 11) based on restriction enzyme digest pattern and serum cross-neutralization test [3]. The most notable diseases associated with FAdVs-infection in chicken are the inclusion body hepatitis (IBH), the hydropericardium syndrome (HPS) and the gizzard erosions (GE) [4–6]. All 12 serotypes of FAdVs have been associated with outbreaks of IBH. Classical IBH affects mostly 3–5 weeks old chickens and is characterized by mortality approaching 10% and hepatic necrosis with microscopic eosinophilic or basophilic intranuclear inclusion bodies in hepatocytes [7, 8]. Most HPS are caused by FAdV serotype 4 (FAdV-4), characterized by accumulation of transparent or straw-colored fluid in the pericardial sac, nephritis, and hepatitis with a high mortality of 30–70% [9, 10]. GE is also induced by several serotypes of FAdVs and frequently found in slaughtered broiler chickens [11–13].

In recent years, the clinical cases of IBH and HPS have been increasing all over the world, resulting in considerable economic losses in many countries, such as USA [14], India [15], Canada [8], Hungary [16], Korea [17] and Japan [7]. A few confirmed outbreaks of IBH have been previously reported in China [18]. However, the number of IBH or HPS cases submitted to and diagnosed by the Diagnostic Center of Livestock and Poultry Epidemic Diseases at China Agricultural University has been increasing since 2012. Most cases were observed in broilers of 3 to 5 weeks of age, and occasionally in layers and breeder pullets aged 10 to 20 weeks. Mortality reported during these outbreaks varied from a slight increase to over 30% in severe cases and FAdVs were isolated from all the cases, but the molecular and pathogenic characterization of these viruses involved in IBH or HPS outbreaks were still unknown.

In the present study, three FAdV strains isolated from recent outbreaks which showed higher mortality rate in the affected flocks were characterized by phylogenetic assays and pathogenic analysis. We determined the complete nucleotide sequence of viral genome, investigated their genomic organization and relationship with other FAdVs, and selected two different isolates for further observation of their pathogenicity in SPF chickens.

Citation: Zhao J, Zhong Q, Zhao Y, Hu Y-x, Zhang G-z (2015) Pathogenicity and Complete Genome Characterization of Fowl Adenoviruses Isolated from Chickens Associated with Inclusion Body Hepatitis and Hydropericardium Syndrome. PLoS ONE 10(7): e0133073. doi:10.1371/journal.pone.0133073

MATERIALS AND METHODS

Animal and Ethics Statement

One hundred and five 3-week-old specific-pathogen-free (SPF) chickens were used to determine the pathogenicity of two FAdV strains. All chickens were kept in isolators at China Agricultural University throughout the experiment and the rearing facilities were approved by Beijing Administration Committee of Laboratory Animals under the leadership of the Beijing Association for Science and Technology, the approval ID is XYXK (Jing) 2013–0013. The study was carried out in strict accordance with the Guidance for the Care and Use of Laboratory Animals formulated by the Ministry of Science and Technology of the People's Republic of China. The protocol (including the possibility of animal death without euthanasia) was specifically considered and approved by the Animal Welfare and Ethical Censor Committee at China Agricultural University. All surgery was performed under sodium pentobarbital anesthesia, and all efforts were made to minimize suffering.

Origin of the Strains

Three FAdV isolates, designated as HBQ12, BJH13 and JSJ13 respectively, were isolated from different geographical areas of China in 2012–2013 from livers of IBH-affected birds which presented with an accumulation of near-amber fluid in the pericardial sac and an enlarged discolored liver with foci of hemorrhage and/or necrosis. The isolates were further identified by polymerase chain reaction (PCR) with a pair of primers (forward: TGCTCGTTGTGGATGGTGAA; reverse: CTCCGTGTTGGGCTGGTC) based on the polymerase gene of the available FAdV nucleotide sequences strain A-2A (GenBank Accession No. AF083975), which specifically amplify a 564-bp fragment of FAdVs. Reactions were performed according to the following protocol: 95°C for 5 min, followed by 30 cycles of 95°C for 45 s, 58°C for 45 s, 72°C for 45 s, and a final elongation step of 10 min at 72°C. PCR products were analyzed by electrophoresis.

The isolates were passaged three times in SPF embryonated chicken eggs via the chorioallantoic membrane (CAM) route following standard method [19]. The CAM from infected embryos was harvested and homogenized up to 10% (w/v) in 0.01 M phosphate-buffered saline (pH 7.4), and clarified by low-speed centrifugation. The supernatant was passed through a 0.45-µm filter. Titration of the virus was carried out by inoculating serial ten-fold dilutions of CAM homogenate into 10-day-old SPF embryonated chicken eggs via the CAM. The median embryo infective dose (EID50) of the FAdV isolate was analyzed using the formula of Reed and Muench [20]. Infected CAM homogenate with a titer of >103.0 EID50/0.2 ml was used as the source of virus for the infection.

Primer Design and DNA Extraction

Based on the available FAdV nucleotide sequences strain A-2A (FAdV-D, GenBank Accession No. AF083975) and ON1 (FAdV-C, GenBank Accession No. GU188428), specific primers were designed to amplify the complete genome sequences of HBQ12, BJH13 and JSJ13, as shown in S1 and S2 Tables. All the primers were synthetized by Sangon Biotech (Shanghai, China). Viral DNAs from infected CAM homogenates were extracted using the DNAVzol (Vigorous, Beijing, China) according to the manufacturer's instructions and then used for the PCR amplification.

PCR and Sequencing

PCR amplification was performed in a thermocycler (Biometra, Goettingen, Germany) using the designed specific primers. Briely, 50 µl PCR reaction mixture containing 1µl (10 pmol) of each primer, 25µl Taq SuperMix (TransGen, Beijing, China), 4µl DNA and 19µl nuclease free water was prepared. Reactions were performed according to the following protocol: 95°C for 5 min, followed by 30 cycles of 95°C for 45 s, 53°C to 60°C for 45 s, 72°C for 1.5 min, and a final elongation step of 10 min at 72°C. PCR products were examined by electrophoresis on a 1.0% (w/v) agarose gel and visualized after Goldview staining. PCR products of the expected length were sequenced directly or purified with a Gel Extraction kit (Omega, Norcross, GA, USA), then cloned into the pEASY-T1 cloning vector (TransGen, Beijing, China) according to the manufacturer's instructions and sequenced at BGI (Beijing, China). The sequencing was performed using the Sanger dideoxy sequencing method with an ABI 3730XL automatic sequencing apparatus. Both strands of the cloned DNA fragments were sequenced and discrepancies between sequences of the two strands were resolved by repeated reactions until a consensus was achieved. The complete sequences of three FAdV isolates were manually assembled using Seqman program in the DNAstar software package (version 5.01, DNAstar, Madison, WI, USA).

Sequence Alignment and Phylogenetic Analysis

The complete genome sequences of the three FAdV isolates were aligned with other available FAdV genome sequences in the GenBank database to determine the nucleotide sequence homologies using the ClustalW multiple alignment algorithm in the MegAlign program of the DNAstar software suite (version 5.01, DNAstar, Madison, WI, USA). A phylogenetic tree was constructed using MEGA4.1 software (Molecular Evolutionary Genetics Analysis, version 4.1) by Neighbor-Joining method (1000 replicates for bootstrap). The evolutionary distances were computed by Pairwise Distance method using the Maximum Composite Likelihood Model [21].

Pathogenicity Assessment of two FAdV Isolates in SPF Chickens

One hundred and five 3-week-old SPF chickens were used to determine the pathogenic potential of two FAdVs. The chickens were randomly divided into three groups, with 35 birds in each group. Chickens were inoculated with either 103.5 EID50 of strain HBQ12 or JSJ13 via an oral route. The control birds received the same volume of phosphate-buffered saline (PBS). All birds were monitored twice daily and scored for clinical signs for 21 days. Clinical signs were given daily clinical scores: 0 for normal, 1 for mild depression, 2 for severe depressed, 3 for paralysis/prostration, and 4 for death. At 3 and 5 days post-infection (dpi), 4 chickens from each group were euthanized with intravenous injection of sodium pentobarbital (200 mg/ml), and heart, liver, spleen, lung, kidney, bursa, cecal tonsil, proventriculus, pancreas, thymus, trachea and small intestine were collected. Some birds died between the intervals of clinical observation and same tissue samples were collected. Tissue samples were placed into 10% formalin for histology or stored at -70°C for DNA detection using PCR method. In addition, 10 serum samples per group collected at 14 and 21 dpi were tested for FAdV-specific antibodies (Abs) by a commercial enzyme-linked immunosorbent assay (ELISA) kit (BioChek, Scarborough, ME, USA).

Viral DNA Detection of Tissue Samples

Collected tissue samples were used to determine tissue distribution of the virus by PCR. Total DNA was extracted from 200 μL of the homogenized tissue samples with DNAVzol Reagent (Vigorous, Beijing, China) according to manufacturer's instructions. PCR was performed as described in section "Origin of the strains".

Histopathology

Samples of liver, kidney, proventriculus and small intestine were fixed in 10% neutral formalin for 48h at room temperature. Tissues were then routinely processed, embedded in paraffin wax, and cut into 5-μm sections. The sections were stained with hematoxylin and eosin (H&E) and examined for lesions associated with FAdV infection using light microscopy. The following scoring scheme was used to assess histopathological lesions in sampled tissues: 0 for no lesion, 1–3 for mild lesions, 4–6 for morderate lesions and 7–10 for severe lesions.

Immunohistochemistry

FAdV antigen was detected in liver by immunohistochemical staining. In brief, 5-μm liver tissue sections were incubated in 10% normal goat serum in PBS for 30 min to block nonspecific binding sites. Slides were further incubated with chicken anti-FAdV hyperimmune serum (1:1000 dilution in 0.01 M PBS, Tiantech, Beijing, China) for 1 h, followed by incubation with a horseradish peroxidase-conjugated rabbit anti-chicken IgG (Solarbio, Beijing, China) for 1 h. The reaction was visualized by the addition of 3,3-diaminobenzidine (Sigma, St. Louis, MO, USA) for 15 min. The sections were counterstained with hematoxylin, air dried, and examined by light microscopy. The numbers of positive cells were quantitatively analyzed under a light microscope at ×40 magnification.

Statistical Analysis

Statistical analysis was performed by two-way analysis of variance (ANOVA) contained in the Prism 5.0 software package (GraphPad software Inc., San Diego, CA, USA), and a P value of <0.05 was considered statistically significant. Results are expressed as means and standard deviations.

RESULTS

Genome Size and Organization of three FADV Isolates

The whole genome nucleotide sequences of HBQ12, BJH13 and JSJ13 isolates are available in the GenBank database under the accession numbers KM096545, KM096546 and KM096544, respectively. The full genome for HBQ12, BJH13 and JSJ13 was found to be 44,081, 43,966 and 43,756 base pairs (bp) in length, respectively. The strain HBQ12 genome contained two regions of repeated sequences TR-1 (the shorter repeat region) and TR-2 (the longer repeat region), as shown in Fig 1. Unlike other FAdV-D strains, TR-1 of HBQ12 contained three identical and contiguous 33 bp-direct repeats. The exact nucleotide sequence of TR-2 was found to consist of 5 identical and contiguous, 135 bp long direct repeats. Strain BJH13 only contained TR-2 and JSJ13 isolate had no repeated sequences of TR-1 or TR-2. Both inverted terminal repeat (ITR) sequences in strain HBQ12 and BJH13 were 72 bp, while in JSJ13 isolate, it is 56 bp in length.

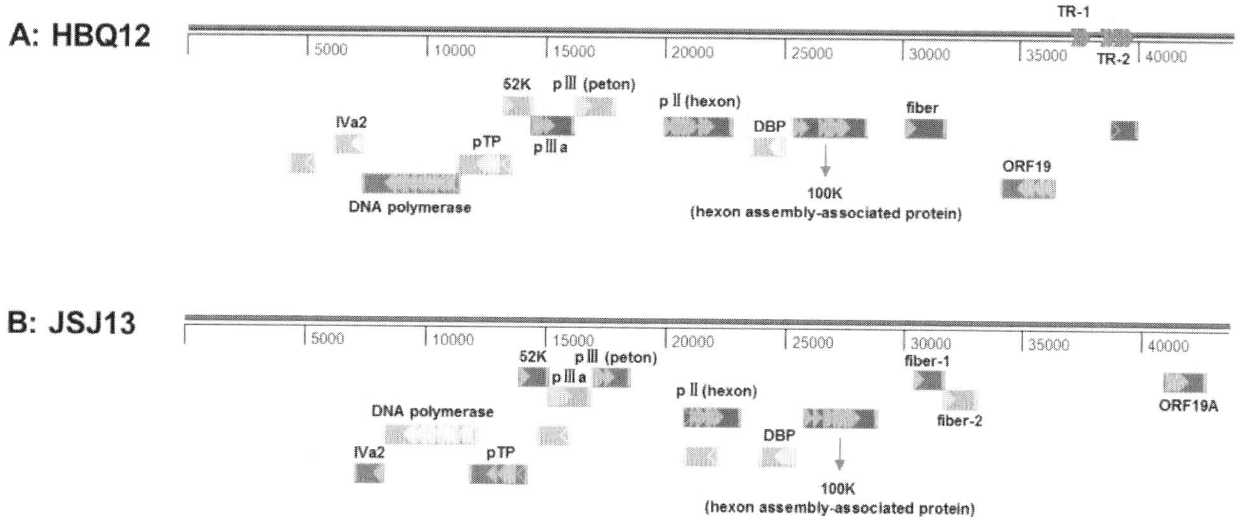

Figure 1: Schematic representation of HBQ12 and JSJ13 genomes, members of the genus Aviadenovirus.

Sequence Alignment and Analysis

The percent sequence identity for available aviadenovirus whole genomes are given in Table 1. Strain HBQ12 and BJH13 were almost identical (99.7%) at the nucleotide level and they showed the highest sequence homology (95.8%) with strain A-2A (FAdV-D, Accession No.AF083975) isolated in the US at nucleotide level. Whereas they showed a low sequence identity (< 61.0%) with the members of other aviadenovirus species. JSJ13 strain was more matched to the KR-5 strain (FAdV-C) isolated in Japan (Accession No. HE608152, 98.4% of identity at the nucleotide level). Sequence identities between JSJ 13 and other species ranged from 38.6% (between JSJ13 and FAdV-A) to 49.7% (between JSJ13 and FAdV-E).

Table 1: Percent nucleotide sequence identities of the whole genomes of aviadenoviruses

Species	Strain	Accession number	Homology		
			HBQ12	BJH13	JSJ13
FAdV-A	CELO	U46933	39.2	39.3	38.7
FAdV-B	340	NC_021221	60.7	61.0	39.9
FAdV-C	ON1	GU188428	39.0	39.1	**98.0**
FAdV-C	KR-5	HE608152	39.1	39.1	**98.4**
FAdV-D	A-2A	AF083975	**95.8**	**95.8**	39.1
FAdV-E	HG	GU734104	56.2	58.1	49.7
FAdV-D	HBQ12	KM096545	/	100	39.4
FAdV-D	BJH13	KM096546	100	/	39.3
FAdV-C	JSJ13	KM096544	39.4	39.3	/

[a]Alignment was performed by using ClustalW. Strain HBQ12, BJH13 and JSJ13 were sequenced in this study and other aviadenovirus strains were obtained from the GenBank database.

Phylogenetic Analysis of Three FADV Strains

According to phylogenetic analysis based on the complete genome (Fig 2A), strain HBQ12 and BJH13 were classified into the same cluster. Both of them belonged to FAdV-D and had a close genetic relationship with strain A-2A (GenBank Accession No. AF083975). Strain JSJ13 was classified into the same cluster (FAdV-C) with ON1 (GenBank Accession No. GU188428), which was reported as the reference strain of serotype 4 of FAdV. The similar evolutionary relationships were obtained from the phylogenetic tree based on the hexon gene (Fig 2B).

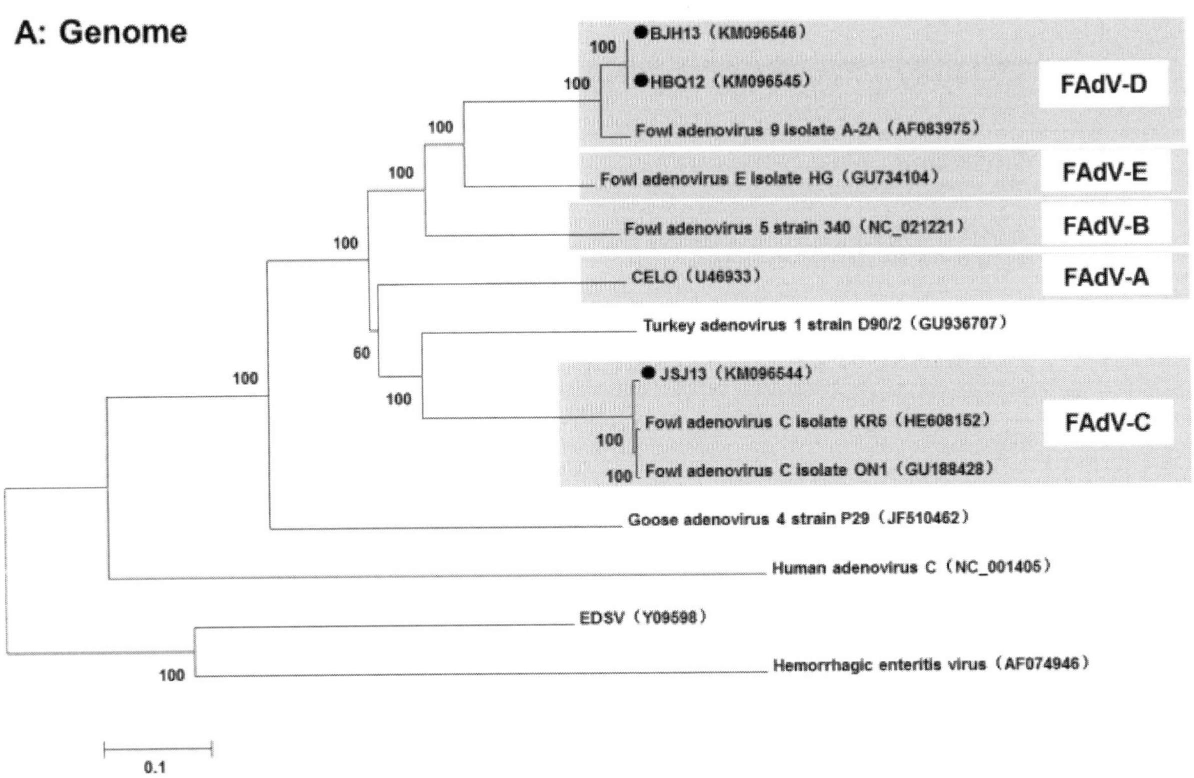

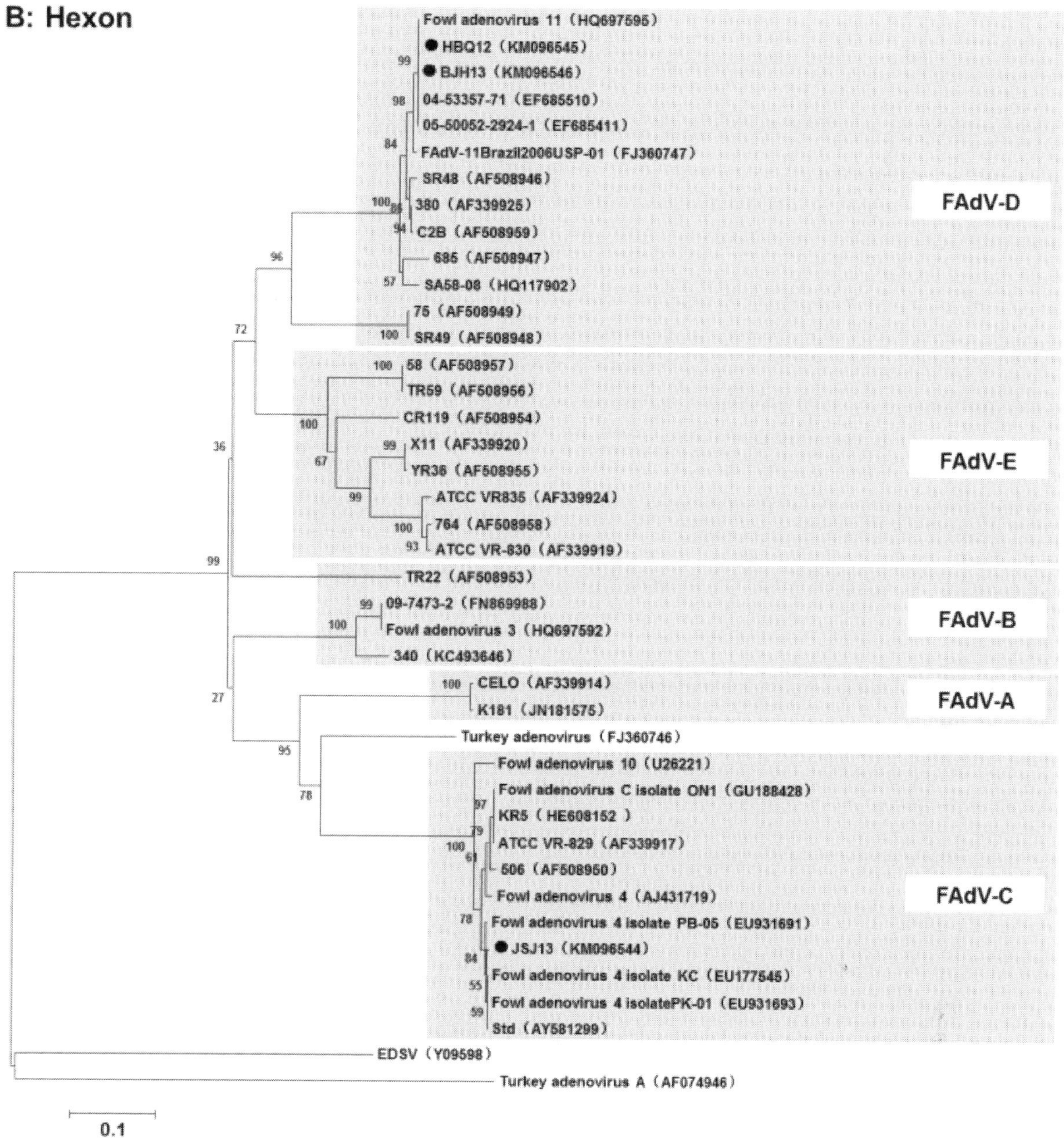

Figure 2: Phylogenetic tree based on the complete genome (A) and hexon gene (B) sequences of avian adenoviruses belonging to genera Aviadenovirus.

Strains HBQ12, BJH13 and JSJ13 were sequenced in this study, other avian adenovirus strains were previously sequenced. Phylogenetic tree was generated by the using Neighbor-Joining method in the MEGA4.1 software. Bootstrap values were indicated for the major nodes. Accession numbers of already published adenovirus sequences were given in the brackets.

Clinical Signs and Gross Pathology

Chickens inoculated with 103.5EID50 of HBQ12 or JSJ13 strain showed depression between 3 and 15 dpi. Flight reactions were significantly reduced. Mortality commenced from 4 dpi and continued for 6 days. During the experimental period, different clinical scores and mortality rates were observed in HBQ12- and JSJ13- infected animals. The clinical scores for HBQ12 infection were less than those of JSJ13 infection at 3 to 8 dpi ($P<0.05$) (Fig 3A). While 3 chicks died as a result of HBQ12 infection, 10 chicks died as a result of JSJ13 infection, and the survival rates were 91.4% and 71.4%, respectively, the survival rate showed significant difference between the JSJ13-infected group and the control group at the end of observation period ($P<0.05$), as shown in Fig 3B. At 3 and 5 days-post-inoculation, mild or severe hepatitis and hydropericardium were observed in necropsy. The pathological characteristics of all dead birds had a high similarity. The major post mortem findings were the swollen, yellow brown coloured liver with necrotic foci and the flabby heart with a severe hydropericardium (Fig 4B and 4C). The swollen and pale kidneys were also observed in most dead birds (Fig 4E and 4F). No significant gross lesions were present in the counterpart tissues of uninoculated control chickens (Fig 4A and 4D).

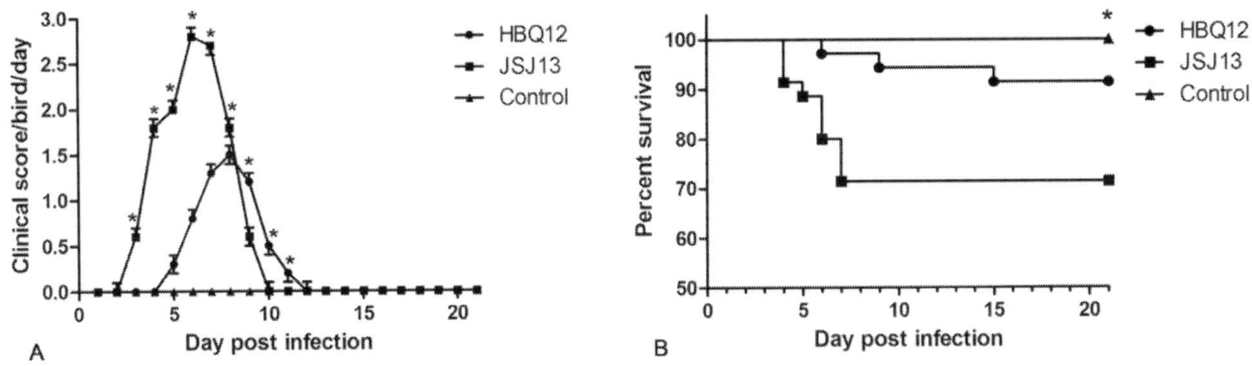

Figure 3: Clinical scores (A) and survival rates (B) of chickens after inoculation with FAdV HBQ12 or JSJ13 strain.

Clinical scoring: 0 for normal, 1 for mild depression, 2 for severe depressed, 3 for paralysis/prostration, and 4 for death. The mean scores per group per day are shown. The error bars indicate standard deviations. Asterisks (*) mark the days in which the clinical scores were significantly different between groups ($P < 0.05$). The percentage of birds that survived in the infected groups were significantly lower than in the control group ($P < 0.05$, log rank analysis).

http://dx.doi.org/10.1371/journal.pone.0133073.g003

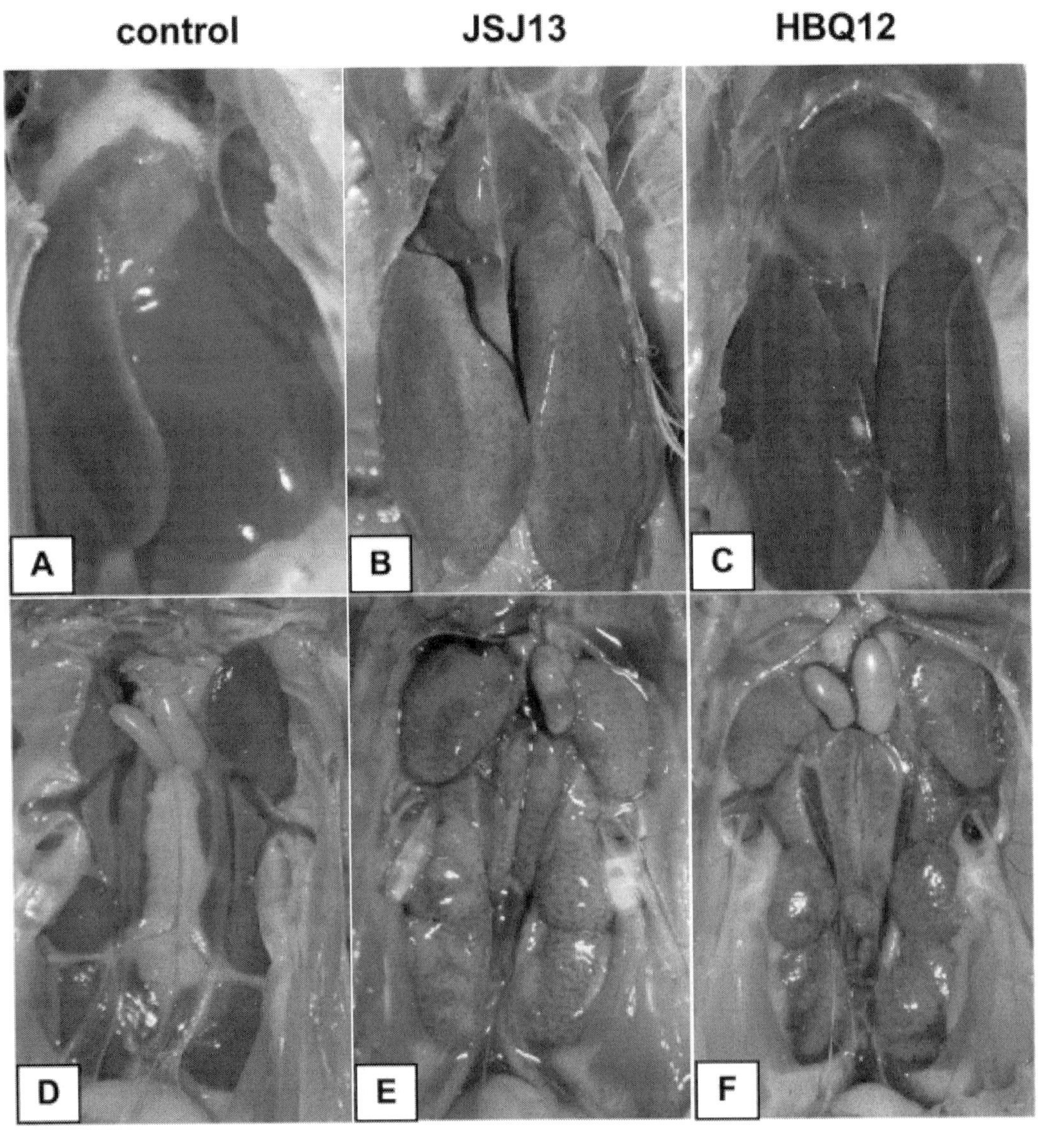

Figure 4: Gross lesions in liver, heart, and kidney from dead chickens at 6 dpi.

(B&C) Swollen and friable liver with multifocal areas of necrosis and petechial haemorrhage and hydropericardium. (E&F) Mild or severe enlargement in the kidney of the infected chickens.

Histology and Immunohistochemistry

All chickens infected with HBQ12 or JSJ13 presented massive histological liver lesions with large fat vacuoles and necrotic focal areas, and basophilic intranuclear inclusion body could be seen in the hepatic cell as indicated by solid arrows (Fig 5A and 5E). Dropout and necrosis of the mucosal epithelia was observed in the proventriculus (Fig 5B and 5F). Protein casts in renal tubules and extensive congestion in renal interstitium were observed in the kidney (Fig 5C and 5G). The pathological changes in small intestine included the villus missing, drop out and necrosis of epithelial cells, subserous congestion or dilatation, as well as congestion of blood capillary in mucosa lamina propria (Fig 5D and 5H). No significant histologic lesions were present in the uninoculated control chickens (Fig 5I, 5J, 5K and 5L). Compared with the control group, the histopathological lesions of four tissues in infected group were obvious ($P<0.05$). The birds infected with JSJ13 caused greater histopathological scores than those of HBQ12 infection ($P<0.05$) (Fig 6). The positive signals for viral antigen were detected extensively in the liver cells by using immunohistochemical staining (Fig 7A). In two infection groups, a large number of positive cells were observed ($P<0.05$) on 6 dpi compared with the control group, but the difference was not significant between two infected groups ($P>0.05$) (Fig 7B).

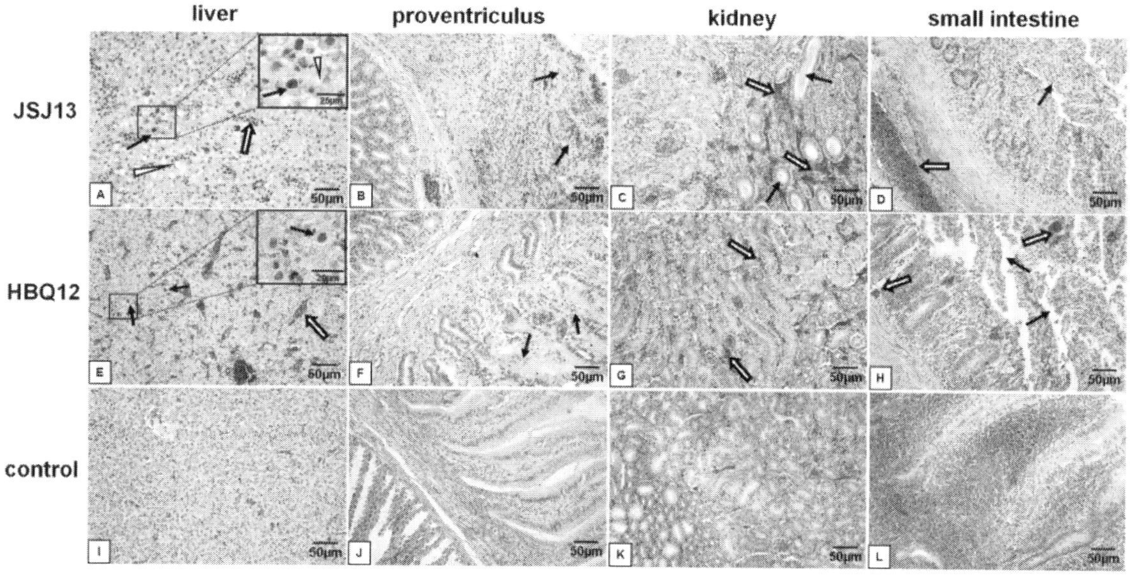

Figure 5: Histopathology in tissues from dead chickens infected with FAdV JSJ13 or HBQ12 (H&E) at 6 dpi.

Liver: solid arrows indicate viral inclusion bodies in liver cells, open arrows indicate congestion in hepatic sinusoid and the triangles indicate hepatocyte steatosis. Proventriculus: solid arrows indicate dropout and necrosis of the mucosal epithelia in the proventriculus. Kidney: solid arrows indicate protein casts in renal tubules and open arrows indicate extensive congestion in renal interstitium. Small intestine: solid arrows indicate villus missing, drop out and necrosis of epithelial cells, open arrows indicate subserous congestion or dilatation and congestion of blood capillary in mucosa lamina proria. Scale bar = 50μm.

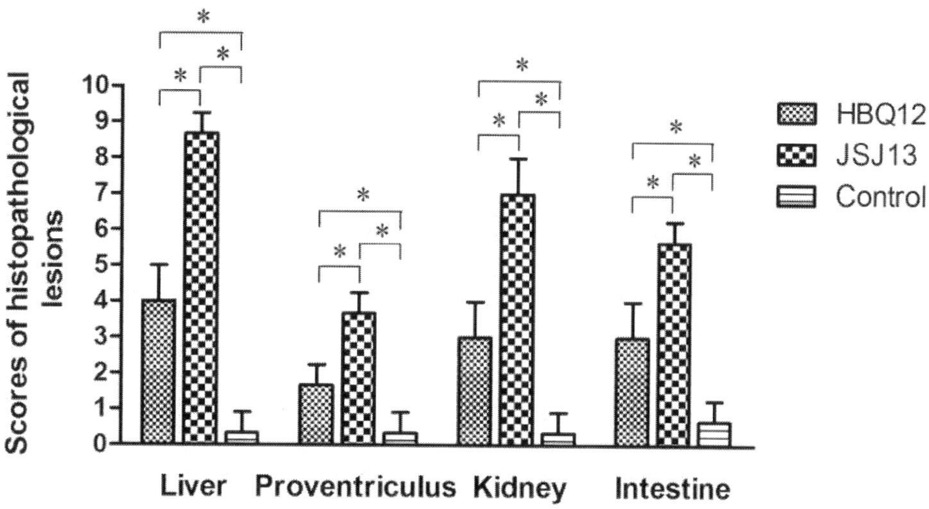

Figure 6: Scores of histopathological lesions in sampled tissues of chickens after inoculation with FAdV HBQ12 or JSJ13 strain at 6 dpi (n = 5 per group).

Lesion scores: 0 for no lesion, 1–3 for mild lesions, 4–6 for morderate lesions and 7–10 for severe lesions. The mean lesion scores per group are shown. The error bars indicate standard deviations. Asterisks (*) mark significant differences between groups (P <0.05).

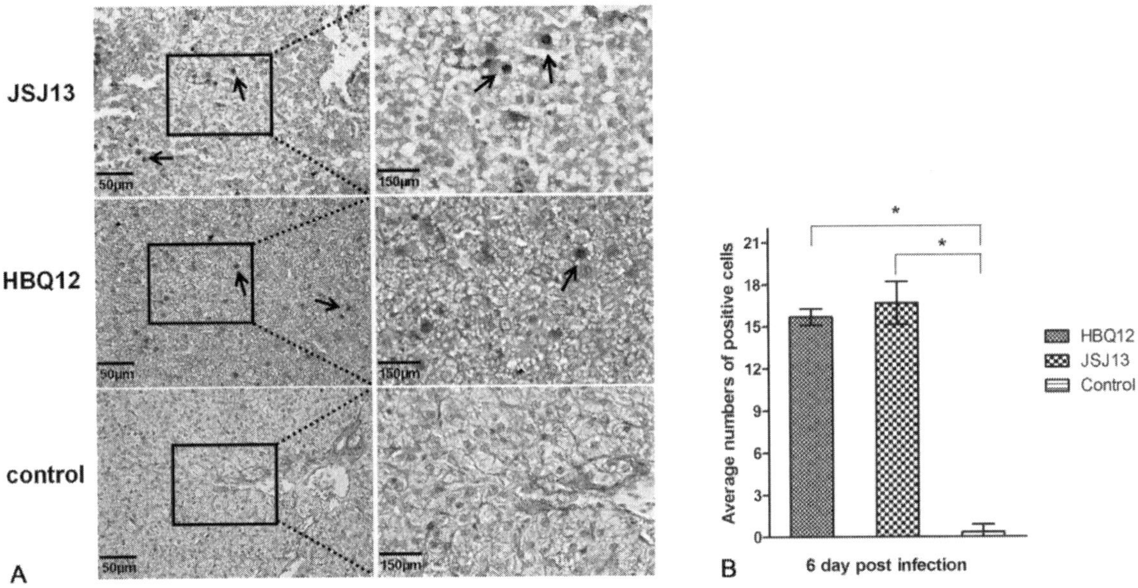

Figure 7: Immunohistochemical analysis.

(A) Detection of FAdV antigens in liver tissues after infection with FAdV HBQ12 or JSJ13 strain at 6 dpi. Solid arrows indicate positive signals for viral antigen detected extensively in the liver cells. Scale bar = 50μm in left column figures and 150μm in right column ones. (B) A summary of the number of positive cells is shown, asterisks (*) mark significant differences between groups ($P < 0.05$, $n = 5$). The error bars indicate standard deviation.

Viral DNA Detection in Tissues

Viral DNA in sampled tissues was detected by PCR and the results were summarized in Table 2. In group HBQ12, viral DNA could only be detected in liver and cecal tonsil collected at 3 and 5 dpi. Whereas total 29/36 (80.6%) of tissue samples collected from dead chickens possessed the detectable viral DNA. The similar results were also observed in group JSJ13, in which total 107/120 (89.2%) of tissue samples from dead chickens were positive for viral DNA, while at 3 and 5 dpi only liver and cecal tonsil in high ratio, occasionally spleen, kidney, pancreas and small intestine, possessed viral DNAratio ratio ratio. No viral DNA was detected in any tissues from chickens in the control group.

Table 2: Tissue distribution of two FAdV isolates in inoculated SPF chickens[a]

Group	Birds	Detection in viral DNA from collected tissues[b]											
		Heart	Liver	Spleen	Lung	Kidney	Bursa	Cecum tonsil	Proventriculus	Small intestine	Pancreas	Thymus	Trachea
HBQ12	Killed (3dpi)	0/4	0/4	0/4	0/4	0/4	0/4	1/4	0/4	0/4	0/4	0/4	0/4
	Killed (5dpi)	0/4	1/4	0/4	0/4	0/4	0/4	2/4	0/4	0/4	0/4	0/4	0/4
	Dead	2/3	3/3	3/3	3/3	3/3	2/3	3/3	2/3	3/3	2/3	2/3	1/3
JSJ13	Killed (3dpi)	0/4	0/4	0/4	0/4	0/4	0/4	2/4	0/4	0/4	0/4	0/4	0/4
	Killed (5dpi)	0/4	4/4	1/4	0/4	1/4	0/4	3/4	0/4	2/4	1/4	0/4	0/4
	Dead	10/10	10/10	10/10	10/10	10/10	7/10	7/10	9/10	10/10	7/10	9/10	8/10
Control	Killed (3dpi)	0/4	0/4	0/4	0/4	0/4	0/4	0/4	0/4	0/4	0/4	0/4	0/4
	Killed (5dpi)	0/4	0/4	0/4	0/4	0/4	0/4	0/4	0/4	0/4	0/4	0/4	0/4

[a]Three-week-old SPF chickens were inoculated with FAdV HBQ12 or JSJ13 ($10^{3.5}$ EID_{50}/bird) via the oral route.

[b]Tissue samples from the euthanized birds at 3 and 5 dpi and dead chickens were examined by PCR method. Data are no. of positive samples/no. of tested samples.

Antibody Response

The Ab response against two FAdV strains was shown in Fig 8. FAdV-specific Ab was not detected in any birds of the control group at all time points. Ab response to viral proteins appeared at 14 and 21 dpi with significant differences between inoculated groups and negative control ($P<0.05$). Chickens infected with strain HBQ12 had higher titers than those infected with strain JSJ13.

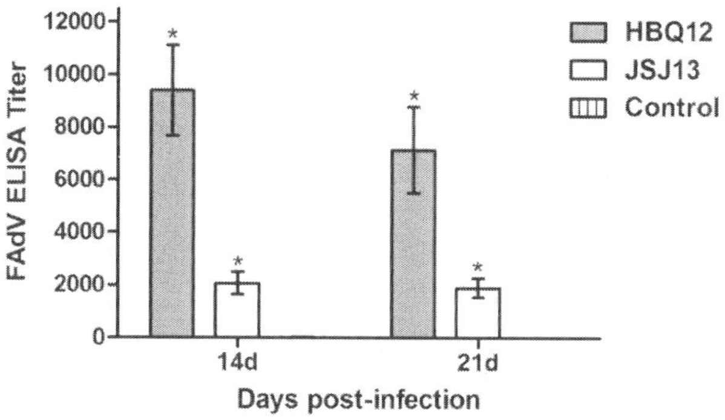

Figure 8: Antibody response to viral proteins in chickens inoculated with HBQ12 or JSJ13 by an oral route as measured by ELISA.

Ten sera samples from each group were randomly collected at 14 and 21dpi. The grey and white bars represent HBQ12- and JSJ13-inoculated chickens, respectively. The bar representing the negative control group (vertical stripes) is not visible as antibodies were not detected in this group. The error bars correspond to 95% confidence intervals. Asterisks (*) mark significant differences between the mock infected group and the groups infected with virus.

DISCUSSION

In recent years, increasing clinical cases of FAdV infections have been concerned and many FAdV strains related to IBH, HPS or GE have been isolated from most cases in many countries [7, 15–17]. Some of them are highly pathogenic to chickens and result in considerable economic losses [22–25]. In China, the outbreaks of IBH or HPS displayed an increasing trend since 2012. Most cases have been observed in broilers of 3 to 5 weeks of age, and occasionally in layers and breeder pullets aged 10 to 20 weeks. Mortality reported during these outbreaks varied from a slight increase to over 30% in severe cases and FAdV strains were isolated from all these cases. In the present study, three FAdV strains isolated from diseased chickens in China were phylogenetic and pathogenic analyzed. Among three FAdV isolates, strain HBQ12 and BJH13 were identified and classified as serotype 11 by sequence alignment and phylogenetic analysis, whereas strain JSJ13 was serotyped as FAdV-4. This is the first report of FAdV-4 strain circulating in China.

So far, the complete genome sequences available from all FAdV species were very limited: FAdV-A (1 strains, GenBank Accession No. U46933), FAdV-B (1 strain, GenBank Accession No. NC_021221), FAdV-C (2 strains, GenBank Accession No. HE608152 and GU188428), FAdV-D (1 strain, GenBank Accession No. AF083975) and FAdV-E (1 strain, GenBank Accession No. GU734104) [26–30]. A wide range of phylogenetic analysis in the whole genome levels would be very necessary and far-reaching for the deep understanding of the evolutionary relationship of aviadenovirus isolates. Given the whole genome of FAdVs haven't been previously reported in China, the present study analyzed firstly the complete genome characterization of three Chinese FAdV prevalent strains. The complete genome of these FAdV isolates was 44,081, 43,966 and 43,756 bp, respectively, and possessed a similar organization with other aviadenoviruses. Both of HBQ12 and BJH13 belonged to FAdV-D and had reduced TR-1 and TR-2 sequence (3 identical repeats for TR-1 and 5 identical repeats for TR-2 in HBQ12, absent TR-1 and 5 identical repeats for TR-2 in BJH13), compared with other isolates that had been sequenced previously, which contained 5 identical repeats for TR-1 and 13 identical repeats for TR-2 [31]. As for strain JSJ13, no TR-1 or TR-2 sequence was found in the genome, which was consistent with sequenced KR-5 (GenBank Accession No. HE608152) and ON1 (GenBank Accession No. GU188428). However, the biological effects of these changes in virus genome are still unclear.

The pathogenicity of FAdV strains varied from subclinical infections to severe diseases [6, 10, 22–26, 30]. The pathogenicity of HBQ12 and JSJ13 strains were determined by observing the clinical signs, gross and histological lesions in chickens. The results demonstrated that both isolates had obvious pathogenicity in SPF chickens and led to 8.6% and 28.6% of mortality, respectively. Strain JSJ13 belonging to FAdV-4 showed higher pathogenicity in chickens than those of HBQ12 (FAdV-11) confirmed by clinical and histopathological scores. The results were consistent with several previous studies that parts of serotype 4 adenoviruses were highly pathogenic strains [5, 23, 25]. The varying degree of mortality correlates to the pathogenicity of the virus, different serotypes of infected strains and susceptibility of the chickens.

Most reported HPS were caused by FAdV serotype 4 (FAdV-4) belonging to FAdV-C, characterized by accumulation of transparent or straw-colored fluid in the pericardial sac, nephritis, and hepatitis with a high mortality of 30–70% [5, 9, 10, 23, 25]. In this study, obvious IBH and HPS were both observed in chickens infected with strain JSJ13 (FAdV-4) and HBQ12 (FAdV-11), suggesting that certain strains of other serotypes except for FAdV-4 are capable to induce HPS via oral route. What is more, maybe some strains of different serotype are able to reproduce IBH and HPS simultaneously.

In addition to the good management practices and strict biosecurity measures, the use of efficacious vaccines may help in prevention of FAdV infections. In some countries, inactivated oil-emulsion vaccine or new generation vaccine against FAdVs has been developed, and study has demonstrated that the inoculation of these vaccines is efficient to protect against IBH or HPS [9, 32]. Maternally derived antibodies to FAdV are known to protect chicks from infection, thereby the vaccination of breeders that produce an adequate level of maternally derived antibodies in progeny is an effective approach to protect young chicks from FAdV, prior to the development of age-related resistance [33]. The commercial vaccines against FAdVs have not yet been developed in China due to a lack of knowledge and understanding of the relevant diseases. Therefore, further genetic and pathogenic analysis of FAdVs is necessary and would provide useful information for the development of the efficient vaccine against IBH or HPS.

ACKNOWLEDGMENTS

This study was supported by the China Agriculture Research System Poultry-related Science and Technology Innovation Team of Peking (CARS-PSTP).

Author Contributions

Conceived and designed the experiments: JZ GZZ. Performed the experiments: JZ QZ YXH. Analyzed the data: JZ YZ YXH GZZ. Contributed reagents/materials/analysis tools: JZ YXH GZZ. Wrote the paper: JZ YXH YZ. Checked and revised the manuscript: GZZ.

REFERENCES

1. Harrach B, Kaján G (2011) Aviadenovirus. In: Tidona C, Darai G, editors. The Springer Index of Viruses: Springer New York.
2. Harrach B, Benkö M, Both GW, Brown M, Davison AJ, Echavarría M, et al. (2012) Family—Adenoviridae. In King AMQ, Adams MJ, Carstens EB, Lefkowitz EJ (Eds.), Virus Taxonomy (9th ed. , pp. 125–141). San Diego: Elsevier Academic Press.
3. Hess M (2000) Detection and differentiation of avian adenoviruses: a review. Avian Pathol 29:195–206. doi: 10. 1080/03079450050045440. pmid:19184805
4. Domanska-Blicharz K, Tomczyk G, Smietanka K, Kozaczynski W, Minta Z (2011) Molecular characterization of fowl adenoviruses isolated from chickens with gizzard erosions. Poult Sci 90: 983–989. doi: 10. 3382/ps. 2010-01214. pmid:21489943
5. Nakamura K, Mase M, Yamaguchi S, Shibahara T, Yuasa N (1999) Pathologic study of specific-pathogen-free chicks and hens inoculated with adenovirus isolated from hydropericardium syndrome. Avian Dis 43: 414–423. pmid:10494409 doi: 10. 2307/1592638
6. Wells RJ, Harrigan K (1974) A fatal adenovirus infection of broiler chickens: inclusion body hepatitis. Vet Rec 94: 481–482. pmid:4211755 doi: 10. 1136/vr. 94. 21. 481
7. Mase M, Nakamura K, Minami F (2012) Fowl adenoviruses isolated from chickens with inclusion body hepatitis in Japan, 2009–2010. The J Vet Med Sci 74: 1087–1089. doi: 10. 1292/jvms. 11-0443
8. Ojkic D, Martin E, Swinton J, Vaillancourt JP, Boulianne M, Gomis S (2008) Genotyping of Canadian isolates of fowl adenoviruses. Avian Pathol 37: 95–100. doi: 10. 1080/03079450701805324. pmid:18202956
9. Schachner A, Marek A, Jaskulska B, Bilic I, Hess M (2014) Recombinant FAdV-4 fiber-2 protein protects chickens against hepatitis-hydropericardium syndrome (HHS). Vaccine 32: 1086–1092. doi: 10. 1016/j. vaccine. 2013. 12. 056. pmid:24397897
10. Kim JN, Byun SH, Kim MJ, Kim Jj, Sung HW, Mo IP (2008) Outbreaks of hydropericardium syndrome and molecular characterization of Korean fowl adenoviral isolates. Avian Dis 52: 526–530. pmid:18939647 doi: 10. 1637/8178-112207-case
11. Gjevre AG, Kaldhusdal M, Eriksen GS (2013) Gizzard erosion and ulceration syndrome in chickens and turkeys: a review of causal or predisposing factors. Avian Pathol 42: 297–303. doi: 10. 1080/03079457. 2013. 817665. pmid:23859215
12. Marek A, Schulz E, Hess C, Hess M (2010) Comparison of the fibers of Fowl adenovirus A serotype 1 isolates from chickens with gizzard erosions in Europe and apathogenic reference strains. J Vet Diagn Invest 22: 937–941. pmid:21088178 doi: 10. 1177/104063871002200613
13. Okuda Y, Ono M, Shibata I, Sato S (2004) Pathogenicity of serotype 8 fowl adenovirus isolated from gizzard erosions of slaughtered broiler chickens. J Vet Med Sci 66:1561–1516. pmid:15644608 doi: 10. 1292/jvms. 66. 1561
14. Mendelson C, Nothelfer HB, Monreal G (1995) Identification and characterization of an avian adenovirus isolated from a 'spiking mortality syndrome' field outbreak in broilers on the Delmarva Peninsula, USA. See comment in PubMed Commons below Avian Pathol 24: 693–706. pmid:18645825 doi: 10. 1080/03079459508419108
15. Mittal D, Jindal N, Tiwari AK, Khokhar RS (2014) Characterization of fowl adenoviruses associated with hydropericardium syndrome and inclusion body hepatitis in broiler chickens. Virus Disease 25: 114–119. doi: 10. 1007/s13337-013-0183-7. pmid:24426318
16. Kajan GL, Kecskemeti S, Harrach B, Benko M (2013) Molecular typing of fowl adenoviruses, isolated in Hungary recently, reveals high diversity. Vet Microbiol 167: 357–363. doi: 10. 1016/j. vetmic. 2013. 09. 025. pmid:24139719
17. Choi KS, Kye SJ, Kim JY, Jeon WJ, Lee EK, Park KY, et al. (2012) Epidemiological investigation of outbreaks of fowl adenovirus infection in commercial chickens in Korea. Poult Sci 91: 2502–2506. pmid:22991534 doi: 10. 3382/ps. 2012 02296
18. Li H, Yin Y, Guo Y, Hu Y, Xu H, Wang X, et al. (2010) Isolation and PCR identification of 12 strains of inclusion body hepatitis virus from clinical broilers. Chinese Vet Sci 40: 722–727.
19. Senne DA (1989) Virus propagation in embryonating eggs. Purchage HG, Arp LH, Domermuth CH, Pearson JE (Eds.). A Laboratory Manual for the Isolation and Identification of Avian Pathogens (3rd Edit.): American Association of Avian Pathologists, Kennett Square, Pennsylvania, pp. 176–181.
20. Reed LJ, Muench H (1938) A simple method of estimating fifty per cent endpoinds. Am J Epidemiol 27:493–497.
21. Tamura K, Dudley J, Nei M, Kumar S (2007) MEGA4: Molecular Evolutionary Genetics Analysis (MEGA) software version 4. 0. Mol Biol Evol 24:1596–1599. pmid:17488738 doi: 10. 1093/molbev/msm092
22. Dahiya S, Srivastava RN, Hess M, Gulati BR (2002) Fowl adenovirus serotype 4 associated with outbreaks of infectious hydropericardium in Haryana, India. Avian Dis 46: 230–233. pmid:11922341 doi: 10. 1637/0005-2086(2002)046[0230:fasawo]2. 0. co;2

23. Schonewille E, Singh A, Göbel TW, Gerner W, Saalmüller A, Hess M (2008) Fowl adenovirus (FAdV) serotype 4 causes depletion of B and T cells in lymphoid organs in specific pathogen-free chickens following experimental infection. Vet Immunol Immunopathol 121:130–139. pmid:17996948 doi: 10. 1016/j. vetimm. 2007. 09. 017

24. Mazaheri A, Prusas C, Voss M, Hess M (1998) Some strains of serotype 4 fowl adenoviruses cause inclusion body hepatitis and hydropericardium syndrome in chickens. Avian Pathol 27: 269–276. doi: 10. 1080/03079459808419335. pmid:18483997

25. Steer PA, Sandy JR, O'Rourke D, Scott PC, Browning GF, Noormohammadi AH (2015) Chronological analysis of gross and histological lesions induced by field strains of fowl adenovirus serotypes 1, 8b and 11 in one-day-old chickens. Avian Pathol 44: 106–113. doi: 10. 1080/03079457. 2015. 1007919. pmid:25609454

26. Ojkic D, Nagy E (2000) The complete nucleotide sequence of fowl adenovirus type 8. J Gen Virol 81:1833–1837. pmid:10859390

27. Chiocca S, Kurzbauer R, Schaffner G, Baker A, Mautner V, Cotten M (1996) The complete DNA sequence and genomic organization of the avian adenovirus CELO. J Virol 70: 2939–2949. pmid:8627769

28. Marek A, Kosiol C, Harrach B, Kajan GL, Schlotterer C, Hess M (2013) The first whole genome sequence of a Fowl adenovirus B strain enables interspecies comparisons within the genus Aviadenovirus. Vet Microbiol 166: 250–256. doi: 10. 1016/j. vetmic. 2013. 05. 017. pmid:23830773

29. Grgic H, Poljak Z, Sharif S, Nagy E (2013) Pathogenicity and cytokine gene expression pattern of a serotype 4 fowl adenovirus isolate. PloS one8: e77601. doi: 10. 1371/journal. pone. 0077601. pmid:24204888

30. Grgić H, Krell PJ, Nagy E (2014) Comparison of fiber gene sequences of inclusion body hepatitis (IBH) and non-IBH strains of serotype 8 and 11 fowl adenoviruses. Virus Genes 48:74–80. doi: 10. 1007/s11262-013-0995-y. pmid:24142408

31. Ojkic D, Nagy E (2001) The long repeat region is dispensable for fowl adenovirus replication in vitro. Virology 283: 197–206. pmid:11336545 doi: 10. 1006/viro. 2000. 0890

32. Alvarado IR, Villegas P, El-Attrache J, Jensen E, Rosales G, Perozo F, et al. (2007) Genetic characterization, pathogenicity, and protection studies with an avian adenovirus isolate associated with inclusion body hepatitis. Avian Dis 51: 27–32. pmid:17461263 doi: 10. 1637/0005-2086(2007)051[0027:gcpaps]2. 0. co;2

33. HarrachToro H, Gonzalez C, Cerda L, Morales MA, Dooner P, Salamero M (2002) Prevention of inclusion body hepatitis/hydropericardium syndrome in progeny chickens by vaccination of breeders with fowl adenovirus and chicken anemia virus. Avian Dis 46: 547–554. pmid:12243517 doi: 10.1637/0005-2086(2002)046[0547:poibhh]2.0.co;2

CHAPTER 12

PATHOGENICITY AND CYTOKINE GENE EXPRESSION PATTERN OF A SEROTYPE 4 FOWL ADENOVIRUS ISOLATE

Helena Grgić, Zvonimir Poljak, Shayan Sharif, Éva Nagy

INTRODUCTION

Fowl adenovirus serotype 4 (FAdV-4) plays a major role in the aetiology of a very important disease in broiler chickens called hydropericardium-hepatitis syndrome (HHS). It was first seen in Pakistan in 1987 [23] and since then outbreaks have also been reported in India, the Middle East, Japan, Mexico, Peru, Ecuador, and Chile [3,10,20,40] causing significant losses to the poultry industry. However, no HHS has been reported in Canada. The disease is prevalent in 3- to 6-week-old broiler chickens. The affected birds may not exhibit clinical signs other than high mortality, up to 75% [18]. The predominant gross lesion is hydropericardium, characterized by accumulation of a gelatinous material in the pericardial sac. The heart appears misshapen and flabby, with its apex floating in the pericardial sac [4,26]. At histological examination, the most consistent findings in the liver are small multifocal areas of necrosis and mononuclear cell infiltration, including basophilic intranuclear inclusion bodies in hepatocytes surrounded by a clear halo or filling the entire nucleus [27]. HHS associated losses could be reduced by vaccination. Inactivated liver homogenates from infected chickens, and inactivated virus propagated in cell culture and/or embryonated eggs based vaccines are available, however they provide only variable protection [6,25]. Recently, Schonewille et al. (2010) reported live FAdV-4 vaccine, but with lack of neutralizing antibodies.

Although there is no history of HHS in Canada, in 2004 a serotype 4 fowl adenovirus was isolated for the first time from a broiler breeder flock with no clinical signs of HHS. The complete nucleotide sequence of this virus (FAdV-4 ON1) was determined by Griffin and Nagy [18]. FAdVs are excellent vaccine vectors, and both FAdV-8 [38,22] and FAdV-9 [7,8,28] have been developed as vectors. This newly identified FAdV-4 ON1 could be considered not only as a live vaccine virus against HHS but as a vector virus as well.

The goals of this study were to 1) determine the pathogenicity of the FAdV-4 ON1 isolate in specific pathogen free (SPF) chickens, 2) investigate virus dissemination in tissues, virus titers in cloacal swabs, and antibody response, and 3) examine the dynamics of interferon (IFN)-γ, interleukin (IL)-10, IL-18, and IL-8 gene expression after FAdV-4 ON1 infection.

MATERIALS AND METHODS

Virus

The FAdV-4 was isolated in 2004 by the Animal Health Laboratory at the University of Guelph. The sample came from a Canadian broiler breeder flock showing no clinical signs of HHS. The virus was plaque purified, and the designated virus (FAdV-4 ON1) was propagated and titrated in a continuous chicken hepatoma cell line (CH-SAH, Solvay Animal Health,

Citation: Grgić H, Poljak Z, Sharif S, Nagy É (2013) Pathogenicity and Cytokine Gene Expression Pattern of a Serotype 4 Fowl Adenovirus Isolate. PLoS ONE 8(10): e77601. doi:10.1371/journal.pone.0077601.

Mendota Heights, Minnesota, USA) as described [5]. The complete genome sequence of the virus was determined [18] and the GenBank/EMBL/DDBJ accession number is GU188428.

Chickens and Housing

White Leghorn SPF chicken eggs were obtained from the Canadian Food Inspection Agency, Ottawa, and they were hatched at the Arkell Poultry Station of the University of Guelph. The birds were housed at the Isolation Unit according to the Animal Care Committee of the University of the Guelph in accordance with the *Guide to the Care and Use of Experimental Animals of The Canadian Council on Animal Care.*

Pathogenicity Assessment

A trial was conducted to evaluate the pathogenicity of the FAdV-4 ON1 isolate according to the University of Guelph Animal Care Committee regulations. The protocol was approved by the University of Guelph Animal Care Committee (Animal Utilization Protocol number 08R126). One hundred chicks were randomly divided into four groups: Group I (FAdV-4) -oral, Group II (FAdV-4) -intramuscular (im), Group III (Control) -oral, and Group IV (Control) -im. Groups I and II were composed of 30 chicks each, and the negative control groups (III and IV) were composed of 20 chicks each. The chicks were tagged with wing bands at 9 days of age and blood samples were collected. Since HHS is not reported in newly hatched birds, ten-day-old chickens in Groups I and II were inoculated with $2x10^8$ plaque forming units (pfu) of virus per chick. The control chicks (mock-infected) received phosphate buffered saline (PBS). Groups I and II were re-inoculated at 14 days of age with the same dose of virus, in order to provide a sufficiently high dose of the virus that could induce clinical signs and to mimic the continuous infection that occurs in a barn. The chicks were observed daily for clinical signs. Three chicks from the control and four chicks from the treatment groups were randomly selected for necropsy at 0, 3, 5, 7, 14, 21, and 28 days post-infection (d.p.i.). The chickens were euthanized with carbon dioxide gas according to the University of Guelph Animal Care Committee regulations, necropsied, and examined for the presence of gross lesions. Tissues from thymus, lung, heart, kidney, liver, bursa of Fabricius, and cecal tonsils were collected. Each organ was sectioned into two portions: one part was placed into 10% formalin, and the other one was stored at -80°C for quantitative polymerase chain reaction (qPCR). Cloacal swabs were collected at 0, 3, 5, 10, 14, 21, 28 d.p.i. and their virus titers were determined by a plaque assay. Blood samples were collected and tested by ELISA and virus neutralization test.

Serologic Testing

The serum samples collected at 0, 7, 14, 21, and 28 d.p.i. were tested for FAdV-specific antibodies (Abs) by ELISA as described [29]. Briefly, microtiter plates (Becton-Dickinson, Franklin Lakes, New Jersey, USA) were coated with FAdV-4 antigen treated with *N*-lauroyl sarcosine at a concentration of 200 ng/well and blocked with 3% bovine serum albumin. The bound antibody was detected by alkaline-phosphatase-labeled goat IgG conjugate against chicken immunoglobulin (Kirkegaard and Perry Laboratories, Gaithersburg, Maryland, USA) and Sigma Fast *p*-nitrophenyl phosphate (Sigma Chemical Company, St. Louis, Missouri, USA).

The results were expressed by calculating the sample-to-positive (S/P) ratio for each sample from mean optical density (OD_{405}) readings.

The serum samples were also assayed for virus neutralizing antibodies [32]. Briefly, 100 pfu of the virus in 100 μl of cell culture medium were mixed with equal volume of 1:50 dilutions of Abs. The mixture was incubated for 1 h, and the sample was then added to CH-SAH cells. After a 1 h adsorption period, the cells were overlaid with a 0.6% agarose. The plaques were visualized by addition of a 2nd overlay after 72 h of incubation at 37°C in 5% CO_2 containing 0.02% neutral red. A sample that inhibited the development of more than 50% of the plaques compared with the negative-control serum sample was considered positive.

Experimental Design for Determination of Cytokine mRNA Expression

Forty ten-day-old chicks were randomly divided into two groups. Group I was composed of 25 birds and Group II (negative control) contained 15 birds. Each bird in Group I received $2x10^8$pfu/ml of the virus im and the control chicks received PBS. Five chicks from the infected and three chicks from the control group were selected randomly for necropsy at 1, 3, 5, 7, and 10 d.p.i. The rationale for selecting these time points was the assumption that one immune measure at one time point is rarely sufficient. Moreover, FAdV infection results in viraemia by 24 hours p.i. with two peaks, the first at day 2 and the second at day 7 p.i. [36]. The chicks were euthanized and the spleen, cecal tonsil, and liver were collected and submerged in RNAlater (Qiagen, Mississauga, ON, Canada) and stored at -80°C until RNA extraction.

DNA Extraction and Quantification of Viral DNA in Tissues

Total DNA was extracted with DNeasy® tissue kit (QIAGEN InC., Mississauga) and the viral load in the tissues was determined by qPCR with SYBR Green as an intercalating dye. The FAdV-4 ON1 ORF14 gene was used as an indicator for the presence of viral DNA. The forward and reverse primers were 5'-AGTGTGTATGTGCGTTGGGTAG-3' and 5'-CATTGTCATAACGATGGTGTAG-3', respectively. The qPCR reaction was carried out as described [16].

RNA Extraction and Real-Time Quantitative PCR for Determination of Cytokine Gene Expression

Total RNA was extracted from tissues by Trizol reagent (Invitrogen, Canada Inc., Burlington, ON, Canada) as described by the manufacturer. The tissues were homogenized in 1 ml Trizol followed by chloroform extraction, isopropanol precipitation, and a 75% ethanol wash. The RNA pellet was dissolved in 20 μl of RNAse-free water. The concentration of the RNA was measured by a NanoVue spectrophotometer. cDNA was synthesized from total RNA with SuperScript™ II Reverse Trancriptase (Invitrogen) according to the manufacturer's instructions, and the random primers were from Invitrogen. The cDNAs were stored at -80°C until use.

A real-time quantitative PCR (RT-qPCR) assay was employed to quantify the expression of β-actin and cytokine genes in cDNA samples. The reaction was prepared in a total volume of 20 μl of Qiagen Quantitect SYBR Green PCR kit (Qiagen) containing hot start Taq polymerase and cDNA. Each assay was performed with a no-template control and in triplicates of the standard in a LightCycler (LC480) instrument (Roche Diagnostics GmbH, Mannheim, Germany).

The sequences and other features of specific primers for chicken IFN-γ, IL-18, IL-10, IL-8, and β-actin have been described [1,2] and they are given in Table 1. The primers were synthesized by Laboratory Services, University of Guelph. The PCR conditions for different segments of each cycle were optimized for all target genes.

Table 1: The sequences of the primers utilized in study

Target	Sequences
IL-18	Forward 5'-GAAACGTCAATAGCCAGTTGC-3'
	Reverse 5'-TCCCATGCTCTTTCTCACAACA-3'
IL-10	Forward 5'-AGCAGATCAAGGAGACGTTC-3'
	Reverse 5'-ATCAGCAGGTACTCCTCGAT-3'
IL-8	Forward 5'-TATCACTGGAGAGTCTCGCTGTC-3'
	Reverse 5'-TCAGTCCTCCTGCAGTGTATACC-3'
IFN-γ	Forward 5' - ACACTGACAAGTAAAGCCGCACA-3'
	Reverse 5' -AGTCGTTCATCGGGAGCTTGGC-3'
β-actin	Forward 5'-CAACACAGTGCTGTCTGGTGGTA-3'
	Reverse 5'-ATCGTACTCCTGCTTGCTGATCC-3'

Since the most variability in cytokine gene expression following FAdV-8 infection was in the liver [15], mRNA expression in liver was determined for IFN-γ, IL-18, IL-10, and IL-8, while for cecal tonsil and spleen, mRNA expression was determined only for IFN-γ and IL-18.

Statistical Analysis

To evaluate if viral copy number differed among tissues, a Kruskal-Wallis test was performed. Data from the oral and im groups were subjected separately to the Kruskal-Wallis test to evaluate the differences in number of viral copies among tissues. When statistically significant *P*-values were identified, pair-wise comparisons were performed between different tissues using the Mann-Whitney U test. *P*-values of <0.05 were considered to be statistically significant.

Real-time quantitative PCR efficiency was calculated as described [35], and mRNA expression was calculated relative to β-actin gene expression [31]. The difference in cytokine expression between groups was assessed by a Mann-Whitney U test and comparisons were considered significant at $P<0.05$.

RESULTS

Clinical Signs, Gross Pathology and Histology

Clinical signs indicative of HHS and characterized by depression, chickens huddling in corners with ruffled feathers, and chickens exhibiting the characteristic posture, with the chest and beak resting on the ground, were not seen in any chicks. Distinctive gross lesions, hydropericardium, and pale, swollen and friable liver were absent. The characteristic histological signs of HHS described as intranuclear inclusion bodies in hepatocytes were also not detected in any of the infected chickens.

Viral Genome Copy Number in Tissues

Viral genome copy numbers in liver, bursa of Fabricius, and cecal tonsil were determined by qPCR and the results are summarized in Table 2. No viral DNA was detected in any tissues from any chickens before inoculation or in the mock-infected chickens. The cecal tonsil was the organ with the highest number of viral copies, followed by the liver and then the bursa of Fabricius, irrespective of the inoculation route. For the orally inoculated birds, viral copy number was higher in cecal tonsil than in liver (P=0.0207), and bursa of Fabricius (P=0.0001), whereas viral copy numbers in the liver ranked only marginally higher than viral copy numbers in the bursa of Fabricius (P=0.0759). For the im inoculated birds, viral copy number was higher in cecal tonsil than in the bursa (P=0.0154), whereas differences among other organs were not significant (P>0.25).

Table 2: Number of positive tissues and viral genome copy numbers in tissues of chickens inoculated with FAdV-4

Days p.i.			**Inoculation route**				
	Tissue		**Oral**			**Intramuscular**	
	Liver		3.4[a]	3/4[b]		7.5	2/4
3	Bursa		5.6	2/4		5	1/4
	Cecal tonsil		26	4/4		530	4/4
	Liver		140	3/4		8.3	4/4
5	Bursa		6.1	4/4		12	2/4
	Cecal tonsil		570	4/4		130	4/4
	Liver		2.2	2/4		2.3	2/4
7	Bursa		0	0/4		6	2/4
	Cecal tonsil		24	1/4		3.5	3/4
	Liver		2.8	4/4		5.9	2/4
14	Bursa		4.9	3/4		4.6	2/4
	Cecal tonsil		12	4/4		8.6	4/4
	Liver		11	4/4		24	4/4
21	Bursa		23	3/4		310	4/4
	Cecal tonsil		6400	4/4		360	3/4
	Liver		650	4/4		350	4/4
28	Bursa		1.4	2/4		74	1/4
	Cecal tonsil		170	4/4		650	4/4

Virus Shedding

No virus was detected in the cloacal swabs of any chickens before inoculation or in the mock-infected groups at any time (data not shown). The virus titers in the cloacal swabs of chickens inoculated with FAdV-4 ON1 are given in Table 3.

Table 3: Virus titers in the cloacal swabs of chickens inoculated with FAdV-4

Days		**Inoculation route**		
post-infection		**Mean titers (pfu/ml) in swabs**		
		Chickens shedding the virus (%)[a]		
		Oral	**Intramuscular**	***P*-value**
3		9.9×10^{3}	1.4×10^{3}	≤0.001
		(93)	(25)	
5		1.2×10^{4}	1.2×10^{3}	≤0.001
		(83)	(32)	
10		1.3×10^{2}	6.8×10^{2}	0.49
		(13)	(20)	
14		3.9×10^{1}	2.5×10^{1}	0.26
		(27)	(13)	
21		3.3×10^{1}	2.5×10^{1}	0.25
		(20)	(10)	
28		0	1.2×10^{1}	0.34
		(0)	(10)	

The virus titers were determined by plaque assay.

P values < 0.05 are statistically significant for comparison between oral and im on each day.

a 26 and 22 chickens at days 3 and 5, respectively; 18 and 14 chickens at days 10 and 14, respectively; 10 and 6 chickens at days 21 and 28, respectively.

The difference in titers between the groups of orally and im inoculated chickens was statistically significant (P<0.001) at 3 and 5 d.p.i., when tested by the Mann-Whitney U test. The oral group had higher ranks. Chickens in the im group shed the virus throughout the experiment (28 d.p.i), while virus shedding in the oral group was recorded until 21 d.p.i.

Antibody Response

FAdV specific Ab was not detected in any birds before inoculation nor in the mock-infected birds. The Ab response against FAdV-4, shown in Figure 1, appeared at 7 d.p.i. and the difference between inoculated groups and negative controls was statistically significant (P<0.001) at 14, 21 and 28 d.p.i. The im inoculated chickens had higher values than birds inoculated orally (P<0.001). When the same sera were tested against another serotype, FAdV-9, the results were similar, but the Ab levels were slightly lower (data not shown).

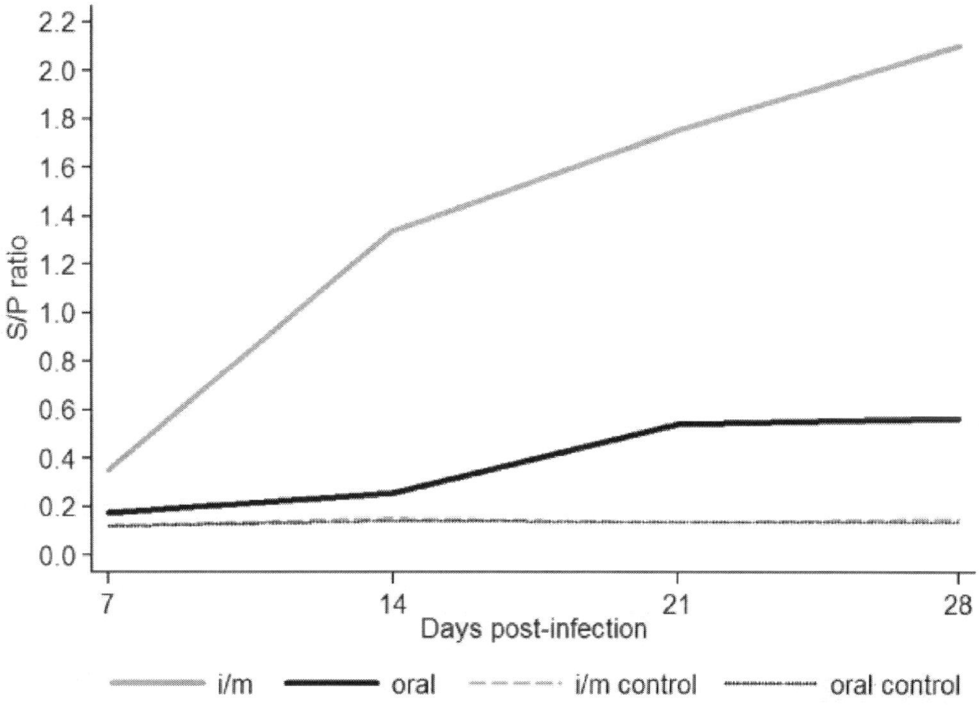

Figure 1: Antibody response to viral proteins in chickens inoculated with FAdV-4 by the oral and intramuscular routes and mock infected as measured by S/P ratios.

Serum samples were collected at 7, 14, 21 and 28 days post-infection. Gray and black lines represent im and orally inoculated chickens, respectively. Dashed gray line and black dotted line represent mock-infected chickens. Statisticaly significant differences among mock infected and infected birds have been detected at 14, 21, and 28 d.p.i.

VIRUS NEUTRALIZATION ASSAY

No Neutralizing Antibodies were Found in any Birds Prior to the Challenge.

In the virus neutralization test, all serum samples collected from both the oral and im groups after 14 days of infection completely neutralized the FAdV-4 ON1 in 1:50 dilution. Finally, negative control birds were negative throughout the whole study.

Expression of Cytokine Genes

Expression of mRNA of cytokine genes after im inoculation of chickens with FAdV-4 ON1 was measured in the cecal tonsil, liver, and spleen.

In the cecal tonsils, there were no significant differences (*P*>0.05) in the expression of IFN-γ and IL-18 mRNA when compared with the values for the uninfected controls (Figure 2).

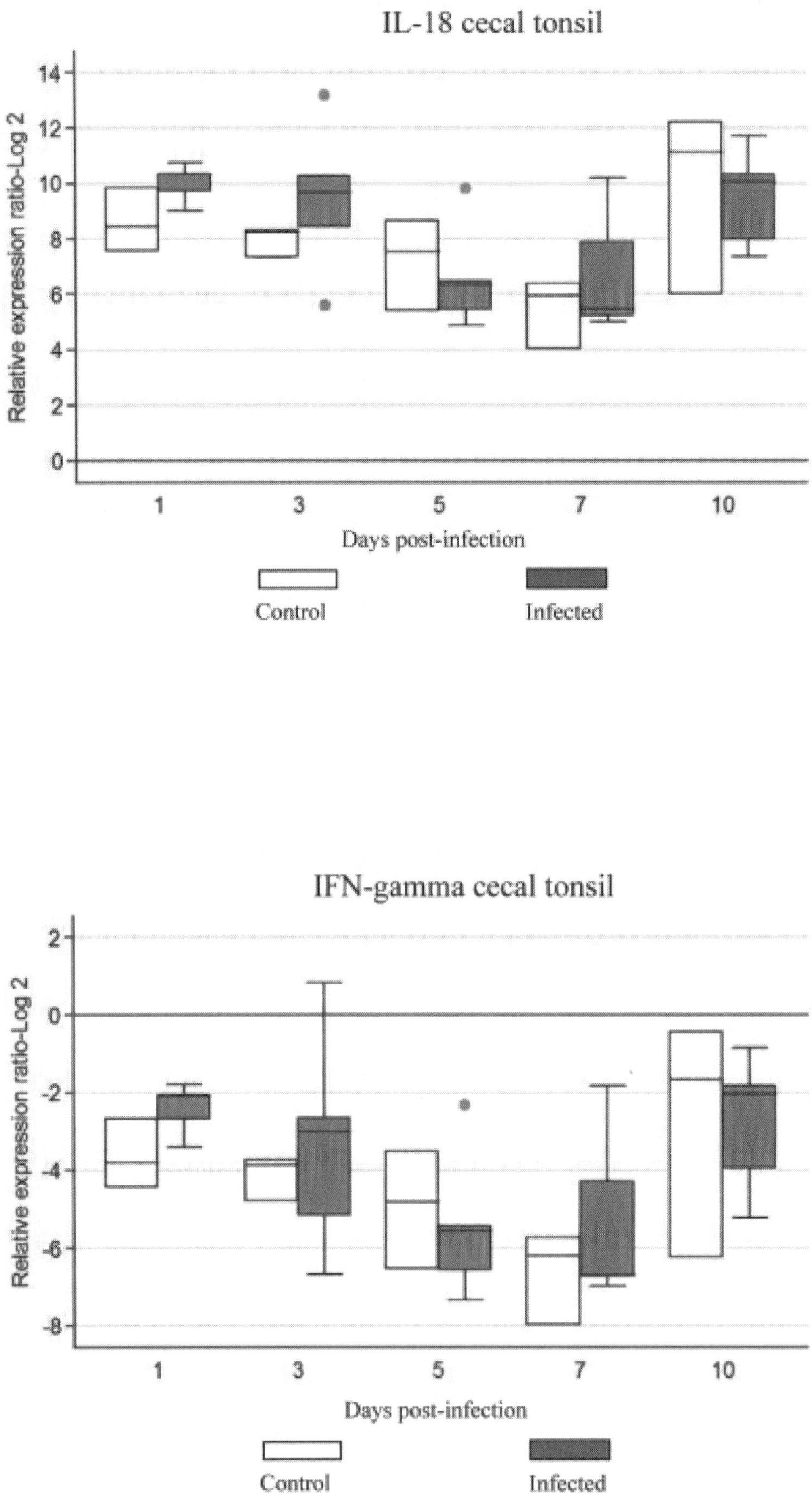

Figure 2: Cytokine mRNA expression in cecal tonsil of chickens infected with FAdV-4.

The groups were: uninfected (negative control), and FAdV-4 infected chickens using im inoculation. Target and reference gene expression was quantified by real-time RT-PCR using SYBR Green. Target gene expression is presented relative to β-actin expression and normalized to a calibrator. The results are diagrammed as whisker boxes with medians. Boxes represent interquartile ranges and whiskers indicate extreme values. There was no statistically significant differences detected (P>0.05) by Mann-Whitney test. ● symbols represent values which were identified as outliers.

The expression of IFN-γ, IL-18, IL-10, and IL-8 mRNA is illustrated in Figure 3. There was a statistically significant increase ($P<0.05$) in IFN-γ gene expression of infected chickens at 3 d.p.i. compared to the uninfected group. The expression pattern of IL-10 was significantly higher at 3 d.p.i. in the infected group of chickens compared to the uninfected group ($P<0.05$). The expression of IL-8 and IL-18 was not significantly different ($P>0.05$) between infected and uninfected groups.

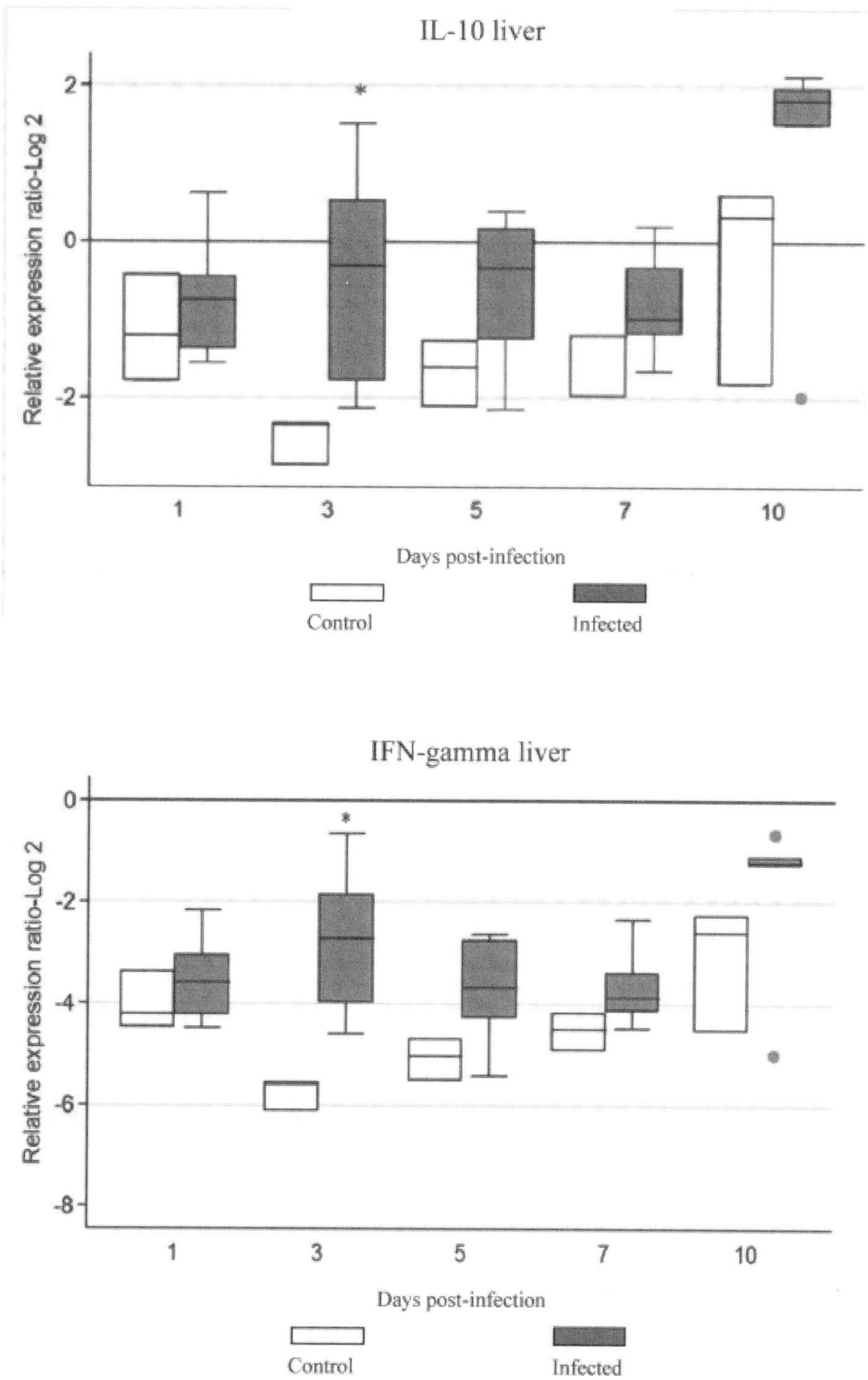

Figure 3: Cytokine mRNA expression in liver of chickens infected with FAdV-4.

The groups were: uninfected (negative control), and FAdV-4 infected chickens using im inoculation. Target and reference gene expression was quantified by real-time RT-PCR using SYBR Green. Target gene expression is presented relative to β-actin expression and normalized to a calibrator. The results are diagrammed as whisker boxes with medians. Boxes represent interquartile ranges and whiskers indicate extreme values. The difference in cytokine expression between groups was assessed by Mann-Whitney test and comparisons were considered significant at $P \leq 0.05$ (*).● symbols represent values which were identified as outliers.

In the spleen, both IFN-γ and IL-18 mRNA were detected at all-time-points (Figure 4). A statistically significant ($P<0.05$) down-regulation of both IFN-γ and IL-18 in the spleen was noted in the infected group at 10 d.p.i.

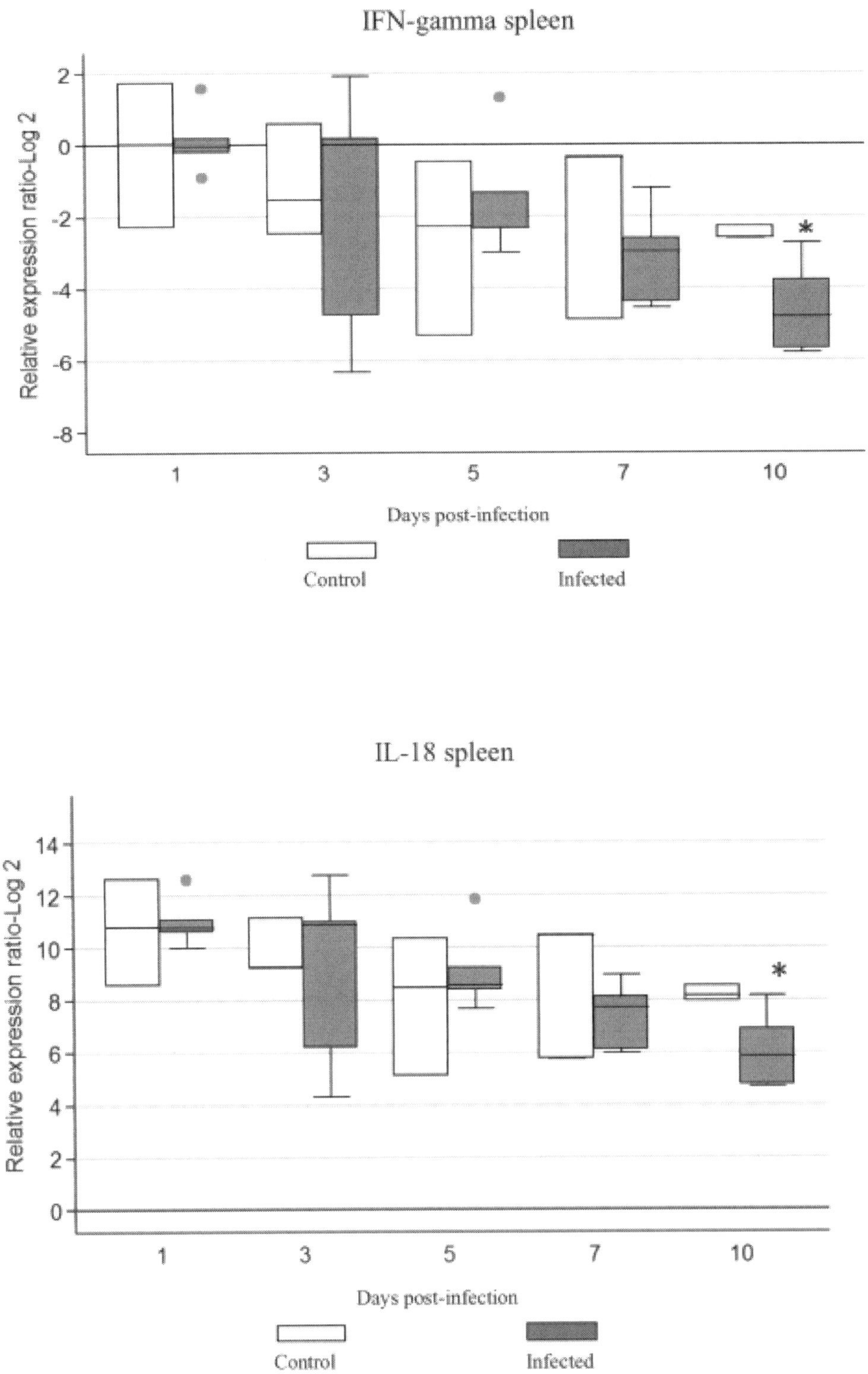

Fig.4.

Figure 4: Cytokine mRNA expression in spleen of chickens infected with FAdV-4.

The groups were: Uninfected (negative control), and FAdV-4 infected chickens using im inoculation. Target and reference gene expression was quantified by real-time RT-PCR using SYBR Green. Target gene expression is presented relative to β-actin expression and normalized to a calibrator. The results are diagrammed as whisker boxes with medians. Boxes represent interquartile ranges and whiskers indicate extreme values. The difference in cytokine expression between groups was assessed by Mann-Whitney test and comparisons were considered significant at $P \leq 0.05$ (*).● symbols represent values which were identified as outliers.

DISCUSSION

There were no clinical signs, or gross and histopathological lesions in any of the inoculated birds by either inoculation route, even though they were inoculated with a high virus dose, suggesting it could be considered non-pathogenic.

The lack of pathological changes was not due to lack of virus replication, since virus was found in tissues and cloacal swabs of the FAdV-4 infected birds; although in numerically lower quantities than previously reported for FAdV-8 and FAdV-9 [16,11]. In both the orally and im inoculated groups, the highest numbers of viral genome copies were detected in the cecal tonsil, in agreement with our work on FAdV-8 [16]. Ojkic and Nagy [29] reported that replication and distribution of another serotype virus (FAdV-9) in tissues are influenced by the route of inoculation. Since FAdV-4 is transmitted horizontally by the oral-fecal route [9], the im and the natural oral routes of infection were compared. In general, the orally inoculated group had a higher number of viral genome copies in contrast to our work on FAdV-8 [16], where the im-inoculated chicks had a higher number of viral genome copies. The reason for this difference might be due to the different serotypes used in these two studies. FAdV-4 ON1 was isolated from cloacal swabs until the end of the study (28 d.p.i.) in the im-inoculated group, while for the oral group, shedding was recorded only up to 21 d.p.i. Orally inoculated chickens had a significantly higher virus titer in feces than the im-inoculated group at all time points. Similar patterns of shedding up to 28 d.p.i. were also reported for FAdV-8 [16]. In further studies different breeds and sources of chickens should be considered in order to increase external validity of results.

Ab response was significantly higher in FAdV-4 im-inoculated chicks than in the orally inoculated chicks. FAdV-8 and FAdV-9 also induced higher antibody levels when the birds were infected via the im route [16,29]. Since sera using a heterologous FAdV serotype, FAdV-9, were positive with ELISA, and although the S/P ratios were lower, these results confirm already publish data that heterologous virus can be used as an antigen to detect adenovirus Abs against other serotypes [33,13]. Importantly, we demonstrated that FAdV-4 ON1 was immunogenic and induced neutralizing antibodies in chickens. Therefore this isolate could be used as a live vaccine virus.

The most notable features related to the genome of the FAdV-4 ON1 are its size and the presence of two fibers. According to previous research [18], FAdV-4 ON1 with 45667 bp in length is the largest AdV genome reported to date. The two fiber genes located adjacent to one another in the central genomic region showed a similar β-strand arrangement to FAdV-1 fiber 1 and fiber 2 [19,12]. Authors speculated that the FAdV-4 fiber 2 might bind a receptor other than CAR and determine the tissue tropism of FAdV-4, possibly leading to the unique clinical features associated with infection with virulent FAdV-4 [18].

Cytokine gene expression associated with T helper (Th)1, pro-inflammatory, and regulatory activities, namely IFN-γ, IL-18, IL-10 and IL-8, was investigated in the liver, spleen and cecal tonsils, and gene activation at early times after infection was shown. The expression of cytokines has been studied following FAdV-8 infection [15], however, this is the first report on the cytokine repertoire associated with FAdV-4 infection. Enhanced expression of IFN-γ in the liver was noted, with a significant increase in expression at 3 d.p.i. IFN-γ is considered an essential cytokine in host defense against a variety of pathogens. For example IFN-γ inhibits Marek's disease virus (MDV) replication by inducing nitric oxide synthesis [41], reduces replication of coccidian parasites [24], and inhibits Rous sarcoma virus *src* oncogene-induced tumor growth [34]. IFN-γ also inhibits transgene expression from adenoviral vectors *in vitro* by a transcription-related mechanism [39].

The IL-10 in the liver was significantly up-regulated at 3 d.p.i. in the infected group, when compared to that of the negative control birds. Our finding that IL-10 was up-regulated following FAdV-4 infection is in agreement with a previous report [15], and may indicate an immune evasive mechanism used by the virus to skew the immune response, supporting persistence of the virus [17]. IL-10 has both immunosuppressive and immunostimulating effects on cells of the adaptive immune system [20]. Simultaneous expression of IFN-γ and IL-10 also has been reported after infection with porcine reproductive and respiratory syndrome virus [21]. More recently Parvizi et al. [30] reported persistent up-regulation of IFN-γ and IL-10 in a MDV-susceptible line of chickens. In humans, up-regulation of both IFN-γ and IL-10 is directly related to IL-12 stimulation of $CD4^+$ and $CD8^+$ T cells [14].

In summary, the lack of clinical signs, and gross and histopathological lesions indicated that FAdV-4 ON1 was a non-pathogenic virus. However, virus replication was demonstrated in tissues and it induced a strong Ab response. The prolonged virus shedding could be advantageous in vaccinated poultry flocks, ensuring a long and uniform exposure to the virus. Infection with this virus was associated with increased expression of IFN-γ and IL-10 in the liver and decreased expression of IFN-γ and IL-18 in the spleen, providing useful information about important mediators involved in immunity against virus infection. Our study demonstrated that the FAdV-4 ON1 potentially could be used as a live vaccine against HHS and developed as vaccine vector. However, further research is needed to confirm protective nature of this vaccine candidate.

ACKNOWLEDGMENTS

We thank the Isolation Unit personnel and Mrs. Sheila Watson for their professional animal care and excellent assistance.

AUTHOR CONTRIBUTIONS

Conceived and designed the experiments: HG. Performed the experiments: HG. Analyzed the data: HG ZP. Contributed reagents/materials/analysis tools: ZP SS EN. Wrote the manuscript: HG ZP SS EN.

REFERENCES

1. Abdul-Careem MF, Hunter BD, Parvizi P, Haghighi HR, Thanthrige-Don N et al. (2007) Cytokine gene expression patterns associated with immunization against Marek's disease in chickens. Vaccine 25: 424-432. doi:10. 1016/j. vaccine. 2006. 08. 006. PubMed: 17070626.
2. Abdul-Careem MF, Hunter BD, Sarson AJ, Mayameei A, Zhou H et al. (2006) Marek's disease virus-induced transient paralysis is associated with cytokine gene expression in the nervous system. Viral Immunol 19: 167-176. doi:10. 1089/vim. 2006. 19. 167. PubMed: 16817759.
3. Abe T, Nakamura K, Tojo H, Mase H, Shibahara T et al. (1998) Histology, immunohistochemistry, and ultrastructure of hydropericardium syndrome in adult broiler breeders and broiler chicks. Avian Dis 42: 606-612. doi:10. 2307/1592690. PubMed:9777164.
4. Adair BM, Fitzgerald SD (2008) Adenovirus infections. In: Saif YM, Fadly AM, Glisson JR, McDougald LR, Nolan LK, Swayne DE editors. Disease of poultry. Ames: Blackwell Publishing Professional. pp. 251-291.
5. Alexander HS, Huber P, Cao JX, Krell PJ, Nagy É (1998) Growth characteristics of fowl adenovirus type 8 in a chicken hepatoma cell line. J Virol Methods 74: 9-14. doi:10. 1016/S0166-0934(98)00062-7. PubMed: 9763123.
6. Anjum AD (1990) Experimental transmission of hydropericardium syndrome and protection against it in commercial broiler chickens. Avian Pathol 19: 655-660. doi:10. 1080/03079459008418721. PubMed: 18679979.
7. Corredor JC, Nagy É (2010) A region at the left end of the fowl adenovirus 9 genome that is non-essential in vitro has consequences in vivo. J Gen Virol 91: 51-58. doi:10. 1099/vir. 0. 013839-0. PubMed: 19759237.
8. Corredor JC, Nagy É (2010) The non-essential left end region of the fowl adenovirus. p. 9.
9. Cowen B (1992) Inclusion body hepatitis-anaemia and hydropericardium syndrome: aetiology and control. Worlds Poult Sci J 48: 247-253. doi:10. 1079/WPS19920019.
10. Cheema AH, Ahmad J, Afzal A (1989) An adenovirus infection of poultry in Pakistan. Rev Sci Tech Off Int Epiz 8: 789-795.
11. Deng L, Sharif S, Nagy E (2013) Oral inoculation of chickens with a candidate fowl adenovirus 9 vector. Clin Vaccine Immunol. doi:10. 1128/CVI 00187-13.
12. El Bakkouri M, Seiradake E, Cusack S, Ruigrok RWH, Schoehn G (2008) Structure of the C-terminal domain of the fowl adenovirus type 1 short fibre. Virology 378: 169-176. doi:10. 1016/j. virol. 2008. 05. 011. PubMed: 18561970.
13. Erny K, Pallister J, Sheppard M (1995) Immunological and molecular comparison of fowl adenovirus serotypes 4 and 10. Arch Virol 140: 491-501. doi:10. 1007/BF01718426. PubMed: 7733822.
14. Gerosa F, Paganin C, Peritt D, Paiola F, Scupoli MT et al. (1996). Interleukine-12 primes human CD4 and CD8 T cell clones for high production of both interferon-gamma and interleukine-10. J Exp Med 183:2559-69. doi: 10. 1084/jem. 183. 6. 2559
15. Grgić H, Sharif S, Haghighi HR, Nagy É (2013) Cytokine patterns associated with a serotype 8 fowl adenovirus infection. Viral Immunol 26: 143-149. doi:10. 1089/vim. 2012. 0078. PubMed: 23537431.
16. Grgić H, Yang D-H, Nagy É (2011) Pathogenicity and complete genome sequence of fowl adenovirus serotype 8 isolate. Virus Res 156: 91-97. doi:10. 1016/j. virusres. 2011. 01. 002. PubMed: 21237223.
17. Grgić H, Philippe C, Ojkić D, Nagy É (2006) Study of vertical transmission of fowl adenoviruses. Can J Vet Res 70: 230-233. PubMed: 16850947.
18. Griffin BD, Nagy É (2011) Coding potential and transcript analysis of fowl adenovirus 4: insight into upstream ORFs as common sequence features in adenoviral transcripts. J Gen Virol 92: 1260-1272. doi: 10. 1099/vir. 0. 030064-0
19. Guardado-Calvo P, Llamas-Saiz AL, Fox GC, Langlois P, van Raaij MJ (2007) Structure of the C-terminal head domain of the fowl adenovirus type 1 long fiber. J Gen Virol 88: 2407- 2416. doi:10. 1099/vir. 0. 82845-0. PubMed: 17698649.
20. Hess M, Raue R, Prusas C (1999) Epidemiological studies on fowl adenoviruses isolated from cases of infectious hydropericardium. Avian Pathol 28: 433-439. doi:10. 1080/03079459994443.
21. Johnsen CK, Bøtner A, Kamstrup S, Lind P, Nielsen J (2002) Cytokine mRNA profiles in bronchoalveolar cells of piglets experimentally infected in utero with porcine reproductive and respiratory syndrome virus: association of sustained expression of IFN-gamma and IL- 10 after viral clearance. Viral Immunol 15: 549-556. doi:10. 1089/088282402320914494. PubMed: 12513926.
22. Johnson MA, Pooley C, Ignjatovic J, Tyack SG (2003) A recombinant fowl adenovirus expressing the S1 gene of infectious bronchitis virus protects against challenge with infections bronchitis virus. Vaccine 21: 2730-2736. doi:10. 1016/S0264-410X(03)00227-5. PubMed: 12798610.
23. Khawaja DA, Ahmad S, Rauf MA, Zulfiqar MZ, Mahmood SMI et al. (1988) Isolation of an adenovirus from hydropericardium syndrome in broiler chicks. Pak J Vet Res 1: 2-17.
24. Lillehoj HS, Choi KD (1998) Recombinant chicken interferon-gamma-mediated inhibition of *Eimeria tenella* development *in vitro* and reduction of oocyst production and body weight loss following *Eimeria Acervulina* challenge infection. Avian Dis 42: 307-314. doi:10. 2307/1592481. PubMed: 9645322.
25. Mansoor MK, Hussain I, Arshad M, Muhammad G (2011) Preparation and evaluation of chicken embryo-adapted fowl adenovirus serotype 4 vaccine in broiler chickens. Trop Anim Health Prod 43: 331-338. doi:10. 1007/s11250-010-9694-z.

26. Mazaheri A, Prusas C, Vo M, Hess M (1998) Some strains of serotype 4 fowl adenovirus cause inclusion body hepatitis and hydropericardium syndrome in chickens. Avian Pathol 27: 269-276. doi:10. 1080/03079459808419335. PubMed:18483997.

27. Nakamura K, Mase M, Yamaguchi S, Shibahara T, Yuasa N (1999) Pathologic study of specific pathogen free chicks and hens inoculated with adenovirus isolated from hydropericardium syndrome. Avian Dis 43: 414-423. doi:10. 2307/1592638. PubMed:10494409.

28. Ojkic D, Nagy É (2001) The long repeat region is dispensable for fowl adenovirus replication in vitro. Virology 283: 197-206. doi:10. 1006/viro. 2000. 0890. PubMed:11336545.

29. Ojkic D, Nagy É (2003) Antibody response and virus tissue distribution in chickens inoculated with wild-type and recombinant fowl adenoviruses. Vaccine 22: 42-48. doi:10. 1016/S0264-410X(03)00544-9. PubMed: 14604569.

30. Parvizi P, Read LR, Abdul-Careem MF, Sarson AJ, Lusty C et al. (2009) Cytokine gene expression in splenic $CD4^+$ and $CD8^+$ T cell subsets of genetically resistant and susceptible chickens infected with Marek's disease virus. Vet Immunol Immunopathol 132: 209-217. doi:10. 1016/j. vetimm. 2009. 06. 009. PubMed: 19615758.

31. Pfaffl MW (2001) A new mathematical model for relative quantification in real-time RT- PCR. Nucleic Acids Res 29: 2003-2007. doi:10. 1093/nar/29. 10. 2003. PubMed:11353068.

32. Philippe C, Grgic H, Nagy É (2005) Inclusion body hepatitis in young broiler breeders associated with a serotype 2 adenovirus in Ontario, Canada. J Appl Poult Res 14: 588-593. doi: 10. 1093/japr/14. 3. 588

33. Philippe C, Grgić H, Ojkić D, Nagy É (2007) Serologic monitoring of a broiler breeder flock previously effected by inclusion body hepatitis and testing the progeny for vertical transmission of fowl adenoviruses. Can J Vet Res 71: 98-102. PubMed: 17479772.

34. Plachý J, Weining KC, Kremmer E, Puehler F, Hala K et al. (1999) Protective effects of type I and type II interferons toward Rous sarcoma virus-induced tumors in chickens. Virology 256: 85-91. doi:10. 1006/viro. 1999. 9602. PubMed: 10087229.

35. Rasmussen R (2001) Quantification on the LightCycler. In Meuer S, Wittwer C, Nakagawara K, editors. Rapid Cycle Real-time PCR, Methods and Application. . Heidelberg: Springer Verlag Press . pp. 21-34.

36. Saifuddin M, Wilks CR (1991) Pathogenesis of an acute viral hepatitis: inclusion body hepatitis in the chicken. Arch Virol 116: 33-43. doi:10. 1007/BF01319229. PubMed:1848068.

37. Schonewille E, Jaspers R, Guntram P, Hess M (2010) Specific-pathogen-free chickens vaccinated with a live FAdV-4 vaccine are fully protected against a severe challenge even in the absence of neutralizing antibodies. Avian Dis 54: 905-910. doi:10. 1637/8999-072309-Reg. 1. PubMed: 20608537.

38. Sheppard M, Werner W, Tsatas E, McCoy R, Prowse S et al. (1998) Fowl adenovirus recombinant expressing VP2 of infectious bursal disease virus induces protective immunity against bursal disease. Arch Virol 143: 915-930. doi:10. 1007/s007050050342. PubMed: 9645198.

39. Sung RS, Qin L, Bromberg JS (2001) TNF alpha and IFN gamma induced by innate anti- adenoviral immune responses inhibit adenovirus-mediated transgene expression. Mol Ther 3: 757-767. doi:10. 1006/mthe. 2001. 0318. PubMed: 11356080.

40. Toro H, Prusas R, Raue R, Cerda L, Geisse C et al. (1999) Characterization of fowl adenoviruses from outbreaks of inclusion body hepatitis hydropericardium syndrome in Chile. Avian Dis 43: 262-270. doi:10. 2307/1592616. PubMed: 10396639.

41. Xing Z, Schat KA (2000) Inhibitory effects on nitric oxide and γ interferon on *in vitro*and *in vivo* replication of Marek's disease virus. J Virol 74: 3605-3612. doi:10. 1128/JVI. 74. 8. 3605-3612. 2000. PubMed: 10729136.

CHAPTER 13

RAPID DETECTION OF MYCOPLASMA SYNOVIAE BY LOOP-MEDIATED ISOTHERMAL AMPLIFICATION

Olimpia Kursa, Grzegorz Woz´niakowski, Grzegorz Tomczyk, Anna Sawicka, Zenon Mint

Keywords: *Mycoplasma synoviae,* Loop-mediated isothermal amplification, Rapid detection

INTRODUCTION

Mycoplasma synoviae (MS) infections are common and main cause of chronic subclinical upper respiratory disease, infectious synovitis, and airsacculitis in chickens and turkeys (Kleven2003; Ferguson-Noel and Noormohammadi 2013). Similar to *Mycoplasma gallisepticum* (MG) MS infections have a great economical impact on the commercial poultry industry causing growth retardation, weight loss, and decreased egg production (King et al. 1973; Kleven 2003; Feberwee et al. 2008; Catania et al. 2010; Ferguson-Noel and Noormohammadi 2013). Its exact effect in layer chicken industry is still a current topic of investigations because of a recently found new clinical form called eggshell apex abnormalities described in several countries (Feberwee et al. 2009; Catania et al. 2010; Landman 2014). An effective treatment of its clinical form was successfully treated using tylosin, administered in the drinking water for 5 days (Catania et al. 2010). However, the most preferred method of MS control in poultry flocks is its eradication (Olson et al. 1956; Stipkovits and Kempf 1996; Fiorentin et al. 2003).

Lateral transmission of MS occurs by direct contact and spreads between the birds kept in the same poultry house (Kleven et al. 1975; Ewing et al. 1998; Ferguson-Noel and Noormohammadi2013). Infection is transmitted via the respiratory tract and usually 100 % of birds become infected. In spite of the clinical form of the disease visible as joint lesions is not always observed, the infected birds remain carriers of MS (Ewing et al. 1998; Kleven 2003). MS infections most frequently occur as a chronic subclinical upper respiratory infection (Landman and Feberwee 2004). Meanwhile, not all chicken flocks showing high antibody levels against MS suffer from infectious synovitis (Jordan 1975; Kleven et al. 1975; Ferguson-Noel and Noormohammadi 2013). Some factors including age of birds, flock size and farming conditions may affect severity of MS in poultry flocks. Taking into account these aspects, rapid and efficient diagnostics of MS is an important issue not only for prevention of clinical disease progress but also infection transmission (Landman 2014).

So far, MS diagnosis has been carried out by bacteriological isolation, serological assays, or polymerase chain reaction and its modification (PCR, multiplex PCR or real-time PCR) (Lauerman et al. 1993; Hong et al. 2004; Hess et al. 2007; Feberwee et al. 2008; Hammond et al.2009; Raviv and Kleven 2009; Dijkman et al. 2013, Fraga et al. 2013; Reck et al. 2013). The sensitivity of direct MS culturing from affected joints is low and often yields negative results, especially in the sub-chronic stage of the disease (Salisch et al. 1998). Furthermore, direct MS culturing is generally expensive and time consuming because it takes up to 28 days. In serological methods, sera collected from infected birds are commonly analysed for the presence of antibodies by serum plate agglutination (SPA) test, seldom by hemagglutination-

Citation: Olimpia Kursa, Grzegorz Woźniakowski, Grzegorz Tomczyk, Anna Sawicka, Zenon Minta, Rapid detection of Mycoplasma synoviae by loop-mediated isothermal amplification, DOI: 10.1007/s00203-014-1063-2.

inhibition (HI) test or enzyme-linked immunosorbent assay (ELISA) (Ewing et al. 1998; Feberwee et al.2008).

Although there are many PCR-based methods for detection of MS, they are still laborious, time consuming and require advanced laboratory equipment (Hong et al. 2004; Hammond et al.2009; Raviv and Kleven 2009; Wetzel et al. 2010; Fraga et al. 2013; Reck et al. 2013).

An advance in detection of many different human and animal pathogens was loop-mediated isothermal amplification (LAMP) developed by Notomi et al. (2000). LAMP depends on auto-cycling strand displacement DNA synthesis performed by *Bst* or *Bsm* DNA polymerase using two or three sets of primers complementary to six or eight different regions in the gene of interest (Fu et al. 2011). LAMP uses a set of inner primers (forward FIP and backward BIP primer), outer primer (including F3 and B3) and forward and backward loop primers (FL and BL) which enable faster formation of loop and hairpin-like structures thus faster detection of specific DNA. Amplification is performed under isothermal conditions of 60–65 °C, while the final products form cauliflower-like structures with multiple overhangs (Notomi et al. 2000; Woźniakowski et al. 2011).

LAMP method has been widely applied for the diagnosis of bacteria like *Shigella spp.,Salmonella spp, E. coli* (Fu et al. 2011) as well as viral and fungal disease in human and animals (Saito et al. 2005; Huang et al. 2010; Yang et al. 2010; Woźniakowski et al. 2012; Mair et al.2013; Kakuya et al. 2014). Some of them are economically important poultry pathogens thus LAMP facilitated an improvement in their efficient detection.

The objective of this study was to develop LAMP for the detection of MS without the need of thermal cycler application using water bath or heating block. This is the first report on the development of LAMP for the specific MS identification.

MATERIALS AND METHODS

Strains

Seven reference *Mycoplasma spp.* strains purchased from ATCC collection comprising of *M. synoviae* (ATCC 25204), *M. gallisepticum* (ATCC 19610), *M. meleagridis* (ATCC 25284), *M. iowae* (ATCC 33552), *M. anatis* (ATCC 25524), *M. anseris* (ATCC 49234) and *M. cloacale*(ATCC 35276) were used. Additionally, 18 field isolates of MS collected during the years 2010–2014 were included. These strains were isolated and identified at the Department of Poultry Diseases at the National Veterinary Research Institute (NVRI, Pulawy, Poland).

DNA Extraction

Extraction of genomic DNA from MS reference strains and field isolates was conducted using the QIAamp Mini Kit (Qiagen, Hilden, Germany) following the manufacturer's instructions. The DNA samples were stored at −70 °C until further analysis.

LAMP

Based on the *vlhA* sequence of SB5654 MS strain (Accession number: KC506824), six complementary primers were designed using Primer Explorer V4 software (NetLaboratory, Tokyo, Japan) and manual assignment. The primer sequences are listed in Table 1. LAMP reactions were conducted in 15 µl of reaction mixtures which contained: 7.5 µl of Isothermal Mastermix (OptiGene, Horsham, West Sussex, United Kingdom), 10 pM of each inner primer FIP and BIP, 2.5 pM of each outer primer F3 and B3, 5 pM of "loop" primers: LoopF and LoopB primer, 1.5 µl of deionised water and 2 µl of DNA template.

Table 1: Sequences of LAMP primers used in the study

Primer	Sequence (5′–3′)	Length (nt)
F3	GGTGATCAAACTCCAGCA	18
B3	TAACCGATCCGCTTAATGC	19
FIP (F1c + F2)	CTCCTGGGTTTCCTGGGTTTCCCTGCTCCAACACCTGG	38
BIP (B1c + B2)	TTGACCCTGTAGAGGCTGCTAGCATCTGCTGTTGTAGTTGT	41
LoopF	TGGGTTTCCTGGATTTGGG	19
LoopB	GCTATTGATGCTGCAACAGAAT	22

F3, forward outer primer; B3, forward backward primer; FIP, forward inner primer (F1C + F2); BIP, backward inner primer (B1c + B2); LoopF, forward loop primer; LoopB, backward loop primer

The reaction mixtures were incubated at the temperature range from 56 to 70 °C for a different time (from 30 min to 105 min) and subsequently heated up to 80 °C for 2 min to inactivate the enzyme. After this step, 1 µl of 1:10 dilution of 10,000× concentrated SYBR Green® (Invitrogen, Paisley, UK) was added to each sample. Results of LAMP were observed by naked-eye, under UV light or after gel electrophoresis in 2 % agarose gels stained with ethidium bromide (0.5 µg/ml). The length of LAMP products was compared to 100-bp DNA Ladder Plus GeneRuler™ (Thermo-scientific, Waltham, Massachusetts, USA). The colour of reaction mixtures in negative controls was orange, while the positives showed green colour or greenish fluorescence. The sensitivity of LAMP was tested using nine 10-fold dilutions (10^6–10^{-2} CFU/ml) of DNA extracted from MS reference strain (ATCC 25204) with the entire titre 1 × 10^6 CFU/ml. The analytical specificity was tested using DNA extracted from six different*Mycoplasma spp.* reference strains.

PCR

PCR was conducted using MSLF and MssR primers as previously described (Wetzel et al. 2010). Reaction was carried out in a final volume of 25 μl containing 12.5 μl of Taq PCR Master Mix (EurX, Gdansk, Poland), 10 pM of each MsLF and MSsR primer and 7.5 μl deionised water. The thermal profile was 94 °C for 2 min, followed by 35 cycles of 94 °C for 30 s; 55 °C for 30 s; 72 °C for 60 s and final extension at 72 °C for 7 min. The PCR products were separated in 2 % agarose gel stained with ethidium bromide (0.5 μg/ml). The length of PCR products was compared to 100-bp DNA Ladder Plus GeneRuler™ (Thermo-scientific, Waltham, Massachusetts, USA).

RESULTS

LAMP Conditions

In order to determine the optimal conditions of LAMP, MS reference strain (ATCC 25204) was used as the target template. LAMP assays were incubated under isothermal condition between 56 and 70 °C. The specific colour change, greenish fluorescence and amplicons were observed at 66.3 °C, but in case of field MS strains, the optimal temperature was 63.0 °C. The optimal LAMP duration time was 60 min. Longer incubation up to 105 min had no influence on the final results. Therefore, the final LAMP conditions for all strains was 63 °C for 60 min then 80 °C for 2 min to deactivate the polymerase.

Analytical Sensitivity and Specificity of LAMP

The analytical sensitivity of LAMP showed that the detection limit was 10^{-1} CFU/ml of MS reference strain (ATCC 25204) and was 1,000 times higher in comparison with conducted PCR (Fig. 1). LAMP was specific only for the reference strain of MS. No colour change, fluorescence nor ladder-like pattern was observed in LAMP samples containing other species of *Mycoplasma*(Fig. 2). This indicated high analytical specificity of established LAMP assay for the selective detection of MS.

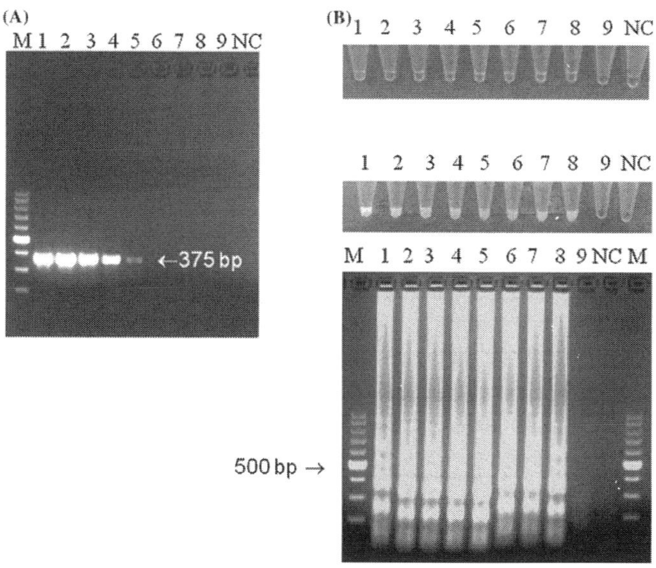

Figure. 1: Analytical sensitivity of LAMP and PCR for MS detection. **a** PCR detection of *vlhA*fragment. The product length is 375 bp and was indicated with an *arrow*. **b** LAMP. Visible colour change in MS-positive samples, greenish fluorescence under UV light and specific ladder-like patter after gel electrophoresis. *DescriptionsM* molecular length marker GeneRuler™ 100 bp DNA Ladder Plus (Thermo-scientific, Waltham, Massachusetts, USA), 10-fold dilutions of reference *M. synoviae* strain (ATCC 25204): *1* 10^6 CFU/ml, *2* 10^5 CFU/ml, *3* 10^4 CFU/ml, *4* 10^3 CFU/ml, *5* 10^2 CFU/ml, *6* 10^1 CFU/ml, *7* 10^0 CFU/ml, *8* 10^{-1} CFU/ml, *9* 10^{-2} CFU/ml, and *NC* negative control—DNA extracted from non-inoculated growth medium.

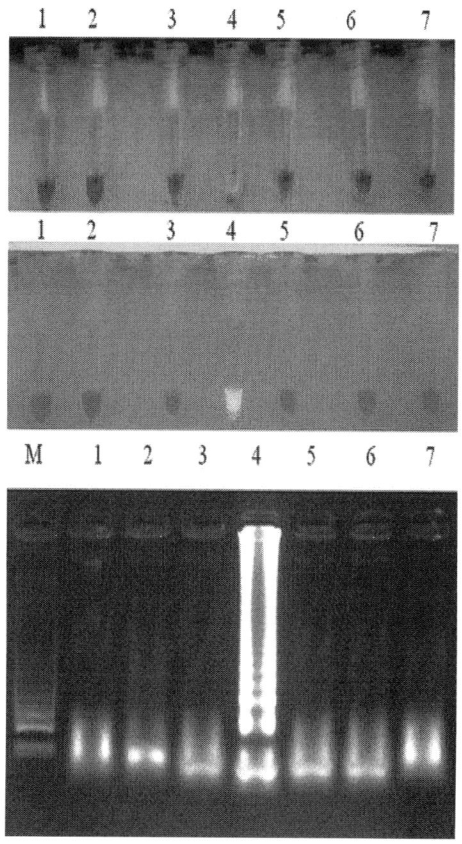

Figure. 2: Analytical specificity of LAMP for MS detection. Visible colour change in MS-positive samples, greenish fluorescence under UV light and specific ladder-like patter after gel electrophoresis. *DescriptionsM* molecular length marker GeneRuler™ 100 bp DNA Ladder Plus (Thermo-scientific, Waltham, Massachusetts, USA), *1 M. gallisepticum*(ATCC 19610), *2 M. meleagridis* (ATCC 25284), *3 M. iowae* (ATCC 33552), *4 M. synoviae*strain (ATCC 25204), *5 M. anatis* (ATCC 25524), *6 M. anseris* (ATCC 49234) and *7 M. cloacale* (ATCC 35276)

Field Samples of MS

All 18 field isolates of MS were positive in conducted MS-specific PCR (data not shown). The similar results were obtained using developed LAMP. The colour change of reaction mixtures and specific ladder-like pattern was detected in reference MS strain as well as in all 18 field isolates (Fig. 3). This indicated on the capability of LAMP to detect both reference and field MS isolates.

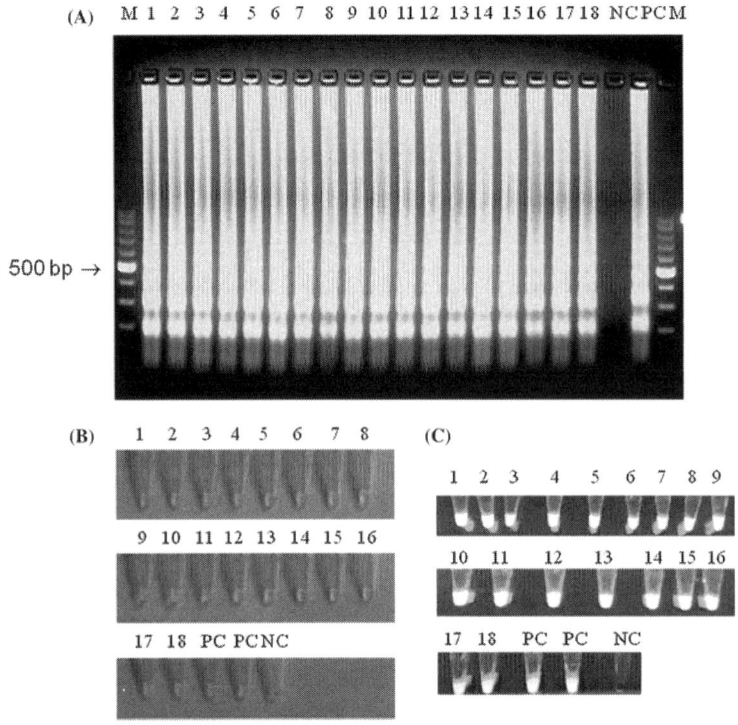

Figure. 3: Analysis of field MS isolates by LAMP. **a** Ladder-like pattern in MS-positive samples after gel electrophoresis, **b** visible colour change in MS-positive samples, **c** greenish fluorescence under UV light. *DescriptionsM* molecular length marker GeneRuler™ 100 bp DNA Ladder Plus (Thermo-scientific, Waltham, Massachusetts, USA), 1–18 field MS isolates, *NC* negative control—DNA extracted from non-inoculated growth medium,*PC* positive control DNA of reference *M. synoviae* strain (ATCC 25204)

DISCUSSION

Until now, MS culturing methods, serological assays and conventional PCRs or real-time PCRs are useful in detection of MS strains (Lauerman et al. 1993; Hong et al. 2004; Hess et al. 2007; Feberwee et al. 2008; Dijkman et al. 2013). However, culturing is very laborious, time consuming and requires specific growth medium and incubation (Ratliff et al. 2014). PCR is highly specific and sensitive technique but requires at least few hours, and electrophoresis is needed to visualize the results. Indeed, it has been previously presented that multiplex PCR is a valuable tool for the detection or differentiation of MG and MS in chickens (Garcia et al. 1995; Hess et al. 2007; Hammond et al. 2009; Wetzel et al. 2010; Fraga et al. 2013). Moreover, using PCR and PCR–RFLP method targeted on single-copy domain of the N-terminal *vlhA* region, it is possible to differentiate field MS isolates from MS-H vaccine strain (Bayatzadeh et al. 2014). Based on 16S-23S rRNA intergenic spacer region, it was also possible to quantify MS copy number in samples collected from chicken joints using specific qPCR technique (Raviv and Kleven 2009; Dijkman et al. 2013). An attractive alternative to traditional methods of MS detection is LAMP which is simple to set and requires no additional thermal cyclers (Woźniakowski et al. 2012). Previously, LAMP has been applied for detection of *M. hyorhinis inspecimens* of bronchoalveolar lavage fluid (BALF) from pigs (Du et al. 2013). The analytical sensitivity of LAMP reached ten DNA copies per ml that was 1,000 times higher than in case of conventional PCR. Indeed, in our study, LAMP showed to be very sensitive detecting 10^{-1}CFU/ml of standard MS strain. Other study published by Mair et al. (2013) showed LAMP usefulness for detection of *M. mycoides* subsp. *mycoides* as the technique that may be run on a battery-driven mobile device. The efficacy of LAMP was also confirmed in the study on detection of *M. pneumoniae* (MP) as the lower respiratory tract pathogen in children (Ratliff et al. 2014). Similar example of LAMP application for the detection of *M. wenyonii* in cattle was presented by Song et al. (2012). The conducted study on development of LAMP for MS identification provides an additional evidence of the usefulness of this technique as a rapid and very sensitive method for *Mycoplasma ssp.* detection.

CONCLUSIONS

In conclusion, the developed LAMP showed to be a powerful and robust method for the specific and selective detection of *M. synoviae*. Using specific region of *vlhA* gene, it was possible to selectively amplify DNA of MS without cross-reactions in other samples containing DNA of*Mycoplasma spp.* This method presents an useful alternative for local veterinarians and small veterinary laboratories without an access to PCR technique.

CONFLICT OF INTEREST

The authors declare that they have no competing interests.

REFERENCES

1. Bayatzadeh MA, Pourbakhsh SA, Ashtari A, Abtin AR, Abdoshah M (2014) Molecular typing of Iranian field isolates *Mycoplasma synoviae* and their differentiation from the live commercial vaccine strain MS-H using vlhA gene. Br Poult Sci 55:148–156
2. Catania S, Bilato D, Gobbo F, Granato A, Terregino C, Iob L, Nicholas RA (2010) Treatment of eggshell abnormalities and reduced egg production caused

by *Mycoplasma synoviae* infection. Avian Dis 54:961–964

3. Dijkman R, Feberwee A, Landman WJ (2013) Validation of a previously developed quantitative polymerase chain reaction for the detection and quantification of *Mycoplasma synoviae* in chicken joint specimens. Avian Pathol 42:100–107
4. Du GM, Liu MJ, Wu YZ, Xiong QY, Bai FF, Feng ZX, Shao GQ (2013) Development of a loop-mediated isothermal amplification assay for rapid detection of *Mycoplasma hyorhinis*. Clin Lab 59:1363–1371
5. Ewing L, Cookson KC, Phillips RA, Turner KR, Kleven SH (1998) Experimental infection and transmissibility of *Mycoplasma synoviae* with delayed serological response in chickens. Avian Dis 42:230–238
6. Feberwee A, de Vries TS, Landman WJ (2008) Seroprevalence of *Mycoplasma synoviae* in Dutch commercial poultry farms. Avian Pathol 37:629–633
7. Feberwee A, de Wit JJ, Landman WJ (2009) Induction of eggshell apex abnormalities by*Mycoplasma synoviae:* field and experimental studies. Avian Pathol 38:77–85
8. Ferguson-Noel N, Noormohammadi AH (2013) *Mycoplasma synoviae* infection. In: Swayne DE, Glisson JR, McDougald LR, Nolan LK, Suarez DL, Nair VL (eds) Diseases of poultry, 13th edn. Wiley, Ames, pp 900–906
9. Fiorentin L, Soncini RA, da Costa JL, Mores MA, Trevisol IM, Toda M, Vieira ND (2003) Apparent eradication *of Mycoplasma synoviae* in broiler breeders subjected to intensive antibiotic treatment directed to control *Escherichia coli*. Avian Pathol 32:213–216
10. Fraga AP, de Vargas T, Ikuta N, Fonseca AS, Celmer AJ, Marques EK, Lunge VR (2013) A multiplex real-time PCR for detection of *Mycoplasma gallisepticum* and *Mycoplasma synoviae*in clinical samples from Brazilian commercial poultry flocks. Braz J Microbiol 44:505–510
11. Fu S, Qu G, Guo S, Ma L, Zhang N, Zhang S, Gao S, Shen Z (2011) Applications of loop-mediated isothermal DNA amplification. Appl Biochem Biotechnol 163:845–850
12. Garcia M, Jackwood MW, Levisohn S, Kleven SH (1995) Detection of *Mycoplasma gallisepticum*, *M. synoviae*, and *M. iowae* by multi-species polymerase chain reaction and restriction fragment length polymorphism. Avian Dis 39:606–616
13. Hammond PP, Ramírez AS, Morrow CJ, Bradbury JM (2009) Development and evaluation of an improved diagnostic PCR for *Mycoplasma synoviae* using primers located in the haemagglutinin encoding gene *vlhA* and its value for strain typing. Vet Microbiol 136:61–68
14. Hess M, Neubauer C, Hackl R (2007) Interlaboratory comparison of ability to detect nucleic acid of *Mycoplasma gallisepticum* and *Mycoplasma synoviae* by polymerase chain reaction. Avian Pathol 36:127–133
15. Hong Y, Garcia M, Leiting V, Bencina D, Dufour-Zavala L, Zavala G, Kleven SH (2004) Specific detection and typing of *Mycoplasma synoviae* strains in poultry with PCR and DNA sequence analysis targeting the hemagglutinin encoding gene *vlhA*. Avian Dis 48:606–616
16. Huang CH, Lai GH, Lee MS, Lin WH, Lien YY, Hsueh SC, Kao JY, Chang WT, Lu TC, Lin WN, Chen HJ, Lee MS (2010) Development and evaluation of a loop-mediated isothermal amplification assay for rapid detection of chicken anemia virus. J Appl Microbiol 108:917–924
17. Jordan FTW (1975) Avian Mycoplasma and pathogenicity—a review. Avian Pathol 4:165–174
18. Kakuya F, Kinebuchi T, Fujiyasu H, Tanaka R, Kano H (2014) Genetic point-of-care diagnosis of*Mycoplasma pneumoniae* infection using LAMP assay. Pediatr Int 56:547–552
19. King DD, Kleven SH, Wenger DM, Anderson DP (1973) Field studies with *Mycoplasma synoviae*. Avian Dis 17:722–726
20. Kleven SH (2003) *Mycoplasma synoviae*. In: Saif YM (ed) Diseases of poultry, 11th edn. Iowa State University Press, Ames, pp 756–765
21. Kleven SH, Fletcher OJ, Davis RB (1975) Influence of strain of *Mycoplasma synoviae* and route of infection on development of synovitis or airsacculitis in broilers. Avian Dis 19:126–135
22. Landman WJ (2014) Is *Mycoplasma synoviae* outrunning *Mycoplasma gallisepticum*? A viewpoint from the Netherlands. Avian Pathol 43:2–8
23. Landman WJM, Feberwee A (2004) Aerosol-induced *Mycoplasma synoviae* arthritis: the synergistic effect of infectious bronchitis virus infection. Avian Pathol 33:591–598
24. Lauerman LH, Hoerr FJ, Sharpton AR, Shah SM, van Santen VL (1993) Development and application of a polymerase chain reaction assay for *Mycoplasma synoviae*. Avian Dis 37:829–834
25. Mair G, Vilei EM, Wade A, Frey J, Unger H (2013) Isothermal loop-mediated amplification (LAMP) for diagnosis of contagious bovine pleuro-pneumonia. BMC Vet Res 9:108
26. Notomi T, Okayama H, Masubuchi H, Yonekawa T, Watanabe K, Amino N (2000) Loop-mediated isothermal amplification of DNA. Nucleic Acids Res 28:E63
27. Olson NO, Shelton DC, Bletner JK, Munro DA, Anderson GC (1956) Studies of infectious synovitis in chickens. Am J Vet Res 17:747–754
28. Ratliff AE, Duffy LB, Waites KB (2014) Comparison of the illumigene *Mycoplasma* DNA amplification assay and culture for detection of *Mycoplasma pneumoniae*. J Clin Microbiol 52:1060–1063
29. Raviv Z, Kleven SH (2009) The development of diagnostic real-time TaqMan PCRs for the four pathogenic avian mycoplasmas. Avian Dis 53:103–107
30. Reck C, Menin A, Canever MF, Milettia LC (2013) Rapid detection of *Mycoplasma synoviae* and avian reovirus in clinical samples of poultry using multiplex PCR. Avian Dis 57:220–224

31. Saito R, Misawa Y, Moriya K, Koike K, Ubukata K, Okamura N (2005) Development and evaluation of a loop-mediated isothermal amplification assay for rapid detection of*Mycoplasma pneumoniae*. J Med Microbiol 54:1037–1041

32. Salisch H, Hinz KH, Graack HD, Ryll MA (1998) A comparison of a commercial PCR-based test to culture methods for detection of *Mycoplasma gallisepticum* and *Mycoplasma synoviae* in concurrently infected chickens. Avian Pathol 27:142–147

33. Song Q, Wang L, Fang R, Khan MK, Zhou Y, Zhao J (2012) Detection of *Mycoplasma wenyonii*in cattle and transmission vectors by the loop-mediated isothermal amplification (LAMP) assay. Trop Anim Health Prod 45:247–250

34. Stipkovits L, Kempf I (1996) Mycoplasmoses in poultry. Rev Sci Tech 15:1495–1525

35. Wetzel AN, Lefevre KM, Raviv Z (2010) Revised *Mycoplasma synoviae vlhA* PCRs. Avian Dis 54:1292–1297

36. Woźniakowski G, Samorek-Salamonowicz E, Kozdruń W (2011) Rapid detection of Marek's disease virus in feather follicles by loop-mediated amplification. Avian Dis 55:462–467

37. Woźniakowski G, Kozdruń W, Samorek-Salamonowicz E (2012) Loop-mediated isothermal amplification for the detection of goose circovirus. Virol J 9:110

38. Yang JL, Rui Y, Cheng AC, Wang MS, Fu LZ, Yang SQ, Zhang SH, Yang L, Xu ZY (2010) A simple and rapid method for detection of goose parvovirus in the field by loop mediated isothermal amplification. Virol J 7:14

CHAPTER 14 A DUPLEX REAL-TIME PCR ASSAY FOR THE DETECTION AND QUANTIFICATION OF AVIAN REOVIRUS AND MYCOPLASMA SYNOVIAE

Li Huang, Zhixun Xie, Liji Xie, Xianwen Deng, Zhiqin Xie, Sisi Luo, Jiaoling Huang, Tingting Zeng and Jiaxun Feng

Keywords: Duplex real-time PCR assay, Avian reovirus, *Mycoplasma synoviae*

BACKGROUND

Infectious arthritis in broilers represents an economic and health problem, resulting in severe losses due to retarded growth and downgrading at the slaughterhouse. The most common agents associated with cases of infectious arthritis in poultry are avian reovirus (ARV) and *Mycoplasma synoviae* (MS). ARV belongs to the Orthoreovirus genus, one of nine genera of the Reoviridae family [1,2]. ARV infection is associated with several disease syndromes and especially viral arthritis/tenosynovitis in chickens [3,4]. Meanwhile, MS is a common pathogen found in turkeys and chickens that causes diseases of the respiratory tract, urogenital tract and joints and impairs growth [5,6]. Mixed infections of ARV and MS have occurred in poultry flocks worldwide and have similar clinical signs, including severe immunosuppression, arthritis, depression, retarded growth, weight loss and decreased egg production. Bradbury [7] and Reck [8] also found that in chickens, a synergistic relationship exists between ARV and MS, which causes much more severe clinical signs and pathological lesions than the additive effects of these two pathogens alone do. The main feature of possible economic importance in ARV and MS infection is the incidence of decreased egg production and fertility, sternal bursitis leading to carcass downgrading and leg abnormalities related to condemnation of broilers. As the elimination of lesioned carcasses at the slaughterhouse is important [3,9], the rapid and efficient detection and diagnosis of ARV and MS are essential prerequisites for the effective control and prevention of these avian pathogens in poultry flocks.

The current methods for ARV and MS detection include serological assays; pathogen isolation and identification; and molecular detection methods, such as single PCR and multiplex PCR [10-13]. However, these assays are laborious and time consuming, have limited specificity and sensitivity, and require post-amplification procedures. Real-time PCR assays for the specific identification of a target sequence by fluorescent probes can overcome these limitations and provide distinct advantages, such as a shorter detection time, improved sensitivity and specificity, simplified closed-tube procedures and the potential for pathogen screening and surveillance in commercial poultry flocks [14-16].

Therefore, the present study developed and validated a duplex real-time PCR assay for the differential diagnosis and quantitative detection of ARV and MS.

MATERIALS AND METHODS

Ethics Statement

This study was performed in strict accordance with the recommendations in the Guide for the Care and Use of Laboratory Animals of the Guangxi Veterinary Research Institute. Additionally, the Animal Care and Use Committee

Citation: Li Huang, Zhixun Xie, Liji Xie, Xianwen Deng, Zhiqin Xie, Sisi Luo, Jiaoling Huang, Tingting Zeng and Jiaxun Feng, A duplex real-time PCR assay for the detection and quantification of avian reovirus and Mycoplasma synoviae, DOI: 10.1186/s12985-015-0255-y.

of the Guangxi Veterinary Research Institute approved all procedures involving the use of animals, and all efforts were made to minimize animal suffering.

Pathogens and Construction of Recombinant Plasmids

DNA was first extracted from MS samples as described previously [17], and total RNA was extracted from ARV samples using TRIzol reagent (Life Technologies, Carlsbad, CA, USA) following the manufacturer's instructions. Next, cDNA was generated as described previously [18] and used as a template for a duplex real-time PCR assay. Target gene fragments from the ARV σC gene or the MS phase-variable surface lipoprotein hemagglutinin (*v*lhA) gene were then amplified with the primers listed in Table 1 and inserted into the pMD18-T vector (TaKaRa, Dalian, China). Subsequently, the constructed plasmids were transformed into DH5α*Escherichia coli*. The recombinant plasmids carrying each target gene were confirmed by sequencing and were used as positive standards for ARV and MS. The copy number of each positive-standard plasmid was calculated as described previously [19].

Table 1: Specific primers used to clone ARV and MS specific genes

Primer name	Primer sequence	Amplicon length	Target gene
ARV C703F	5'-TGTGGATCCATGGCGGGTCT-CAAT-3'	981 bp	σC gene
ARV C703R	5'-CCGGAATTCTAAGGTGTCGAT-GCC-3'		
MS *v*lhA F2	5'-CTGTTATAGCAATTTCATGTG-GTG-3'	283 bp	phase-variable surface lipoprotein hemagglutinin (*v*lhA)
MS *v*lhA R2	5'-TGTTGTAGTTGCTTCAACTT-GTCT-3'		

Oligonucleotide Primers and DNA Probes for Duplex Real-Time PCR

DNASTAR software (DNASTAR Inc., Madison, WI, USA) was used to confirm the highly conserved regions in the ARV and MS genomes, and Primer Express 3.0 software (Applied Biosystems, Foster City, CA, USA) was used to design the primers and probes listed in Table 2 for ARV and MS, based on their highly conserved regions. The cross-reactivity of the oligonucleotides was assessed by BLAST analysis. Both sets of primers and probes were synthesized by TaKaRa (Dalian, China).

Table 2: Primers and probes used for the duplex real-time PCR assay

Primer/probe name	Primer/probe sequence	Target gene	Amplicon length
ARV F	5'-CGTTCCCTGTGGACGTATCA-3'	σC	69 bp
ARV R	5'-GAGTACACCCCATACGCTTGGT-3'		
ARV P	5'-(FAM) TCACCCGCGATTCTGCGACTCAT (Eclipse)-3'		
MS-F	5'-ATAGCAATTTCATGTGGTGATCAA-3'	*v*lhA	143 bp
MS-R	5'-TGGATTTGGGTTTTGAGGATTA-3'		
MS-P	5'-(ROX) CAGCACCTGAACCAACACCTG-GAA (Eclipse)-3'		

Duplex Real-Time PCR Assay for Simultaneous MS and ARV Detection

The duplex real-time PCR was performed in a 20-μl volume. The reaction mixture included 1× real-time PCR Premix (Perfect Real Time PCR Kit, TaKaRa, Dalian, China); 0.3 μM ARV F, ARV R and ARV P primers; 0.3 μM MS F, MS R and MS P primers; and 2 μl of positive-plasmid template. Sterilized H_2O was added to bring the final volume to 20 μl. The protocol for the reaction was 95°C for 30 sec; 45 cycles of 95°C for 10 sec and 60°C for 30 sec; and, finally, 40°C for 5 sec. The fluorescence was measured at the end of each 60°C incubation. The data analysis was performed using Light Cycler 2.0 system software (Roche, Molecular Biochemical, Mannheim, Germany).

Conventional RT-PCR and PCR

Conventional RT-PCR for ARV amplification and conventional PCR for MS amplification were performed. The PCR mixture contained 2× Premix Taq (TaKaRa, Dalian, China), 0.4 μM forward primer or reverse primer, 2 μl of template and sterilized H_2O to bring the final reaction volume to 25 μl. The conditions for PCR were 95°C for 5 min; 72°C for 7 min; and three-step cycling 35 times at 95°C for 30 sec, 60°C for 30 sec and 72°C for 30 sec. The PCR product was run on a 2% agarose gel at 80 V for 45 min and visualized on a molecular imager Gel Doc XR+ imaging system with Image Lab software (Bio-Rad, Life Science Research, Hercules, CA, USA).

Specificity and Sensitivity of the Duplex Real-Time PCR Assay

To assess the specificity of the assay, DNA from *Mycoplasma gallisepticum* (MG), *Mycoplasma iowae* (MI) and *Mycoplasma meleagridis* (MM) were extracted as described previously [17]. Additionally, cDNA was generated from total RNA that was extracted from cases of newcastle disease virus (NDV), infectious bursal disease virus (IBDV), avian infectious bronchitis virus (AIBV), the H9 subtype of the avian influenza virus (AIV), Marek's disease virus (MDV), reticuloendotheliosis virus (REV), and avian leukosis virus (ALV) using TRIzol reagent (Life Technologies, Carlsbad, CA, USA) following the manufacturer's instructions. The DNA and cDNA were mixed together in equal concentrations as the templates and were subjected to the optimized duplex real-time PCR assay to detect ARV and MS. The sensitivity of the duplex real-time PCR assay was determined using serial 10-fold dilutions (10^1-10^8 copies/μl) of positive-plasmid combinations carrying the MS and ARV target genes as templates. These results were compared with the results of conventional PCR. To generate a standard curve for ARV and MS, the threshold cycle (Ct) of these standard dilutions was plotted against the log value of the copy number of the corresponding standard plasmid.

Reproducibility and Interference Tests of the Duplex Real-Time PCR Assay

To assess the intra- and inter-assay reproducibility, three samples with the same concentration (10^8 copies/μl) of the MS or ARV target gene were assessed using the duplex real-time PCR assay. The same experiments were repeated in triplicate every two days for seven days. The reproducibility was then analyzed based on the standard deviation (SD) and the coefficient of variability (CV) of the Ct average. To determine the reaction efficiency interference, different concentrations of positive plasmids carrying the ARV or MS target gene ($10^8$and 10^1 copies/μl, respectively) were analyzed using the duplex real-time PCR assay.

Duplex Real-Time PCR Analysis of Field Samples

All field samples, such as joints and joint contents, were collected from chicks and broilers exhibiting clinical signs of MS or ARV infections and were used to validate the duplex real-time PCR assay. The results were compared with those of traditional diagnostic methods, such as pathogen isolation and identification and serological tests.

RESULTS

Specificity Test

The specificity of the duplex real-time PCR assay was verified by examining DNA/cDNA from different samples infected with different pathogens. As shown in Figure 1 and Table 3, the duplex real-time PCR assay was able to detect and differentiate ARV and MS independently and simultaneously. In contrast, the other avian pathogens (NDV, IBDV, AIBV, AIV, MDV, REV, ALV, MG, MI and MM) were not detected using the duplex real-time PCR assay. When samples were co-infected with both ARV and MS, unique amplification curves were simultaneously produced in the 530 nm and 610 nm channels, whereas a single amplification curve was observed in the 530 nm or 610 nm channel when samples were infected with only ARV or MS, respectively. Thus, the specificity of the duplex real-time PCR assay was 100%, with no detectable fluorescent signals for other avian pathogens or negative controls.

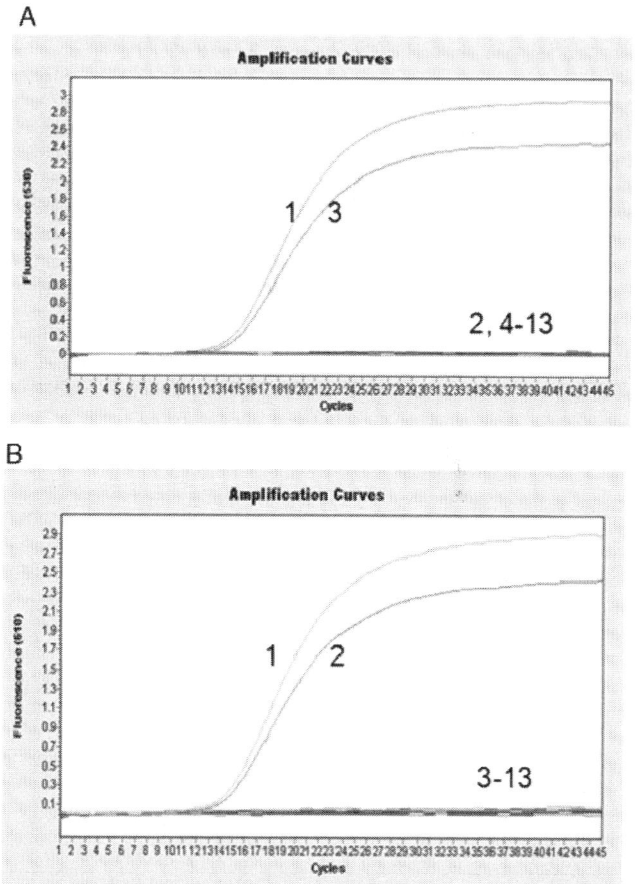

Figure 1: Specificity of the duplex real-time PCR assay for ARV and MS. (A) Specificity of the duplex real-time PCR assay for ARV. 1. MS+ARV, 2. MS, 3. ARV, 4. IBDV, 5. NDV, 6. AIBV, 7. MDV, 8. H9 subtype of AIV, 9. REV, 10. ALV, 11. MG, 12. MI, 13. MM, 14. negative control. **(B)** Specificity of the duplex real-time PCR assay for MS. 1. MS+ARV, 2. MS, 3. ARV, 4. IBDV, 5. NDV, 6. AIBV, 7. MDV, 8. H9 subtype of AIV, 9. REV, 10. ALV, 11. MG, 12. MI, 13. MM, 14. negative control.

Table 3: Pathogens used and Ct values of the duplex real-time PCR assay's specificity

Pathogen	Number of samples	Ct values of duplex real-time PCR assay	
		MS	ARV
NDV	1	Negative	Negative
H9 subtype of AIV	1	Negative	Negative
AIBV	1	Negative	Negative
IBDV	1	Negative	Negative
MG	1	Negative	Negative
MM	1	Negative	Negative
MI	1	Negative	Negative
MDV	1	Negative	Negative
pMD18-T-ARV	1	Negative	15.58
pMD18-T-MS	1	15.31	Negative
REV	1	Negative	Negative
ALV	1	Negative	Negative

Sensitivity Test

The sensitivity of the duplex real-time PCR assay was verified by testing 10^1-10^8 copies/μl of recombinant plasmids carrying the ARV or MS target gene. In Figure 2, the ARV amplification curves are shown in the 530 nm channel (Figure 2A), and the MS amplification curves are shown in the 610 nm channel (Figure 2B). Moreover, the standard curves for ARV and MS are shown in Figure 3, and the Ct values are listed in Table 4. The results revealed that even with a template amount as low as 2×10^1 copies, the ARV or MS target gene was still detectable. In contrast, the detection limit of the conventional PCR template was 2×10^3 copies for ARV and MS (Figure 4), which is 100 times lower than that of the duplex real-time PCR assay. Thus, the duplex real-time PCR assay is highly sensitive.

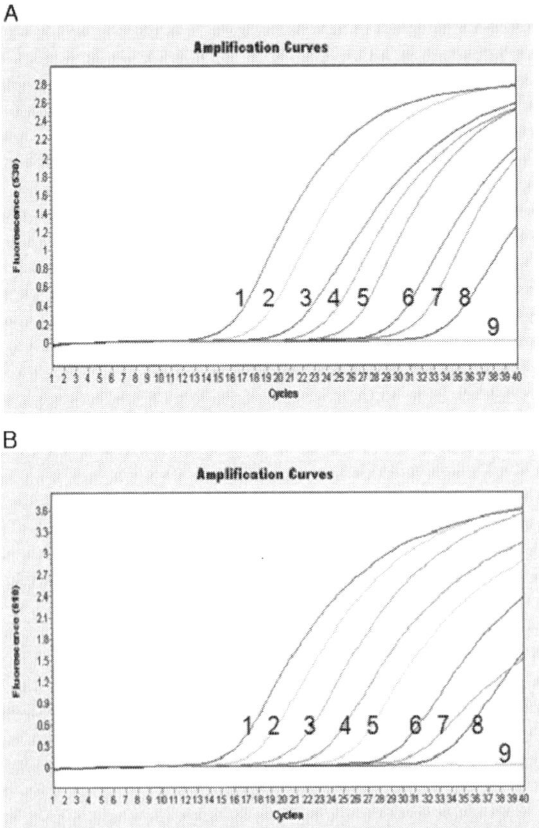

Figure 2: Sensitivity of the duplex real-time PCR assay for ARV and MS. (A) Sensitivity of the duplex real-time PCR assay for ARV. 1. 2×10^8 copies, 2. 2×10^7 copies, 3. 2×10^6 copies, 4. 2×10^5 copies, 5. 2×10^4 copies, 6. 2×10^3 copies, 7. 2×10^2 copies, 8. 2×10^1 copies, 9. negative control. (B) Sensitivity of the duplex real-time PCR assay for MS. 1. 2×10^8copies, 2. 2×10^7 copies, 3. 2×10^6 copies, 4. 2×10^5 copies, 5. 2×10^4 copies, 6. 2×10^3 copies, 7. 2×10^2 copies, 8. 2×10^1 copies, 9. negative control.

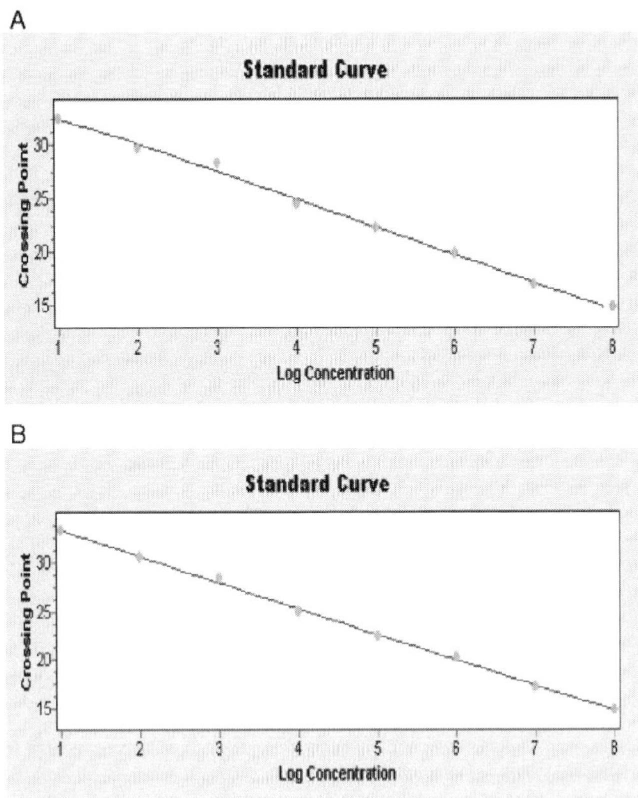

Figure 3: ARV and MS standard curves. (A) ARV standard curve. (B) MS standard curve.

Table 4: Ct values from the serial dilution of positive plasmids

Tenfold dilution	**2×10^8copies**	**2×10^7copies**	**2×10^6copies**	**2×10^5copies**	**2×10^4copies**	**2×10^3cop-ies**	**2×10^2copies**	**2×10^1copies**
ARV (Ct)	14.86	17.21	20.2	22.49	24.9	28.36	30.45	33.18
MS (Ct)	14.79	16.83	19.77	22.18	24.49	28.24	29.72	32.41

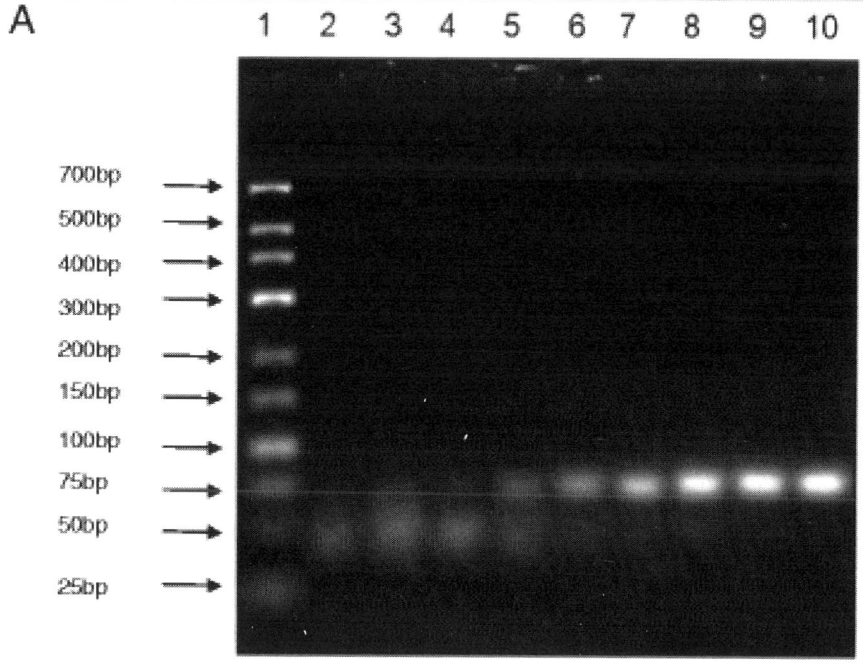

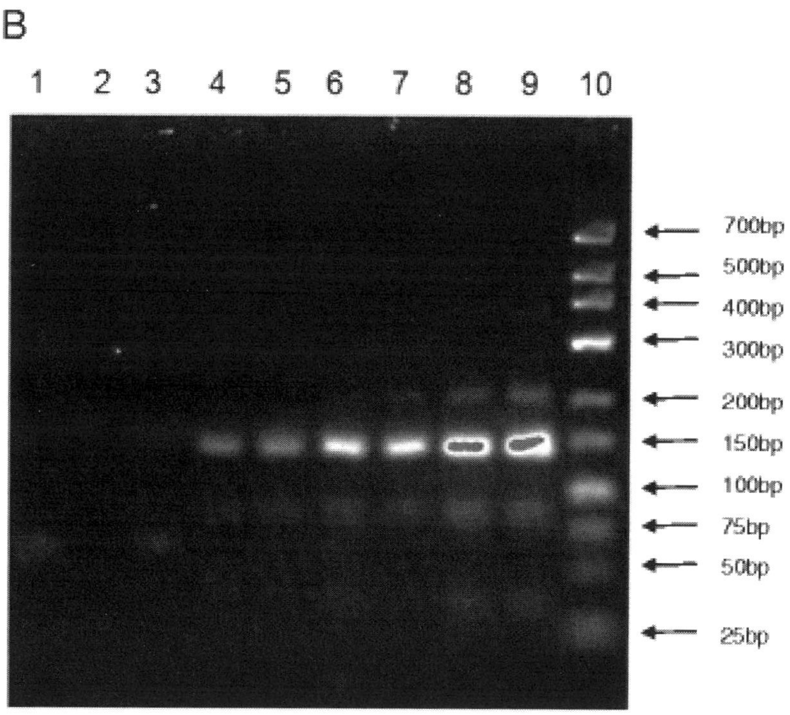

Figure 4: Sensitivity of conventional PCR for ARV and MS. (A) Sensitivity of conventional PCR for ARV. 1. low ladder, 2. negative control, 3. 2×10^1 copies, 4. 2×10^2 copies, 5. 2×10^3 copies, 6. 2×10^4 copies, 7. 2×10^5 copies, 8. 2×10^6copies, 9. 2×10^7 copies, 10. 2×10^8 copies. (B) Sensitivity of conventional PCR for MS. 1. negative control, 2. 2×10^1copies, 3. 2×10^2 copies, 4. 2×10^3 copies, 5. 2×10^4 copies, 6. 2×10^5 copies, 7. 2×10^6 copies, 8. 2×10^7 copies, 9. 2×10^8 copies, 10. low ladder.

Reproducibility and Interference Tests

The reaction of reproducibility was determined by testing three samples of the same concentration at the same time points and was assessed using the SD and CV of the Ct values for each sample. The intra-assay reproducibility results are shown in Figure 5, and the inter-assay reproducibility results are listed in Table 5. The CV values were 1.61% for ARV and 1.89% for MS. These data indicate that the findings produced by the duplex real-time PCR assay are reproducible.

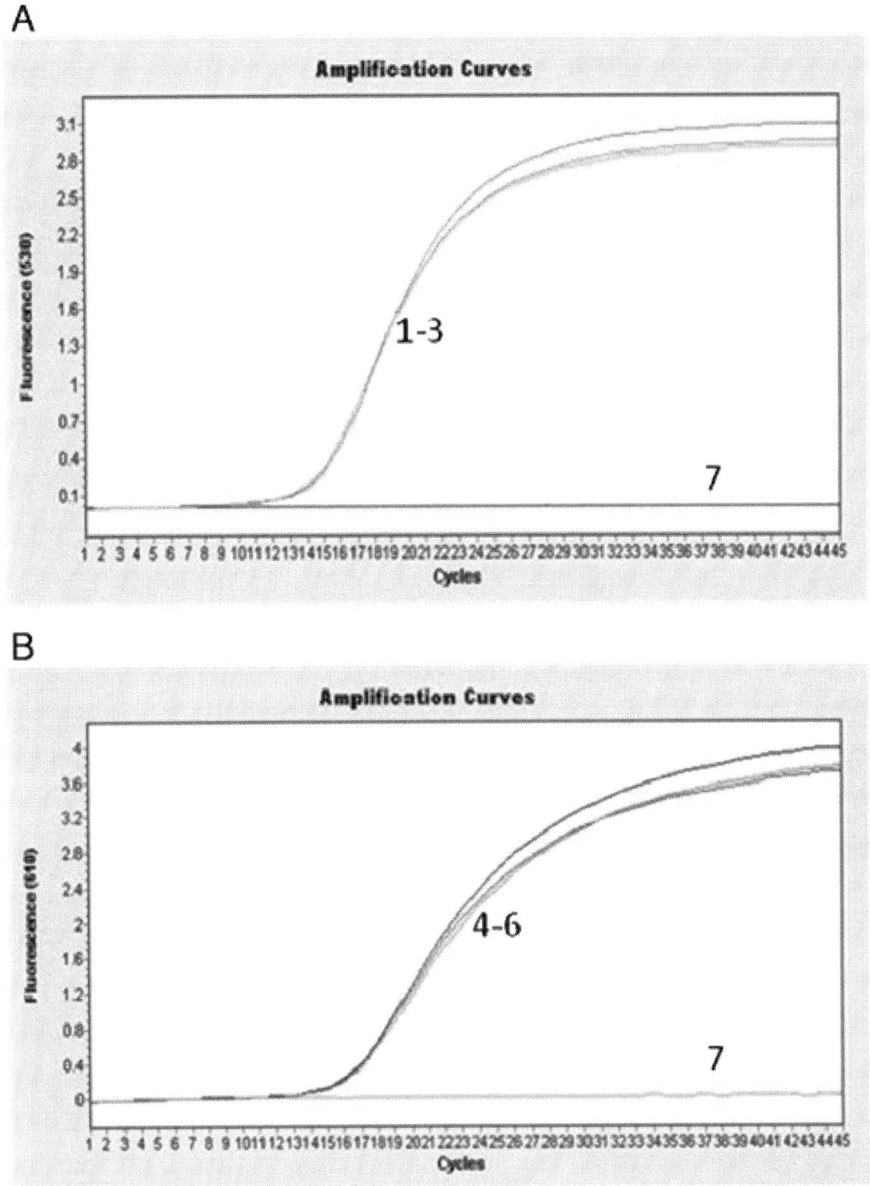

Figure 5: Reproducibility of the duplex real-time PCR assay for ARV (A) and MS (B). 1–3. ARV, 4–6. MS, 7. negative control.

Table 5: Reproducibility of the duplex real-time PCR assay for ARV and MS

Pathogen	Ct values of same samples at different time points				
	Day 1	Day 4	Day 7	SD	CV
ARV	15.23/1 × 10^8	15.4/1 × 10^8	15.36/1 × 10^8	0.24	1.61%
MS	14.96/1 × 10^8	14.7/1 × 10^8	15.09/1 × 10^8	0.28	1.89%

Because the presence of other templates may affect the amplification efficiency of a PCR assay, we tested the influence of mixtures of different concentrations on reaction sensitivity. In particular, a combination of a high concentration (10^8 copies/μl) and a low concentration (10^2 copies/μl) of positive plasmids carrying the ARV or MS target gene was used in an interference test (Table 6). The results indicated that there were no systematic deviations in the amplification curves of the mixed templates compared with those of the single template; moreover, the CV value was less than 3% (data not shown). The results suggest that the newly developed duplex real-time PCR assay is stable.

Table 6: Samples used in the interference test

Pathogen	Sample 1	Sample 2	Sample 3	Sample 4	Sample 5	Sample 6
ARV	2×10^8 copies	2×10^2 copies	2×10^8 copies	--	2×10^2 copies	--
MS	2×10^2 copies	2×10^8 copies	--	2×10^8 copies	--	2×10^2 copies

Field Samples

The detection results of the duplex real-time PCR assay for 40 field samples are listed in Table 7 and the results were confirmed by pathogen isolation and identification or serological tests. The ARV detection rate was 7.5%, and the MS detection rate was 5%. Additionally, the range of Ct values for ARV was 15.29-34.42, and the range of Ct values for MS was 13.53-30.68. Thus, the results of this new assay were comparable with the results of other detection approaches.

Table 7: Detection of field samples using the duplex real-time PCR assay

	ARV	MS
Positive samples/total samples	3/40	2/40
Positive rate	7.5%	5%

DISCUSSION

Both MS and ARV can cause similar clinical signs and lesions and may be present as co-infections in chickens and other avian species, which can lead to huge economic losses [20]. In this paper, we developed a duplex real-time PCR assay and described its use for the rapid, sensitive and accurate quantitative detection of ARV and MS.

The primary advantage of this duplex real-time PCR assay is the simultaneous detection and differentiation of ARV and MS. By using unique primer and probe sets within the highly conserved gene regions of ARV and MS, this duplex real-time PCR assay is readily able to detect and differentiate these pathogens via one reaction. Furthermore, this assay is optional and can be utilized as a single-target assay or combined into duplex assays, without impacting the quality of the results. Specifically, duplexing reduces the expense of reagents and the required time for analysis, and the single-target assay makes this assay adaptable to circumstances that may not require the simultaneous detection of these two pathogens for diagnostic purposes. These advantages greatly facilitate clinical application, which is an important criterion for the usefulness of a diagnostic assay for the early surveillance and prevention of diseases [21].

For a method of pathogen detection to be used as a clinical diagnostic tool, sensitivity is a key criterion [22,23]. Using the newly developed assay, as few as 2 x 10^1 copies could be detected for both ARV and MS, which was more sensitive than the results of a duplex real-time PCR assay reported by Sprygin [24] and the results of the multiplex PCR performed by Reck [11]. Moreover, during detection with mixed samples (other non-targeted pathogens) and field samples, the specificity of this new assay was comparable with that of traditional methods, such as pathogen isolation and identification and serological tests. Therefore, this duplex real-time PCR assay with higher sensitivity rates could be promising as a tool for rapid clinical differentiation and diagnosis at the early stage of ARV and/or MS infection.

Another distinct feature of this duplex real-time PCR assay is the short turn-around time for the results. In the present study, the results for ARV and MS infections were obtained within 2 h with this duplex real-time PCR assay, which is very important for rapid diagnosis, especially during emergent disease outbreaks. Furthermore, the obtained results could be directly visualized on a computer connected to the real-time PCR station. Compared with the conventional diagnostic approaches for ARV and MS infections (and even single and multiplex PCRs [25,26]), this assay does not require additional unique equipment or specialized labor. This method also minimizes post-amplification procedures, such as electrophoresis and UV visualization, which are time consuming. As compared to recently developed isothermal methods for ARV or MS detection, including loop-mediated isothermal amplification [27,28] and cross-priming amplification [29], for which there is no need for expensive equipment except a water bath, the main drawback of the duplex real-time PCR assay is the absolute need for the thermal cycle. However, the method capability of simultaneous detection for ARV and MS highlights its importance and great value for the rapid detection of ARV and MS infections in the laboratory.

Considering the high cost of probe synthesis and the possibility of different genotypes as well as variant or vaccine strains of ARV or MS, the development of new technology or novel reagents for probe synthesis and the design of more primers based on more highly conserved regions of the ARV and MS genomes would be necessary to investigate further modification and optimization of this new assay.

CONCLUSION

In this study, we developed a rapid, specific and sensitive duplex real-time PCR assay for the simultaneous detection of ARV and MS. Based on its speed and sensitivity, this newly developed assay could be useful not only for the clinical diagnosis of ARV and MS infections but also for the control and prevention of these infections.

ACKNOWLEDGMENTS

This study was supported by the National Natural Science Foundation of China (31160512), China Postdoctoral Science Foundation (2014M552536XB), Guangxi Science and Technology Projects (1222003-2-4, 2013GXNSFBA019120 and 2014GXNSFCA118006), and Guangxi Government Senior Scientist Foundation (2011B020) (Guangxi, China).

COMPETING INTERESTS

The authors declare that they have no competing interests.

AUTHORS' CONTRIBUTIONS

LH and ZX designed the experiments. LX, XD, ZX, SL, TZ and JF prepared the RNA/DNA samples. LH designed the primers and optimized the conditions of the duplex real-time PCR assay. LH, LX, XD, ZX, SL and TZ carried out the experiments shown in Figures 1, 2, 3, 4 and 5 and in Tables 3, 4, 5 and 7. LH performed the data analysis. LH wrote the manuscript. All authors reviewed and approved the final version of the manuscript.

REFERENCES

1. Teng L, Xie Z, Xie L, Liu J, Pang Y, Deng X, et al. Complete genome sequences of an avian orthoreovirus isolated from guangxi, china. Genome Announc. 2013;1(4):e00495–13.
2. Teng L, Xie Z, Xie L, Liu J, Pang Y, Deng X, et al. Sequencing and phylogenetic analysis of an avian reovirus genome. Virus Genes. 2014;48:381–6.
3. Jones RC. Avian reovirus infections. Rev Sci Tech. 2000;19:614–25.
4. van der Heide L. The history of avian reovirus. Avian Dis. 2000;44:638–41.
5. Lockaby SB, Hoerr FJ, Lauerman LH, Kleven SH. Pathogenicity of Mycoplasma synoviae in broiler chickens. Vet Pathol. 1998;35:178–90.
6. Hinz KH, Blome C, Ryll M. Virulence of Mycoplasma synoviae strains in experimentally infected broiler chickens. Berl Munch Tierarztl Wochenschr. 2003;116:59–66.
7. Bradbury JM, Garuti A. Dual infection with Mycoplasma synoviae and a tenosynovitis-inducing reovirus in chickens. Avian Pathogy. 1978;7(3):407–19.
8. Reck C, Menin ÁM, Pilati C, Milettia LC. Clinical and histologic lesions of mixed infection with Avian orthoreovirus and Mycoplasma synoviae in broilers. Pesqui Vet Bras. 2012;32:687–91.
9. Landman WJ, Feberwee A. Aerosol-induced Mycoplasma synoviae arthritis: the synergistic effect of infectious bronchitis virus infection. Avian Pathol. 2004;33:591–8.
10. Ortiz A, Kleven SH. Serological detection of Mycoplasma synoviae infection in turkeys. Avian Dis. 1992;36:749–52.
11. Reck C, Menin A, Canever MF, Milettia LC. Rapid detection of Mycoplasma synoviae and avian reovirus in clinical samples of poultry using multiplex PCR. Avian Dis. 2013;57:220–4.
12. Noormohammadi AH, Markham PF, Markham JF, Whithear KG, Browning GF. Mycoplasma synoviae surface protein MSPB as a recombinant antigen in an indirect ELISA. Microbiology. 1999;145(Pt 8):2087–94.
13. Ben Abdelmoumen Mardassi B, Ben Mohamed R, Gueriri I, Boughattas S, Mlik B. Duplex PCR to differentiate between Mycoplasma synoviae and Mycoplasma gallisepticum on the basis of conserved species-specific sequences of their hemagglutinin genes. J Clin Microbiol. 2005;43:948–58.
14. Troxler S, Marek A, Prokofieva I, Bilic I, Hess M. TaqMan real-time reverse transcription-PCR assay for universal detection and quantification of avian hepatitis E virus from clinical samples in the presence of a heterologous internal control RNA. J Clin Microbiol. 2011;49:1339–46.
15. Wittwer CT, Herrmann MG, Gundry CN, Elenitoba-Johnson KS. Real-time multiplex PCR assays. Methods. 2001;25:430–42.
16. Mackay IM, Arden KE, Nitsche A. Real-time PCR in virology. Nucleic Acids Res. 2002;30:1292–305.
17. Xie Z, Luo S, Xie L, Liu J, Pang Y, Deng X, et al. Simultaneous typing of nine avian respiratory pathogens using a novel GeXP analyzer-based multiplex PCR assay. J Virol Methods. 2014;207:188–95.
18. Xie Z, Xie L, Pang Y, Lu Z, Xie Z, Sun J, et al. Development of a real-time multiplex PCR assay for detection of viral pathogens of penaeid shrimp. Arch Virol. 2008;153:2245–51.
19. Vaitomaa J, Rantala A, Halinen K, Rouhiainen L, Tallberg P, Mokelke L, et al. Quantitative real-time PCR for determination of microcystin synthetase e copy numbers for microcystis and anabaena in lakes. Appl Environ Microbiol. 2003;69:7289–97.
20. Montebugnoli L, Venturi M, Gissi DB, Cervellati F. Clinical and histologic healing of lichenoid oral lesions following amalgam removal: a prospective study. Oral Surg Oral Med Oral Pathol Oral Radiol. 2012;113:766–72.
21. Heidenreich A, Bellmunt J, Bolla M, Joniau S, Mason M, Matveev V, et al. EAU guidelines on prostate cancer. Part 1: screening, diagnosis, and treatment of clinically localised disease. Eur Urol. 2011;59:61–71.
22. Heidenreich A, Bellmunt J, Bolla M, Joniau S, Mason M, Matveev V, et al. EAU guidelines on prostate cancer. Part I: screening, diagnosis, and treatment of clinically localised disease. Actas Urol Esp. 2011;35:501–14.
23. Wu H, Rao P, Jiang Y, Opriessnig T, Yang Z. A sensitive multiplex real-time PCR panel for rapid diagnosis

of viruses associated with porcine respiratory and reproductive disorders. Mol Cell Probes. 2014;28(5-6):264–70.

24. Sprygin AV, Andreychuk DB, Kolotilov AN, Volkov MS, Runina IA, Mudrak NS, et al. Development of a duplex real-time TaqMan PCR assay with an internal control for the detection of Mycoplasma gallisepticum and Mycoplasma synoviae in clinical samples from commercial and backyard poultry. Avian Pathol. 2010;39:99–109.

25. Jindal N, Chander Y, Patnayak DP, Mor SK, Ziegler AF, Goyal SM. A multiplex RT-PCR for the detection of astrovirus, rotavirus, and reovirus in turkeys. Avian Dis. 2012;56:592–6.

26. Xie Z, Fadl AA, Girshick T, Khan MI. Amplification of avian reovirus RNA using the reverse transcriptase-polymerase chain reaction. Avian Dis. 1997;41:654–60.

27. Kursa O, Wozniakowski G, Tomczyk G, Sawicka A, Minta Z. Rapid detection of Mycoplasma synoviae by loop-mediated isothermal amplification. Arch Microbiol 2014. doi:10.1007/s00203-014-1063-2

28. Xie Z, Peng Y, Luo S, Wang Y, Liu J, Pang Y, et al. Development of a reverse transcription loop-mediated isothermal amplification assay for visual detection of avian reovirus. Avian Pathol. 2012;41:311–6.

29. Wozniakowski G, Niczyporuk JS, Samorek-Salamonowicz E, Gawel A. The development and evaluation of cross-priming amplification (CPA) for the detection of avian reovirus (ARV). J Appl Microbiol 2014. doi:10.1111/jam.12705.

CHAPTER 15

DEVELOPMENT AND EVALUATION OF A DAS-ELISA FOR RAPID DETECTION OF TEMBUSU VIRUS USING MONOCLONAL ANTIBODIES AGAINST THE ENVELOPE PROTEIN

Hao Chen, Quanbin Ou, Yi Tang, Xuhui Gao, Lili Wu, Cong Xue, Chunmei Yu, Jingteng Cui, Youxiang Diao

INTRODUCTION

A new infectious disease has been causing a sudden fall in the egg production of ducks from Jiangsu, Fujian, Anhui, Shandong, Hebei and other provinces in China since April 2010 [1]–[3]. The main pathological changes observed consistently in the affected ducks are ovarian hyperemia, hemorrhage, degeneration, distortion, macrophage and lymphocyte infiltration, and hyperplasia. Morbidity by this virus is usually high (up to 90%)[3], and mortality varies from 5% to 10%. The pathogen was identified by the NS5 gene sequence as TMUV, which belongs to the *Ntaya* virus (NTAV) group in the genus *Flavivirus* of the family*Flaviviridae* [1]. This duck TMUV disease has since caused an enormous loss in many layers and meat-type duck farms in China [2]–[5].

The mature virion of TMUV is 45 nm to 50 nm in diameter, coated with a capsule membrane, and contains a 10, 990 nt single-stranded positive-sense genomic RNA. The translation product of the genome is a polyprotein that contains three structural proteins (capsid, prM and envelope) and eight nonstructural proteins (NS1, NS2A, NS2B, NS3, NS4A, 2K, NS4B and NS5) [5]–[8]. The ‘pr’ peptide is cleaved from the prM structure during the maturation of flaviviruses, and the M protein remains in the mature particle as a transmembrane protein beneath the envelope (E) protein shell [9]. The ‘pr’ peptide was proposed to function as a cap-like structure that protected the fusion peptide on E from undergoing pre-mature fusion before virus release [10]. The E protein is the major surface protein of flaviviruses and is the dominant antigen that induces immunological responses in infected hosts and consequently elicits virus-neutralizing antibodies[11]–[13].

Several serological and molecular biological methods have been established, including reverse-transcription loop-mediated isothermal amplification (RT-LAMP) [14], [15], real-time RT-polymerase chain reaction (PCR) [16], blocking enzyme-linked immunosorbent assay (ELISA) [17]. These methods are not applicable both in directly antigen detection and large-scale molecular epidemiology under field conditions. TMUV has been detected in culex mosquitoes in China, in Thailand and Malaysia, indicating that mosquitoes may be important for transmission and that TMUV may have a large distribution outside of China. A new and effective detecting method is therefore urgently needed. This study aimed to develop a simple, convenient, time-saving, and cost-effective ELISA for the detection of TMUV. Monoclonal antibodies (mAbs) 12B1, 10C6, and 2D2 obtained in our study were able to recognize different linear epitopes on the E protein. A double-antibody sandwich enzyme-linked immunosorbent assay (DAS-ELISA) was established using high-affinity mAbs 12B1 and 2D2 as capture and detection antibodies, respectively. The assay demonstrated high sensitivity and specificity and may serve as an alternative method for detecting duck TUMV with its easy manipulation and low cost.

Citation: Chen H, Ou Q, Tang Y, Gao X, Wu L, et al. (2014) Development and Evaluation of a DAS-ELISA for Rapid Detection of Tembusu Virus Using Monoclonal Antibodies against the Envelope Protein. PLoS ONE 9(5): e96366. doi:10.1371/journal.pone.0096366.

MATERIALS AND METHODS

Ethics Statement

The animal procedures were approved by the Animal Care and Use Committee of Shandong Agricultural University (Permit number: 20010510) and performed in accordance with the "Guidelines for Experimental Animals" of the Ministry of Science and Technology (Beijing, China). All surgery was performed according to recommendations proposed by the European Commission (1997), and all efforts were made to minimize suffering. Mice were housed in a temperature-controlled room with proper darkness-light cycles, fed with a regular diet, and maintained under the care of the Experimental Animal Center, College of Animal Science and Veterinary Medicine, Shandong Agricultural University. The mice were euthanized using sodium pentobarbital and the house sparrows were euthanized in CO_2.

Duck and duckling samples were collected from Gaotang, Linyi, Pingyi and Feicheng counties, respectively. House sparrows were collected in Gaotang County and bred for further research. No specific permissions were required for these locations/activities. There are not endangered or protected species involved in these locations. GPS and companies of these lacations are listed as follows:

- Gaotang County (36°50′14″N, 116°12′38″E), Company: Shandong Yonghui Food Limited Company, E-mail: moc.361@yhdehuil.
- Linyi County (37°10′33″N, 116°49′16″E), Company: Linyi Liuhe Jiusheng Limited Company, E-mail:moc.ehuil@cl-iynilzd.
- Pingyi County (35°28′20″N, 117°44′25″E), Company: Pingyi Liuhe Food Limited Company, E-mail:moc.qq@12332749.
- Feicheng County (36°4′37″N, 116°46′16″E), Company: Shandong Yonghui Food Limited Company, E-mail: moc.621@rhnauygnakeh.

Relevant Materials

Tembusu virus (TMUV), duck plague virus (DPV), H9 subtype avian influenza virus (AIV), Newcastle disease virus (NDV), duck hepatitis A virus-type 1 (DHAV-I) and duck reovirus (DRV) strains were collected in Shandong province and stored in Poultry Diseases Institute, Shandong Agricultural University. Purified recombinant E protein (40.5 KD) expressed in *E. coli* was obtained from Gao[18] and SP2/0 myeloma cell line was stored in our laboratory. BALB/c mice were purchased from the Experimental Animal Center of Shandong (Jinan, China).

Production of MABS to TMUV

Five 6-week-old female BALB/c mice were prepared as immunized animals. The mice were initially inoculated subcutaneously with 100 μg of immunogen (purified E protein) emulsified in an equal volume of complete Freund's adjuvant (Sigma, Missouri, USA). Immunogen emulsified in incomplete Freund's adjuvant (Sigma, Missouri, USA) was subsequently inoculated on the mice three times at 2-week intervals. Antiserum was collected from the lateral tail veins of immunized mice one week after the fourth inoculation and tested by indirect-ELISA to monitor that if the production of antibody was adequate for cell fusion. A final inoculation of immunogen without adjuvant was administered intraperitoneally four days prior to fusion. The mice were euthanized using sodium pentobarbital and sensitized spleen cells were fused with mouse myeloma cell SP2/0 using PEG1500 (Roche, Mannheim, Germany). The hybridomas were selectively cultured for approximately two weeks, and the cell supernatants were screened by ELISA against E protein expressed in *Escherichia coli* (*E.coli*). Positive hybridomas cultures were cloned at least twice by a limiting dilution in a 96-well cell culture plate.

The mAbs were generated by inoculating the selected hybridoma cells into BALB/c mice treated with sterile paraffin oil. Indirect ELISA, using E protein expressed in *E. coli* as coating antigen, was performed to titrate mAbs in the culture supernatants and ascitic fluids. The isotypes of the mAbs were determined by ELISA, using a mouse monoclonal antibody isotyping reagents kit (Sigma, Missouri, USA). The mAbs were purified from the ascitic fluids by ammonium sulfate precipitation. The concentration was determined by spectrophotometry.

Characterization of MABS

The reactions between mAbs and E protein were identified by western blot analysis. The E protein was separated by sodium dodecyle-sulfate polyacrylamide gel electrophoresis (SDS-PAGE) and transferred electrophoretically onto a polyvinylidene difluoride membrane (Roche, Mannheim, Germany). The membrane was blocked by the blocking buffer (PBS containing 0.5% (v/v) Tween-20 (PBST) and 2.5% skim milk powder) overnight at 4°C. Then the membrane was incubated with the purified mAbs diluted 1:1000 in the blocking buffer, and goat anti-mouse IgG (H+L) (BoAoSeng Company, Beijing, China) was used to detect the bound antibodies. A colorimetric reaction was observed using 3, 3′-diaminobenzidine enhanced liquid substrate system (TianGen Corporation, Beijing, China).

The cross-reactivity of mAbs was examined by indirect ELISA. The mAbs were reacted with TMUV, but no reaction with DPV, AIV subtype H9, NDV, DHAV-I, and DRV were observed.

Preparation of Field Samples

A total of 171 field samples from dead ducks suspected of TMUV infection and 20 house sparrows living around the infected duck farms were collected in Shandong province during 2010–2012 (Table 1). These duck farms are located in Gaotang, Linyi, Pingyi and Feicheng counties, respectively. House sparrows were collected in Gaotang County and euthanized in CO_2 in our laboratory for further research. These tissue samples were stored at −80°C. All samples were homogenized in 5 mL of PBS containing penicillin (5, 000 U/mL) and streptomycin (5 mg/mL). The suspensions were subjected to three freeze–thaw cycles and centrifuged at 3, 000 g for 10 min. The supernatant samples were incubated for 15 min at 37°C before testing.

Table 1: Field samples collected from different duck farms in Shandong, China

Species	Type of samples	Age(days)	No. of samples
laying duck [a]	theca folliculi	>150	137
duckling [a]	brain	>10	34
house sparrow [b]	brain	unknown	20
Total			191

[a]Samples were collected from dead ducks with symptoms of reduction in egg production (laying ducks), loss of appetite and encephalitis (ducklings).
[b]These samples were collected from house sparrow euthanized with CO_2 near the TMUV-infected duck farms.
doi:10.1371/journal.pone.0096366.t001

Selection of Pairing Antibodies

mAbs were labeled with horseradish peroxidase (HRP, Sigma, Missouri, USA) by the sodium periodate oxidation method described by Kanpp [19]. The titers of the labeled mAbs were determined by indirect ELISA. The 96-well microtiter plates (Maxisorp Nunc, Denmark) were coated with purified capture mAbs diluted in sodium carbonate buffer. TMUV was added as antigen and the plates were incubated for 30 min at 37°C. Then mAbs conjugated with horseradish peroxidase (detection antibody) were added. The unbound conjugates were washed off after incubation, and 3, 3›, 5, 5›-Tetramethylbenzidine (TMB) substrate solution (TIANGEN, Beijing, China) was added to each well. Incubation was carried out for 30 min and the reaction was stopped by adding 3M H_2SO_4. Plates were read at 450 nm on an automated ELISA plate reader (Bio-Rad, USA). The best pairing antibodies were obtained according to the recorded result.

Development of DAS-ELISA

The optimal concentrations of capture mAb (5, 2.5, 1.25 and 0.625 μg/mL), detection mAb (1:500, 1:1000, 1:1500, 1:2000, 1:2500 and 1:3000), dilution of positive control (purified TMUV) and negative control (SPF duck serum) (1:5, 1:10, 1:20, 1:40, 1:80 and 1:160) were screened by checkerboard titration. DAS-ELISA was performed by the following procedure. Briefly, 96-well microtiter plates were coated with capture mAb (100 μL/well) diluted in 0.01 M PBS (pH 7.2) and incubated at 4°C overnight. The plates were washed three times with PBST (PBS containing 0.05% Tween-20), then applied with blocking buffer (200 μL/well) (1% bovine serum albumin (BSA), 5% skim milk, 5% fetal bovine serum (FBS), and 1% gelatin) at 37°C for 1 h. After washing three times with PBST, clarified field sample suspensions diluted 1:2 in blocking buffer (100 μL/well) were added and incubated at 37°C for 1 h. Afterwards, the plates were washed three times with PBST and 100 μL diluted detection mAb conjugated to HRP was added to each well. After incubation at 37°C for 1 h and washing with PBST three times, TMB substrate solution (100 μL/well) was added to each well, and the plates were incubated in the dark at 37°C for 15 min. 3 M H_2SO_4 of 50 μL volume was added to stop reaction and the OD values were read at 450 nm using an automated ELISA plate reader. The optimal condition was obtained by comparing the positive/negative ratio (P/N) of the samples.

Positive/Negative cut-Off Value

Twenty-four negative sera from SPF ducks were detected by the established DAS-ELISA under the determined optimal conditions. The cut-off value at OD450 to define a virus positive was calculated based on the formula: positive and negative cut-off value = negative sample mean +3 standard deviations (mean+3 SD).

Cross-Reaction Test

The specificity of the DAS-ELISA was evaluated by testing antisera to other avian pathogens including DPV, AIV, NDV, DRV, DHAV-1 and TMUV. The specificity of the method was evaluated based on the results.

Duplicability Test

Each sample was detected in parallel three wells in a 96-well plate coated with one batch of 12B1 (capture mAb) by DAS-ELISA for intra-batch assay (n=10). These samples were detected by DAS-ELISA in 96-well plates coated with different batches of 12B1 for inter-batch assay. All these tests were repeated three times. The intra- and inter-assay coefficients of variation (%CV) were calculated by the following formula: %CV = standard deviations (SD)/ mean OD_{450} of samples ×100%.

RT-PCR Assay

The RT-PCR assay used the recently described primers [20], and amplified a fragment of approximately 211 nucleotides, which corresponded to nucleotides 9340 to 9551 in the TMUV genome. The cycling protocol comprised an initial of denaturation at 94°C for 30 s, annealing at 55°C for 30 s, and extension at 72°C for 60 s, followed by a final elongation step at 72°C for 5 min. The PCR products were analyzed on 2.0% (w/v) agarose gel and stained with ethidium bromide. All the PCR products were cloned into a pMD18-T vector (Takara, Dalian, China) and sequenced (Invitrogen, Beijing, China).

Comparison of DAS-ELISA and RT-PCR

A total of 191 field samples obtained from different duck farms in Shandong province were detected simultaneously by DAS-ELISA and RT-PCR. The sensitivity, specificity and accuracy were calculated by the following formulas: sensitivity= true positive / (true positive + false negative) ×100%, specificity= true negative/(true negative + false positive) ×100%, accuracy = (true positive +true negative)/ (true positive+ false positive + true negative + false negative) ×100%.

RESULTS

Preparation and Characterization of MABS

The purified recombinant E protein was used as the immunogen. Cell culture supernatants were screened by indirect-ELISA for the presence of antibodies against E protein. After cell fusion, cell culture, antibody detection and subcloning, three mAbs designated as 12B1, 10C6 and 2D2 were obtained, and each recognized different linear epitopes on the E protein. The three obtained hybridoma lines were injected intraperitoneally into BALB/c mice to produce ascitic fluids. Their titer in the culture supernatant and ascites was 1:400, 1:400, 1:800 and 1:256000, 1:128000, and 1:512000 for 12B1, 10C6 and 2D2, respectively. Additionally, the isotype of the mAbs was IgG1, IgG1, and IgG2a, respectively. All of the three mAbs types contained κ light chains. The concentration of the purified mAbs was 2.181 mg/mL, 2.409 mg/mL and 2.233 mg/mL for 12B1, 10C6 and 2D2, respectively (Table 2).

Table 2: Properties of monoclonal antibodies against TMUV E protein

MAbs	Isotype	Ascites titer	IgG yield ($mg{\cdot}mL^{-1}$)
12B1	IgG_1, κ chain	1:256000[a]	2.181
10C6	IgG_1, κ chain	1:128000	2.409
2D2	IgG_{2a}, κ chain	1:512000	2.233

[a]MAb titer was the last dilution that yielded an absorption value above 0.3 at 30 min after adding the substrate at room temperature.
doi:10.1371/journal.pone.0096366.t002

Reactivity and Specific Identification of MABS

Western blot was used to test the reactivity and specific identification of mAbs against the TMUV E protein. The corresponding analytic results demonstrated that the three mAbs reacted strongly with the TMUV E protein expressed in *E. coli* (Fig. 1) but did not react with other viruses that can also infect ducks (DPV, AIV, NDV, DRV and DHAV-I). Thus, the three mAbs were shown to be very specific for TMUV.

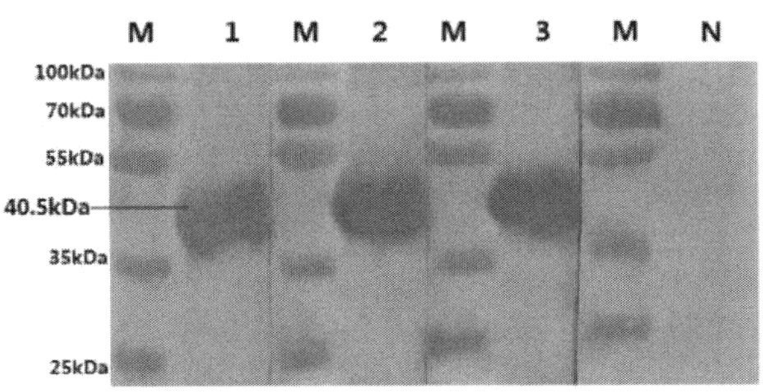

Figure 1: Reactivity of three mAbs with E protein by Western blotting.

Development of DAS-ELISA

mAb (12B1) and mAb (2D2) were determined as the optimal capture and detector antibodies by indirect ELISA, respectively. The 96-well plates were coated with 0.25 μg/well of purified mAb 12B1 (2.5 μg/mL) that was diluted in sodium carbonate buffer (pH 9.6). The plates were incubated at 4°C overnight. Dilutions of the positive and negative control were determined to be 1:1000 and 1:20, respectively. The most suitable blocking buffer concentration was 1% BSA. Using the 24 TMUV-antibody negative serum samples, the average OD value was determined to be 0.14±0.0153 (mean OD value± S.D.), resulting in a cut-off OD value of 0.2. No cross-reactions were detected by the DAS-ELISA using antisera against DPV, AIV, NDV, DRV and DHAV-1, with the OD values ranging from 0.099 to 0.124, demonstrating an excellent specificity of the DAS-ELISA for detection of the TMUV antigen (Table 3).

Table 3: Cross-reaction of DAS-ELISA

Samples	DPV	AIV	NDV	DRV	DHAV-I	TMUV(+)	TMUV(−)
mean OD_{450}[a]	0.122	0.110	0.101	0.120	0.115	1.458	0.099

[a]Each sample was tested in two parallel repetitions by established DAS-ELISA. Samples except TMUV positive sera (OD450 values<0.2) were determined as negative.
doi:10.1371/journal.pone.0096366.t003

Stability of DAS-ELISA

The mean %CV of the intra- and inter-batch duplicability tests was 4.31% and 8.2%, respectively. The duplicability and stability of the developed DAS-ELISA were thus shown to be adequate for TMUV detection.

Detection of TMUV in Field Samples

A total of 194 field samples of laying ducks, ducklings, house sparrows, and mosquitoes were screened for the presence of TMUV using DAS-ELISA and RT-PCR. In DAS-ELISA, 68 of the 194 field samples were determined to be positive, whereas 72 samples were positive by RT-PCR. Among the 194 samples, 67 were positive in both DAS-ELISA and RT-PCR and 118 were negative in both tests (Table 4). DAS-ELISA was found to be 99.1% specific (118/119) and 93.1% sensitive (67/72) relative to RT-PCR. The results of both DAS-ELISA and RT-PCR showed that the accuracy ((67+118)/191) of the two detection methods was 95.8%.

Table 4: Comparison of DAS-ELISA with RT-PCR for field samples detection

DAS-ELISA	RT-PCR(+)	RT-PCR(−)	Total
Positive	67	1	68
Negative	5	118	123
Total	72	119	191

doi:10.1371/journal.pone.0096366.t004

DISCUSSION

TMUV is an important pathogen in ducks in the south and east of China, and the early and rapid detection of TMUV is essential to prevent and control the spread of disease caused by the virus. In this study, three mAbs against TMUV E protein were obtained. A simple, convenient, and rapid DAS-ELISA procedure was established for the large-scale diagnosis of TMUV based on the three high- quality mAbs.

Monoclonal antibody is an important immunological tool. It has an important application in the determination of pathogen biological characteristics and antigenicity, as well as in pathogen detection, treatment, immune mechanism research, among others. Monoclonal antibody-based DAS-ELISA is a time-saving, conveniently operated method that has high sensitivity, specificity and efficiency. This method is also used extensively for the routine diagnosis of multiple pathogens, such as Influenza A virus [21], classical swine fever virus [22], and Peste Des Petits Ruminants Virus [23].

The E protein of the flavivirus is the largest structural protein and the main envelope glycoprotein that constitutes the virus surface protuberance. Under neutral and micro alkaline conditions, envelope protein is exciting in homologous dimer on mature virus [24]. The E protein that binds to cell receptors is the viral component that mediates the targeting of flaviviruses to vital host organs and has thus generated the highest interest in research [25]. The E protein also enables viral penetration during infection through the lipid endosomal membrane [26] and induces neutralizing, haemagglutination-inhibiting and protective antibodies[27]. The advantage of E protein based diagnostics is that the epitopes on the E protein are conserved among flaviviruses in previous studies. This feature helps avoid the overlooking of flavivirus infections from other geographical regions during detection.

In this study, BALB/c mice were immunized with purified TMUV emulsified with Freund›s adjuvant. Three hybridoma cell lines were obtained and designated as 12B1, 10C6 and 2D2. These three mAbs are able to recognize three different linear epitopes on the E protein eliminating the competition of mAbs on the same epitope and minimizing steric hindrance. The DAS-ELISA assay was established by capture antibody 12B1 and detection antibody 2D2. The accuracy of the two detection methods was 95.8%. The developed method for TMUV detection was also shown to have no cross-reactivity with DPV, AIV, NDV, DRV and DHAV-I. These results collectively indicate that the developed DAS-ELISA method can be used for the survey and diagnosis of TMUV.

In summary, three mAbs against TMUV E protein were obtained in our study, and the established DAS-ELISA assay for TMUV was shown to be convenient and easy to perform. The DAS-ELISA assay is suitable for the diagnosis and antigen detection of clinical cases, investigation of duck TMUV.

FUNDING STATEMENT

This study was supported by the National Natural Science Foundation of China (31272583), the China Agriculture Research System (CARS-43-34)and the Modern Agro-industry Technology Research System (nycytx-45-11). The funders had no role in study design, data collection and analysis, decision to publish, or preparation of the manuscript.

REFERENCES

1. Cao Z, Zhang C, Liu Y, Ye W, Han J, et al. (2011) Tembusu virus in ducks, china. Emerg Infect Dis 17: 1873–1875
2. Yan P, Zhao Y, Zhang X, Xu D, Dai X, et al. (2011) An infectious disease of ducks caused by a newly emerged Tembusu virus strain in mainland China. Virology 417: 1–8
3. Su J, Li S, Hu X, Yu X, Wang Y, et al. (2011) Duck egg-drop syndrome caused by BYD virus, a new Tembusu-related flavivirus. PLoS One 6: e18106.
4. Yun T, Zhang D, Ma X, Cao Z, Chen L, et al. (2012) Complete genome sequence of a novel flavivirus, duck tembusu virus, isolated from ducks and geese in china. J Virol 86: 3406–3407
5. Tang Y, Diao Y, Gao X, Yu C, Chen L, et al. (2012) Analysis of the complete genome of Tembusu virus, a flavivirus isolated from ducks in China. Transbound Emerg Dis 59: 336–343
6. Li L, An H, Sun M, Dong J, Yuan J, et al. (2012) Identification and genomic analysis of two duck-origin Tembusu virus strains in southern China. Virus Genes 45: 105–112
7. Wan C, Huang Y, Fu G, Shi S, Cheng L, et al. (2012) Complete genome sequence of avian tembusu-related virus strain WR isolated from White Kaiya ducks in Fujian, China. J Virol 86: 10912.
8. Yu K, Sheng ZZ, Huang B, Ma X, Li Y, et al. (2013) Structural, antigenic, and evolutionary characterizations of the envelope protein of newly emerging Duck Tembusu Virus. PLoS One 8: e71319.
9. Yu IM, Zhang W, Holdaway HA, Li L, Kostyuchenko VA, et al. (2008) Structure of the immature dengue virus at low pH primes proteolytic maturation. Science 319: 1834–1837
10. Perera R, Kuhn RJ (2008) Structural proteomics of dengue virus. Curr Opin Microbiol 11: 369–377

11. Wu KP, Wu CW, Tsao YP, Kuo TW, Lou YC, et al. (2003) Structural basis of a flavivirus recognized by its neutralizing antibody: solution structure of the domain III of the Japanese encephalitis virus envelope protein. J Biol Chem 278: 46007–46013
12. Stiasny K, Kiermayr S, Holzmann H, Heinz FX (2006) Cryptic properties of a cluster of dominant flavivirus cross-reactive antigenic sites. Journal of virology 80: 9557–9568
13. Beasley DW, Barrett AD (2002) Identification of neutralizing epitopes within structural domain III of the West Nile virus envelope protein. Journal of virology 76: 13097–13100
14. Tang Y, Diao Y, Yu C, Gao X, Chen L, et al. (2012) Rapid detection of Tembusu virus by reverse-transcription, loop-mediated isothermal amplification (RT-LAMP). Transbound Emerg Dis 59: 208–213
15. Wang Y, Yuan X, Li Y, Yu K, Yang J, et al. (2011) Rapid detection of newly isolated Tembusu-related Flavivirus by reverse-transcription loop-mediated isothermal amplification assay. Virol J 8: 553.
16. Yu CM, Diao YX, Tang Y, Cui JT, Gao XH, et al. (2012) Fluorescence Quantitative RT-PCR Assay for Detection of Tembusu Virus. Scientia Agricultura Sinica 45: 4492–4500
17. Li X, Li G, Teng Q, Yu L, Wu X, et al. (2012) Development of a blocking ELISA for detection of serum neutralizing antibodies against newly emerged duck Tembusu virus. PLoS One 7: e53026.
18. Gao X (2013) Establishment and Application of Digoxigenin-labeled DNA probe and Differential Indirect ELISAs Detection Methods for TMUV. Tai›an: Shandong Agricultural University.
19. Knapp JS, Holmes KK (1983) Modified oxidation-fermentation medium for detection of acid production from carbohydrates by Neisseria spp. and Branhamella catarrhalis. J Clin Microbiol 18: 56–62
20. Tang Y (2013) The Characterization of Tembusu Virus and Inhibition of Tembusu Virus Replication with Recombinant Adenovirus Expressing shRNA in Vitro [S852]. Tai›an City, China: Shandong Agricultural University.
21. Siebinga JT, de Boer GF (1988) Influenza A viral nucleoprotein detection in isolates from human and various animal species. Arch Virol 100: 75–87
22. Wu CW, Chien MS, Liu TY, Lin GJ, Lee WC, et al. (2011) Characterization of the monoclonal antibody against classical swine fever virus glycoprotein E(rns) and its application to an indirect sandwich ELISA.Appl Microbiol Biotechnol 92: 815–821
23. Singh RP, Sreenivasa BP, Dhar P, Bandyopadhyay SK (2004) A sandwich-ELISA for the diagnosis of Peste des petits ruminants (PPR) infection in small ruminants using anti-nucleocapsid protein monoclonal antibody. Arch Virol 149: 2155–2170
24. Gritsun TS, Holmes EC, Gould EA (1995) Analysis of flavivirus envelope proteins reveals variable domains that reflect their antigenicity and may determine their pathogenesis. Virus Res 35: 307–321
25. Chen Y, Maguire T, Hileman RE, Fromm JR, Esko JD, et al. (1997) Dengue virus infectivity depends on envelope protein binding to target cell heparan sulfate. Nat Med 3: 866–871
26. Gollins SW, Porterfield JS (1986) A new mechanism for the neutralization of enveloped viruses by antiviral antibody. Nature 321: 244–246
27. Heinz FX, Tuma W, Guirakhoo F, Kunz C (1986) A model study of the use of monoclonal antibodies in capture enzyme immunoassays for antigen quantification exploiting the epitope map of tick-borne encephalitis virus. J Biol Stand 14: 133–141

CHAPTER 16 DEVELOPMENT OF VACCINE-INDUCED IMMUNITY AGAINST TRT IN TURKEYS DEPENDS REMARKABLY ON THE LEVEL OF MATERNAL ANTIBODIES AND THE AGE OF BIRDS ON THE DAY OF VACCINATION

Marcin Smialek, Daria Pestka, Bartlomiej Tykalowski, Tomasz Stenzel and Andrzej Koncicki

Keywords: Avian metapneumovirus, Turkeys, Vaccination, Humoral immunity, Cell mediated immunity, Maternally derived antibodies

BACKGROUND

The avian metapneumovirus (aMPV), a member of the family Paramyxoviridae, genus Metapneumovirus [1], is a strongly infectious RNA virus that causes turkey rhinotracheitis (TRT) in flocks of turkeys and the swollen head syndrome (SHS) in chickens. aMPV infections cause massive financial losses in the poultry industry worldwide. aMPV was first identified in South Africa in 1978 [2], and since then, it has spread to numerous countries, excluding Australia and Canada. aMPV has been classified into 4 subtypes (A–D) based on its nucleotide sequence and antigen structure [3-5]. In most immunoprophylaxis programs, one-day-old turkey poults are administered live attenuated vaccines against TRT by coarse spray to provide the earliest possible protection of the upper respiratory tract against aMPV infections [6]. Despite the above, TRT outbreaks in the field are noted very frequently [7-9].

Humoral immunity is strongly stimulated by vaccination or aMPV infection [7,10-13], but antibodies do not play a key role in protection against TRT and should not be considered as indicators of immunity against aMPV infections [7,9,11,12,14-16]. It has been shown, however, that high antibody titers suppress aMPV replication in the upper respiratory tract, thus alleviating the clinical course of TRT [6,11,12,14]. In one-day-old turkey poults, most of which originate from parent flocks vaccinated against TRT, the possible influence of maternal antibodies on the development of vaccine-induced immunity against aMPV is questionable. Cell-mediated immunity (CMI) is increasingly often considered as the decisive factor in protection against TRT. Unfortunately, little is known about local CMI mechanisms in turkeys' upper respiratory tract during infections and after vaccination against TRT. These mechanisms seem to play a particularly important role because aMPV infects the host through mucosal sites in the upper respiratory tract. Liman and Rautenschlein [7] demonstrated a significant increase in the percentage of CD4+ T lymphocytes with concurrent upregulation of IL-6 and/or IFN-γ in the Harderian gland (HG) after vaccination or infection with aMPV/A or B. Conversely, Cha [17] reported an increase in the percentage of CD8+ T cells, but not in CD4+ T cells, in the upper respiratory tract after aMPV/C inoculation. The cited results indicate that CMI actively participates in protection against aMPV infections and that local CMI mechanisms may be related to age and the aMPV subtype. In view of the diverse immunopathogenesis of TRT, the aim of this study was to describe selected parameters of local CMI and humoral immunity in the upper respiratory tract of turkeys immunized against TRT with live attenuated aMPV/A vaccines. The specific goal of the study was to characterize the development of vaccineinduced immunity in variously-aged birds and in turkeys with, or without maternal anti aMPV antibodies on the day of vaccination.

Citation: Marcin Smialek, Daria Pestka, Bartlomiej Tykalowski, Tomasz Stenzel and Andrzej Koncicki, Development of vaccine-induced immunity against TRT in turkeys depends remarkably on the level of maternal antibodies and the age of birds on the day of vaccination, DOI 10.1186/s12917-015-0345-5.

METHODS

The experimental procedures and animal handling procedures were conducted with the approval of the Local Ethic Committee for Animal Experiments in Olsztyn, Poland (resolution No. 37/2011).

Turkeys and Vaccination

The experiments were carried out on 568 commercial Hybrid Converter turkeys of both sexes. The birds were not vaccinated at hatchery, and the absence of aMPV genetic material in samples collected at hatch was confirmed by nested RT-PCR. The MDA+ group comprised 284 turkeys that were positive for maternally derived antibodies and were obtained from a parent flock vaccinated against TRT (3 times with live aMPV/A vaccine and twice with inactivated vaccine). MDA+ birds were provided by the Grelavi S.A. hatchery (a Hendrix Genetics Company) in Kętrzyn, Poland. The MDAgroup consisted of 284 turkeys that were negative for maternally derived antibodies and originated from a parent flock not vaccinated against TRT (reproductive flock from Canada). MDA- birds were provided by the Gerczak Nord-Pol Hatchery in Laseczno, Poland. Turkeys were housed in isolated units maintained at a biosafety PCL 3 facility. Water and feed were given to birds ad libitum. Birds were vaccinated with aMPV/A strain BUT1 #8544 of lyophilized attenuated commercial vaccine. Each bird was inoculated with 10^4 of tissue culture infectious doses (TCID50) per bird oculonasally. Unvaccinated birds received sterile vaccine diluent.

EXPERIMENTAL DESIGN

Experiment I

Experiment I was carried out on 160 one-day-old MDA+ turkeys. The birds were randomly divided into vaccinated (MDA+0/V) and unvaccinated (MDA+0/NV) groups. Before vaccination, samples were collected from 24 randomly selected turkeys for further analysis. The day of vaccination (hatch day) was termed zero day of life (DOL), and the results for samples collected on this day were marked as zero day post-vaccination (DPV). Birds of both groups were raised to 14 DOL, and further samples were collected on 3, 7 and 14 DPV (from 20–24 birds in each group in each sampling). In each sampling, choanal swabs (e-Swab, Copan Diagnostics, Canada) were collected from 5 birds for RT-PCR analysis (swabs were stored at −76°C until further analysis) and blood samples were collected from 15 birds. The birds were euthanized, and HG and trachea (TC, cut from the larynx to the cranial aperture of body cavity) were collected. HG and TC were pooled from 4–6 birds per sample (the number of birds per sample were equal in each sampling), which produced 4 samples of each organ for flow cytometry analysis. Tracheal washings were collected from 10 birds in each group by passing 1 ml of phosphatebuffered saline (PBS, pH 7.2, Sigma Aldrich, Germany) through the lumen of the trachea with a 2 ml syringe. The trachea was massaged gently to intensify the washing effect. Following centrifugation, the supernatant and serum samples were stored at −20°C for further use.

Experiment II

The experimental design was identical to that of Experiment I. The experiment was carried out on 124 MDA+ turkeys raised until 14 dol. On 14 DOL, samples were collected from 20 birds, and the remaining turkeys were divided randomly into vaccinated (MDA+14/V) and unvaccinated (MDA+14/NV) groups. Similarly to Experiment I, the results reported for samples that were collected before vaccination (14 DOL) represent 0 DPV, and further samples were collected on 3 (17 DOL), 7 (21 DOL) and 14 (28 DOL) DPV (from 16–20 birds in each group in each sampling).

Experiment III

Experiment III was carried out on 160 one-day-old MDA- turkeys. The experimental design and schedule, the day of vaccination and the sample collection protocol were identical to those described in Experiment I. Turkeys were divided into two groups of vaccinated (MDA-0/V) and unvaccinated (MDA-0/NV) birds.

Experiment IV

Experiment IV was carried out on 124 fourteen-day-old MDA- turkeys. The experimental design and schedule, the day of vaccination and the sample collection protocol were identical to those described in Experiment II. Turkeys were divided into two groups of vaccinated (MDA-14/V) and unvaccinated (MDA-14/NV) birds.

Isolation and Determination of Mononuclear Cell Counts

HG were homogenized in a manual Dounce tissue grinder in 3 ml of complete cell culture medium (RPMI – 1640, 10% Fetal Bovine Serum (FBS), 1% MEM Nonessential Amino Acids solution, 1% Penicillin – Streptomycin, 1% HEPES, 1% Sodium pyruvate, Sigma Aldrich, Germany) and filtered (70 μm mesh). After centrifugation in culture medium at 450 g for 10 min at 20°C, the supernatant was discarded and cell pellets were resuspended in 3 ml of 40% Percoll density gradient and gently layered on 3 ml of 60% Percoll. Percoll concentrations were obtained by combining 100% Percoll solution (Percoll with 10% Hank's balanced salt solution, Sigma Aldrich, Germany) with the appropriate amount of culture medium. Mononuclear cells were collected from the interphase after density centrifugation (20 min, 1900 g, 20°C) and washed twice in PBS with 5% FBS. TC were cut into 2–3 mm pieces and placed in 100 ml an Erlenmeyer flask with 25 ml of DL-Dithiothreitol solution (RPMI – 1640, 5% FBS, 1% HEPES, 2 mM DLDithiothreitol, Sigma Aldrich, Germany) and incubated at room temperature for 5 min to remove epithelial cells with intraepithelial lymphocytes. TC were filtered (70 μm mesh), and flasks were rinsed with RPMI – 1640 before collagenase treatment. TC were placed in 25 ml of Collagenase type IV solution (RPMI – 1640, 1% HEPES, 1% Penicillin – Streptomycin, 200 Collagen Digestion Units/ml, Sigma Aldrich, Germany) and incubated at 38°C for 35 min to isolate lamina propria lymphocytes. TC were filtered (70 μm mesh) after incubation. Filtrates from both incubations were washed three times in culture medium. Final pellets of intraepithelial lymphocytes were merged with lamina propria lymphocytes from the same sample, they were washed and resuspended in 3 ml of 40% Percoll. The isolation procedure was identical to that involving HG. The obtained HG and TC mononuclear cells were resuspended in 1 ml of PBS, and the absolute lymphocyte counts (ALC)/ml were calculated for each sample with the Vi-cell XR (Beckman Coulter, USA) cell counter and cell viability analyzer.

Flow Cytometry and Dual Platform Analysis

2.5×10^5 (or 1.25×10^5 , depending on the ALC) of viable mononuclear cells from HG or TC were stained with monoclonal Mouse anti-Chicken CD4 - FITC (clone 2–35) and CD8 - PE (clone 11–39) for T lymphocytes, or polyclonal Goat anti-Chicken IgM - FITC for B cells (AbD Serotec, UK), incubated for 30 min on ice and washed twice in PBS. The cells were analyzed with the FACSCanto II (BD, USA) flow cytometer. Data acquisition was performed in FACSDiva Software 6.1.3. (BD, USA). Cells were analyzed and immunophenotyped in the FloJo 7.5.5 (Tree Star, USA). Relative cell counts (RCC) of T and B lymphocytes in HG and TC were established in the above environment. Absolute cell counts (ACC) of T and B cells were calculated by dual platform analysis with the use of the following formula: ACC = (ALC * RCC)/100%. Data were expressed as the mean ACC of $CD4^+$ and $CD8^+$ T lymphocytes or the mean x-fold change in ACC of IgM+ B lymphocytes in vaccinated groups relative to control groups.

Enzyme-Linked Immunosorbent Assay (Elisa)

Sera and tracheal washings were tested for specific IgY anti-aMPV antibodies with the APV Ab ELISA kit (IDEXX Laboratories, USA). Sera of MDA- birds were diluted 100-fold rather than 500-fold (manufacturer's recommendations) to enhance detection of low antibody levels. 100-fold dilution, as a comparative to 500-fold, was also used in analyses of MDA+14/V and NV serum. Tracheal washings were incubated undiluted. Successive steps of the procedure were performed according to the manufacturer's recommendations. ELISA was carried out with the use of the Eppendorf epMotion 5075 LH automated pipetting station (Eppendorf, Germany), BioTek ELx405 automatic plate washer (BioTek, USA) and BioTek ELx800 plate reader. The sample to positive (S/P)-ratio was calculated based on the ODs and used to express the mean (S/P)-ratio +/− SD per group and per DPV.

Nested RT-PCR

Nested RT-PCR was performed in accordance with the protocol described in a previous study [18] with minor modifications. RNA was extracted from choanal swabs with the RNeasy Mini Kit (Qiagen, Germany) according to the manufacturer's recommendations. The concentrations and purity of extracted RNA were evaluated with the NanoDrop 2000 spectrophotometer (ThermoScientific, USA). Samples ranging from 1.8 to 2.0 of the OD 260/280 ratio were used for further analysis. RT was performed with the EnhancedAvian HS RT-PCR Kit (Sigma-Aldrich, USA) according to the manufacturer's recommendations. PCR was performed with the use of the HotStarTaq Plus Master Mix (Qiagen, Germany) and the vapo.protect (Eppendorf, Germany) thermocycler. The first PCR was carried out with 10 µl of HotStarTaq Plus DNA, 0.1 µl of 100 µM primers G1+A and G6, 6.8 µl of RNase-free water and 3 µl of cDNA. Nested PCR was performed with 10 µl of HotStarTaq Plus DNA, 0.1 µl of 100 µM primers G8+A and G5, 2 µl of CoralLoad PCR, 5.8 µl RNase-free water and 2 µl of the amplicon from the first PCR. Primer sequences and product sizes for every PCR are summarized in Table 1. Both PCRs had the following thermal profile: 95°C for 5 min, 30 cycles of 94°C for 1 min, 54°C for 45 s, 72°C for 45 s, and final elongation at 72°C for 10 min. Nested PCR amplicons were analyzed in 2% agarose gel with 0.5 µl/ml of ethidium bromide dye (SigmaAldrich, USA). Electrophoresis results were visualized with GelDoc XR+ (Bio-Rad, USA).

Statistical Analysis

The flow cytometry results were processed by Student's t-test for independent samples in Graphpad Prism 6 software (San Diego, USA) and results of serological examination by non-parametric Mann–Whitney U-test in Statistica PL V10. Differences were considered statistically significant at $p \leq 0.05$ and highly significant at $p \leq 0.01$.

RESULTS

Absolute Lymphocyte Count

No statistical differences in absolute lymphocyte counts in HG and TC were observed between vaccinated and unvaccinated birds in all experiments, excluding ALC in TC in the MDA-14/V group. In this group, ALC in TC on 14 DPV increased significantly in comparison with control group (data not shown).

Table 1: Sequences of the oligonucleotide primers used in nested RT-PCR analysis and expected PCR product size

Primer	Sequence	Product size
G1+A	5'-GGGACAAGTATCTCTATG-3'	444 bp
G6	5'-CTGACAAATTGGTCCTGATT-3'	
G8+A	5'-CACTCACTGTTAGCGTCATA-3'	268 bp
G5	5'-CAAAGArCCAATAAGCCCA -3'	

Flow Cytometry Analysis of T Cells

In Experiment I, a highly significant and significant increase in $CD4^+$ T ACC was noted in the MDA^+0/V group on day 7 and 14 PV in HG and TC, respectively, in comparison with virus-free birds (Figure 1A). In Experiment II, $CD4^+$ T ACC did not increase significantly in vaccinated birds, but a brief increase of this parameter was reported in HG on 7 DPV in the MDA^+14/V group. A significant decrease in $CD4^+$ T ACC was observed in the MDA^+14/V group on 7 DPV in TC in comparison with control (Figure 1B). The noted decrease was shortlived and no significant differences in the value of analyzed parameter were reported between groups in Experiment II on day 14 PV. In Experiment III, significant increase in $CD4^+$ T ACC was noted on 14 DPV in HG in the MDA-0/V group, whereas a significant decrease in this parameter was observed in TC on 7 DPV in comparison with unvaccinated birds (Figure 1C). In Experiment IV, a statistical increase in $CD4^+$ T ACC on day 7 PV and a highly significant increase on day 14 PV were reported in HG of MDA- 14/V birds, whereas a significant decrease in $CD4^+$ T ACC in TC was observed on day 3 PV in comparison with unvaccinated turkeys (Figure 1D). In Experiments I and II, no significant differences in T $CD8^+$ ACC were observed in neither HG nor TC (Figures 2A and B). The number of these cells increased briefly in vaccinated birds (in particular in HG), but the observed changes were not statistically significant. In Experiment III, a highly significant increase in $CD8^+$ T ACC was reported in HG of MDA-0/V birds on days 7 and 14 PV, and a highly significant decrease in the analyzed parameter was observed in TC on 7 DPV in comparison with unvaccinated birds (Figure 2C). The recorded decrease in CD8+ T cell count was short-lived, and on 14 DPV, a highly significant increase in this parameter was reported in TC of vaccinated turkeys in comparison with virus-free birds. The most prominent infiltration of CD8+ T cells induced by the vaccine virus was observed in Experiment IV. In the MDA-14/V group, a highly significant increase in $CD8^+$ T ACC was noted in TC on days 7 and 14 PV and in HG on 7 DPV in comparison with control. T $CD8^+$ cell counts in HG decreased significantly on 14 DPV (Figure 2D). In Experiments III and IV, significant differences were also observed in double-positive $CD4^+$ $CD8^+$ T ACC (data not shown). In Experiment III, a significant decrease in the absolute counts of double-positive T cells was reported in the MDA-0/V group in TC on day 7 PV. In Experiment IV, a highly significant increase in double-positive T ACC was noted in both HG and TC on 14 DPV. No statistical differences in $CD4^+$ $CD8^+$ T

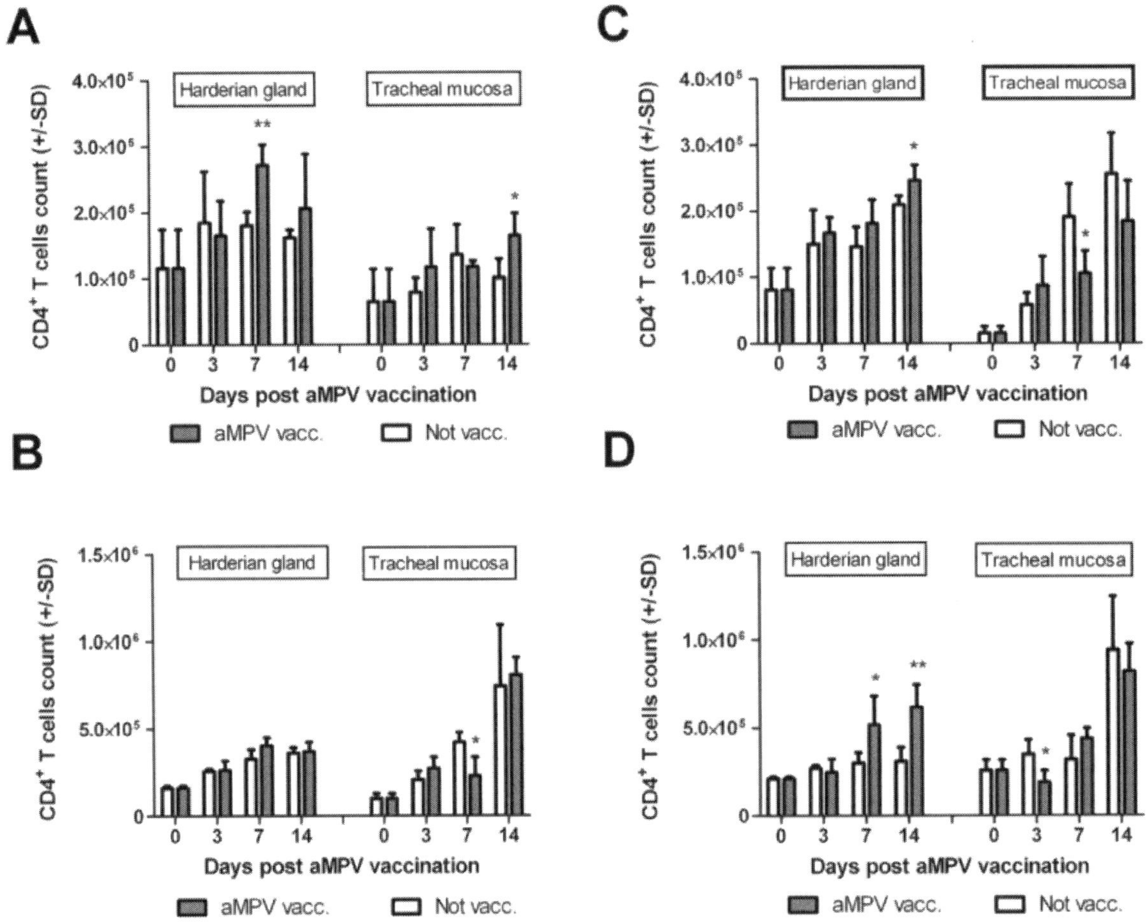

Figure 1: Summary of $CD4^+$ ACC in HG and TC (n = 4) at different days post aMPV/A vaccination. Results for vaccinated (aMPV vacc.) and not vaccinated (Not vacc.) groups in different experiments (A) Experiment I; groups MDA^+0/V and MDA^+0/NV; (B) Experiment II; groups MDA^+14/V and MDA^+14/NV; (C) Experiment III; groups MDA-0/V and MDA-0/NV (D) Experiment IV; groups MDA-14/V and MDA-14/NV. Results for every experiment are presented as mean $CD4^+$ ACC ± SD. */**Significant differences at different DPV (T-test, *as $p < 0,05$, **as $p < 0,01$).

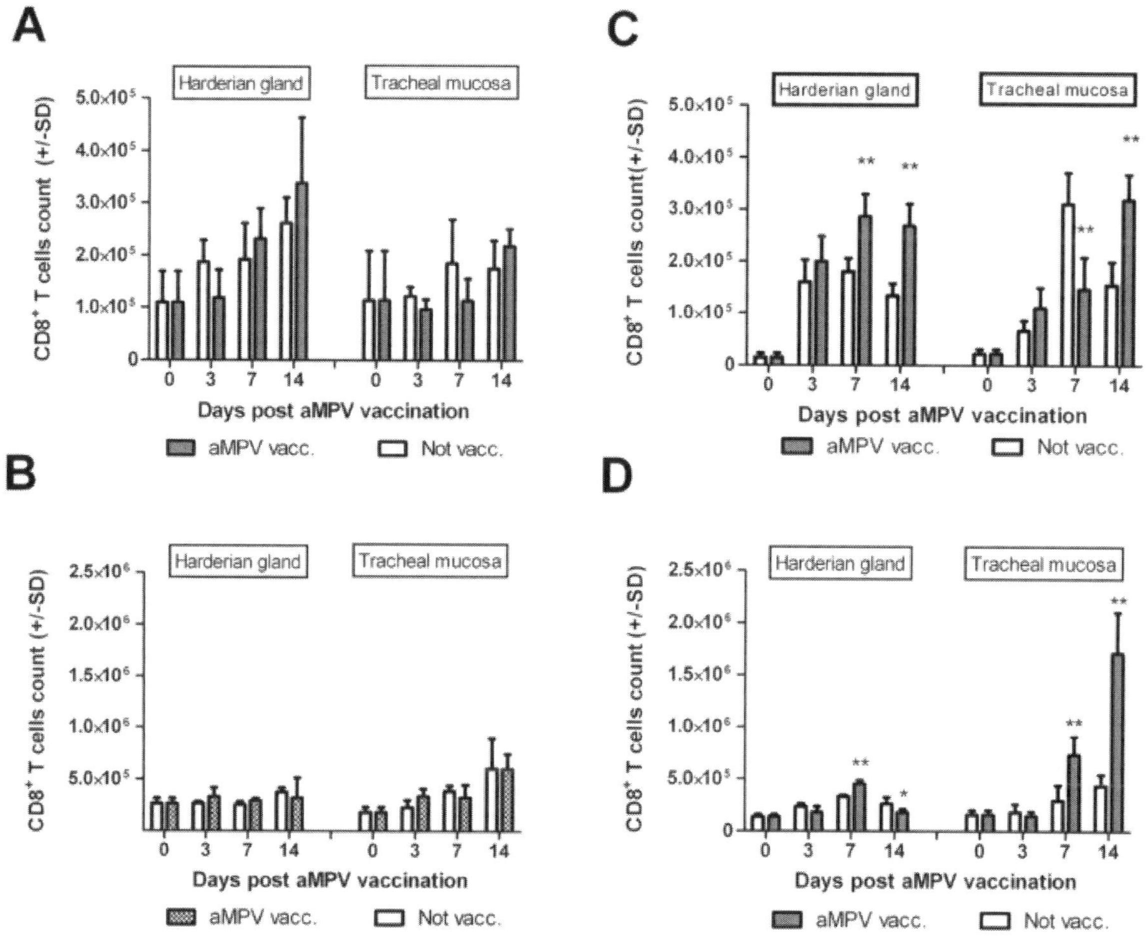

Figure 2: Summary of CD8+ ACC in HG and TC (n = 4) at different days post aMPV/A vaccination. Results for vaccinated (aMPV vacc.) and not vaccinated (Not vacc.) groups in different experiments (A) Experiment I; groups MDA+0/V and MDA+0/NV; (B) Experiment II; groups MDA+14/V and MDA+14/NV; (C) Experiment III; groups MDA-0/V and MDA-0/NV (D) Experiment IV; groups MDA-14/V and MDA-14/NV. Results for every experiment are presented as mean CD8+ ACC ± SD. */**Significant differences at different DPV (T-test, *as $p < 0,05$, **as $p < 0,01$).

ACC were observed between MDA+ vaccinated groups and unvaccinated birds (data not shown).

Flow Cytometry Analysis of B Cells

In Experiment I, a significant decrease in B IgM+ ACC in TC was observed in MDA+0/V on 7 DPV in comparison with the unvaccinated group (Table 2). The above decrease was transient, and no differences in B IgM+ ACC were observed on day 14 PV. In Experiment II, B IgM+ ACC increased significantly in HG of MDA+14/V birds on 7 DPV (Table 2). In Experiments III and IV, vaccination against TRT induced a highly significant increase in B IgM+ cell counts in TC on 14 DPV (Table 2). In Experiment IV, the vaccine virus also caused a significant decrease in IgM+ B cell count in HG of MDA-14/V birds on 14 DPV in comparison with control.

Serology

No statistical differences in the level of specific anti – aMPV antibodies in tracheal washes were observed between vaccinated and not vaccinated birds in any of the experiments (Table 3). In all MDA+ groups on the day of vaccination, in tracheal washes turkeys had detectable maternally derived antibodies (Table 3), and their level decreased with age in large part proportionally to their level in the serum. Only in MDA-14/V group the level of these antibodies increased in tracheal washes on day 14 PV, but this increase was not statistically significant (Table 3). In Experiment I, high level of anti-aMPV antibodies were reported in the serum and tracheal washes on the day of vaccination. In the MDA+0/V group, mean (S/P)-ratios in the serum and tracheal washes decreased faster than in unvaccinated birds. In most of cases, the observed decrease was not statistically significant, except on day 3 PV when mean serum antibody level were significantly lower in vaccinated turkeys than in unvaccinated birds (Table 4). In Experiment II, MDA levels were detectable on vaccination day with the mean (S/P)-ratio (at 500-fold dilution) of 0.41 (data not shown), which is twice the recommended (S/P)-ratio (0.2) cut-off value. In the MDA+14/V group, MDA levels increased significantly on 14 DPV in comparison with unvaccinated birds, but differences were noted only when serum were diluted 100-fold (Table 4), and not 500-fold (data not shown). In Experiment III and IV, anti-aMPV antibodies were not detected in the serum (at 100-fold dilution) or in tracheal washes on the day of vaccination (Tables 3 and 4). Serum level of anti-aMPV IgY antibodies in MDA-0/V increased gradually after vaccination and peaked significantly on 14 DPV (Table 4). Similar results were observed in Experiment IV, where serum antibody levels in MDA- 14/V turkeys began to increase significantly (at 100-

fold serum dilution) on 7 DPV, and increased further on 14 DPV (Table 4).

Table 2: Mean x-fold change of IgM+ B ACC at days post aMPV vaccinationa in HG and TC M

			Mean x-fold change[b]			
Experiment	Group		0	3	7	14
I	MDA+0/V	HG	1	0,72	1,33	0,88
		TC	1	0,83	0,75*	0,88
II	MDA+14/V	HG	1	1,23	1,49*	0,76
		TC	1	1,28	1,32	1,08
III	MDA-0/V	HG	1	1,24	0,85	0,89
		TC	1	1,31	0,91	1,66**
IV	MDA-14/V	HG	1	1,18	0,99	0,74*
		TC	1	0,78	1,17	3,31**

[a]In all experiments vaccinated birds were inoculated oculonassaly with 10^4 $TCID_{50}$ of live attenuated aMPV/A vaccine.
[b]x-fold change = mean IgM+ B ACC of vaccinated group divided by mean IgM +B ACC of not vaccinated group.
*/**Significant difference in mean IgM+ B ACC of vaccinated birds in comparison to the not vaccinated group (T-test, *as $p < 0,05$ and **as $p < 0,01$).

Molecular Biology

The effectiveness of aMPV-A vaccine virus detection varied considerably between experiments. Between day 3 and 7 PV in Experiments I and III, aMPV RNA was detected by nested RT-PCR in 100% (5 out of 5) of choanal swabs in both MDA+0/V and MDA-0/V groups. On 14 DPV, aMPV RNA was detected in 40% (2/5) and 100% (5/5) of samples from groups MDA+0/V and MDA-0/V, respectively (Table 5). In Experiment II, aMPV RNA was barely detectable in the MDA+14/V group. In this group, genetic material of the vaccine virus was identified only on day 3 PV in 40% (2/5) of choanal swabs. aMPV/A RNA was undetectable on days 7 and 14 PV (Table 5). In Experiment IV, no viral RNA was detected at any stage of the experiment in choanal swabs of MDA-14/V birds (Table 5). The viral genome was not identified in unvaccinated birds in any experiments (data not shown).

DISCUSSION

Different TRT vaccination schedules in parent flocks of turkeys are likely to cause highly variable transfer of specific anti-aMPV maternal antibodies to hatching poults. Although these antibodies do not protect turkey poults against TRT, their impact on the development of vaccine-induced immunity against aMPV infections has been investigated by very few studies. The present situation is problematic because most breeder turkey flocks are kept in areas with endemic prevalence of aMPV [19].

Table 3: Tracheal washings anti – aMPV IgY antibody level after aMPV vaccinationa of turkeys

		Mean (S/P)-ratio ± S.D.[b] at days post aMPV/A vaccination			
Experiment	Group	0	3	7	14
I	MDA+0/V	1.152 ± 0.995	0.446 ± 0.476	0.117 ± 0.136	0.097 ± 0.102
	MDA+0/NV	1.152 ± 0.995	0.598 ± 0.695	0.123 ± 0.204	0.128 ± 0.097
III	MDA-0/V	0.00 ± 0.00	0.00 ± 0.00	0.00 ± 0.00	0.005 ± 0.011
	MDA-0/NV	0.00 ± 0.00	0.00 ± 0.00	0.00 ± 0.00	0.00 ± 0.00
II	MDA+14/V	0.128 ± 0.097	0.025 ± 0.036	0.022 ± 0.033	0.023 ± 0.025
	MDA+14/NV	0.128 ± 0.097	0.025 ± 0.039	0.031 ± 0.039	0.025 ± 0.042
IV	MDA-14/V	0.00 ± 0.00	0.00 ± 0.00	0.00 ± 0.00	0.029 ± 0.058
	MDA-14/NV	0.00 ± 0.00	0.00 ± 0.00	0.00 ± 0.00	0.00 ± 0.00

[a]In all experiments vaccinated birds were inoculated oculonassaly with 10^4 $TCID_{50}$ of live attenuated aMPV/A vaccine.
[b]10 samples of undiluted tracheal washings per group were analyzed.

Cook et al. [11,12] did not report any differences in the development of post-vaccination immunity between MDA+ and MDA- turkeys immunized with an attenuated aMPV strain (3B (Att.)) on 1 or 7 DOL. However, the attenuated aMPV strain used by the cited authors was characterized by a relatively high virulence index indicated by post-vaccination clinical scoring system. Contemporary TRT vaccines are less pathogenic, and they do not cause such acute side effects [20,21]. Reevaluating research aiming to compare the development of vaccine-induced immunity to aMPV infections in MDA+ and MDA- birds has not been undertaken to date. Liman and Rautenschlein [7] demonstrated that the percentage of CD4+ T cells subpopulation (but not CD8+ T cells) increased briefly in HG of turkeys 7–14 DPV after vaccination with aMPV/B or after infection with aMPV/A or B, which corroborates our results for MDA+ vaccinated birds. The cited authors suggested that the short-lived protection against TRT offered by vaccination and the frequency of TRT outbreaks could be explained by the transient character of the observed stimulation. Liman and Rautenschlein [7] analyzed turkeys from a breeder flock immunized against TRT, and vaccinated the birds at the age of 33 days when specific anti-aMPV IgY antibodies could still be detected in their serum. Their findings corroborate our observations and indicate that the presence of maternal antibodies somehow protects the upper respiratory tract of turkeys against infiltration by CD8+ T cells regardless of the birds' age. In this context, the presence of specific antibodies could contribute to the induction of aMPV/A immunophagocytosis through opsonization of the vaccine virus. The above observation is additionally validated by faster viral clearance in the upper respiratory tract and a significant decrease in MDA levels in MDA+0/V birds after vaccination. Cha [17] demonstrated that inoculation of two-week old MDA- turkeys with aMPV/C increased the percentage of CD8+ T cells in nasal turbinate mucosa, without inducing any changes in the percentage of CD4+ T cells. Additionally, an increase in absolute counts of CD4+ and CD8+ T cells was reported in HG in MDA- chickens inoculated with aMPV/A or B [9], which is consistent with our results in MDA- vaccinated groups. The observed increase in absolute counts of both T cell subpopulations in HG and CD8+ T lymphocytes in TC could indicate that CD4+ T cells alone are unable to control the vaccine virus replication. Greater stimulation of vaccine-induced humoral immunity in MDA- vaccinated birds could indicate that the CD4+ T cells are unable to maintain such metabolic pressure, which triggers the additional CD8+ T cells infiltration in the upper respiratory tract of vaccinated turkeys.

Table 4: Serum anti – aMPV IgY antibody level after aMPV vaccinationa of turkeys

		Mean S/P-ratio ± S.D.[b] at days post aMPV/A vaccination			
Experiment	Group	0	3	7	14
I	MDA+0/V (1:500)[c]	2.19 ± 1.87	0.9 ± 0.51*	0.57 ± 0.5	0.23 ± 0.16
	MDA+0/NV (1:500)	2.19 ± 1.87	1.9 ± 1.36	0.6 ± 0.62	0.41 ± 0.36
III	MDA-0/V (1:100)	0.001 ± 0.00	0.006 ± 0.01	0.015 ± 0.017	0.179 ± 0.164**
	MDA-0/NV (1:100)	0.001 ± 0.00	0.005 ± 0.005	0.005 ± 0.008	0.00 ± 0.00
II	MDA+14/V (1:100)	0.598 ± 0.492	0.42 ± 0.361	0.21 ± 0.200	0.307 ± 0.217*
	MDA+14/NV (1:100)	0.598 ± 0.492	0.423 ± 0.345	0.228 ± 0.242	0.140 ± 0.116
IV	MDA-14/V (1:100)	0.00 ± 0.00	0.014 ± 0.038	0.034 ± 0.055*	0.123 ± 0.194*
	MDA-14/NV (1:100)	0.00 ± 0.00	0.00 ± 0.00	0.00 ± 0.00	0.00 ± 0.00

[a]In all experiments vaccinated birds were inoculated oculonassaly with 10^4 $TCID_{50}$ of live attenuated aMPV/A vaccine.
[b]15 samples of serum were analyzed per group.
[c]Sample dilution for ELISA procedure.
*/**Significant difference in mean (S/P)-ratio of vaccinated birds in comparison to the not vaccinated group (U-test, *as p < 0,05 and **as p < 0,01).

Table 5: Detection of aMPV genome in choanal swabs of aMPV vaccinated turkeys by nested RT-PCR

		Number of aMPV/A positive birds/total[a] at days post vaccination			
Group	Experiment	0	3	7	14
MDA+0/V	I	0/5	5/5	5/5	2/5
MDA-0/V	III	0/5	5/5	5/5	5/5
MDA+14/V	II	0/5	2/5	0/5	0/5
MDA-14/V	IV	0/5	0/5	0/5	0/5

[a]In all experiments vaccinated birds were inoculated oculonassaly with 10^4 $TCID_{50}$ of live attenuated aMPV/A vaccine.

Studies of other respiratory infections demonstrated that CD8+ T memory cells confer immunity to reinfection by restricting the spread of the virus at the site of its replication [22,23] but in turn, they may be directly responsible for damage to the host's anatomical structures (e.g. mucous membranes) [24]. The question that arises in connection with our results is whether the presence of maternal antibodies interferes with the acquisition of T memory cells after vaccination against TRT. Further research is needed to examine the role of CD4+ and/or CD8+ T memory cells in protection against aMPV infections. As demonstrated earlier, the TRT virus has immunosuppressive activity [25-28]. In our study a significant decrease in CD4+ and/or CD8+ T ACC in tracheal mucosa was noted in vaccinated birds 3–7 DPV in Experiments II - IV. Although, we can not assume that the above changes are indicative of T cell immunosuppression we may speculate that attenuated aMPV strains could also deliver such effects. Further research is necessary to establish whether aMPV vaccine strains possess immunosuppressive activity and if this activity in vivo could depend on specific antibody levels and birds' age on the day of vaccination. The role of double positive CD4+ CD8+ T cells in the host's immune response has not yet been fully elucidated, but in some production lines of hens those cells were found to represent a significant percentage of T lymphocytes in secondary immune structures [29]. Double-positive T cells could also have regulatory functions or could play the role of both T helper cells and cytotoxic T cells [30]. As it turned out in our study, this parameter varied significantly only in MDA- vaccinated turkeys. Although most of the cited authors concluded their studies based on relative T cell count and not ACC, these values are highly proportional. Therefore it can be concluded that differences exist in the development and activation of CMI in the upper respiratory tract of turkeys vaccinated against TRT, and that those variations are correlated mainly with the birds' age and maternal antibody level on the day of vaccination. The cited experiments produced similar results despite the use of different aMPV strains, which seems to suggest that differences in TRT immunopathogenesis are not strainrelated [6,7,17]. In our study, MDA+0/V was the only group of vaccinated birds that did not produce anti-aMPV IgY antibodies in response to vaccination. On day 3 PV, anti-aMPV antibody level in the serum of MDA+0/V birds was significantly lower in comparison with unvaccinated birds. Those findings are partly consistent with results reported by Cook et al. [12] who did not observe an increase in serum levels of specific anti-aMPV antibodies in most MDA+ chickens vaccinated against TRT on the first day of life. In our experiment, a faster decrease in MDA levels in MDA+0/V birds after vaccination could be attributed to immunophagocytosis. MDA+14/V and both MDA- vaccinated groups of turkeys responded to vaccination by producing specific IgY anti-aMPV antibodies that were detected in the serum (but not in tracheal washings) at 100-fold dilution on 14 DPV. In MDA-14/V birds, the level of those antibodies were elevated already on day 7 PV. Those results indicate that the level and the time required to induce humoral immunity after TRT vaccination are determined by age and MDA level on the day of vaccination. The above could impair early assessment of vaccine-induced immunity against TRT based on the results of routine serological monitoring which also has other disadvantages [4,31-33]. The absolute count of IgM^+ B cells is a general, nonspecific indicator of humoral immunity stimulation after vaccination. In this study, the only group of birds in which this parameter did not change or decreased significantly was MDA^+0/V. A significant decrease in IgM^+ B ACC in HG was also reported in the MDA-14/V group on day 14 PV, but in this case, it was probably compensated by intensive IgM^+ B cell infiltration in tracheal mucosa.

Secretory IgA plays the main role in humoral immunity of upper respiratory tract. IgA^+ B cells and specific IgA participates in the local immune response against TRT in the upper respiratory tract [15]. Absolute counts of IgA^+ B cells and the levels of specific IgA were not determined in this study, but it could be hypothesized that, similarly to IgM^+ B cell infiltration and specific IgY production, the production of specific IgA could also depend on the MDA level and the age of birds on the day of vaccination against TRT. Further research is needed to determine the involvement of IgA^+ B cells and the role of IgA in protection against aMPV infections. The observed differences in aMPV/A replication in nasal choanae between the MDA+0/V and MDA-0/V groups could be associated with high antibody levels in MDA+ birds upon vaccination. High levels of the these immunoglobulins probably facilitated and stimulated (upon immunophagocytosis) viral clearance from the upper respiratory tract, as suggested earlier by Cook et al. [11,12]. In both groups vaccinated on 14 DOL, aMPV/A RNA was practically undetectable in choanal swabs. Despite the above, the upper respiratory tract immune system was strongly stimulated in those groups after vaccination. Those results partially corroborate the findings of Rubbenstroth and Rautenschlein [16] who reported very high levels of vaccine-induced immunity against aMPV infections despite the fact the aMPV/A RNA was not regularly detected in nasal turbinates of turkeys vaccinated against TRT on 13 or 14 DOL. Differences between birds vaccinated at different age suggest that the replication of vaccine aMPV/A is limited during the maturation of the immune and/or respiratory system, but this does not influence the development of vaccineinduced immunity. This work has some limitations. Given the fact that MDA^+ and MDA- turkeys had different origin we were unable to perform statistical analysis of assayed parameters between them. However we could trace the diversity of their development after vaccination in each experiment in comparison to adequate not vaccinated control group. To ensure the reproducibility of the results, in each experiment, before the vaccination the birds were confirmed to be free (with PCR technique) from other common respiratory pathogens Mycoplasma spp., Ornithobacterium rhinotracheale, Bordetella avium and Newcastle Disease Virus (data not shown). Experiments I and III as well as II and IV were performed simultaneously. Additionally, birds for experiment I and II as well as III and IV were obtained from the hatcheries at the same time.

CONCLUSION

Previous studies demonstrated that, under laboratory conditions, the live attenuated vaccines are highly efficient in protecting against TRT [7,9,17,21]. On the other hand, despite vaccination, TRT outbreaks are frequently observed in the field [7-9], which may stem not only from aMPV high virulence but, additionally, from differences in post-vaccination immunity development in birds with diversified immunological status and from the impact of other negative environmental and microbiological agents. From the perspective of our study, at least a few pathomechanisms of this adverse situation may be concluded: (I) possible immunosuppressive activity of aMPV vaccine strains causing disorders in T cells activity or T cells activation, (II) possible disorders in antigen specificity acquisition of B and/or T lymphocytes in MDA^+ groups, (III) high infiltration of $CD8^+$ T lymphocytes in TC of MDA-groups, that (most likely) is not neutral to the morphology of tracheal mucosa, as well as (IV) temporary character of $CD4^+$ T lymphocytes increase in the upper respiratory tract of MDA+ birds.

Competing Interests

None of the authors of this paper have a financial or personal relationship with other people or organizations that could inappropriately influence or bias the content of the paper.

ACKNOWLEDGEMENTS

The project has been defrayed with the funds from the National Science Centre granted in terms of decision number DEC-2011/01/N/NZ6/05757.

REFERENCES

1. Pedersen JC, Reynolds DL, Ali A. The sensitivity and specificity of a reversetranscriptasepolymerase chain reaction assay for the avian pneumovirus (Colorado strain). Avian Dis. 2000;44:681–5.
2. Buys SB, Du Preez JH. A preliminary report on the isolation of a virus causing sinusitis in turkeys in South Africa and attempts to attenuate the virus. Turkeys. 1980;28:36.
3. Collins MS, Gough RE, Alexander DJ. Antigenic differentiation of avian pneumovirus isolates using polyclonal antibody and mouse monoclonal antibodies. Avian Pathol. 1993;22:469–79.
4. Cook JKA, Jones BV, Ellis MM, Li J, Cavanagh D. Antigenic differentiation of strains of Turkey rhinotracheitis virus using monoclonal antibodies. Avian Pathol. 1993;22:257–73.
5. Seal BS. Matrix protein gene nucleotide and predicted amino acid sequencedemonstrate that the first US avian pneumovirus isolate is distinct from European strains. Virus Res. 1998;58:45–52.
6. Smialek M, Tykalowski B, Stenzel T, Koncicki A. The perspective of immunoprophylaxis and selected immunological issues in the course of the turkey rhinotracheitis. Pol J Vet Sci. 2012;15:175–80.
7. Liman M, Rautenschlein S. Induction of local and systemic immune reactions following infection of turkeys with avian Metapneumovirus (aMPV) subtypes A and B. Vet Immunol Immunopathol. 2007;115:273–85.

8. Catelli E, Lupini C, Cecchinato M, Ricchizzi E, Brown P, Naylor CJ. Field avian metapneumovirus evolution avoiding vaccine induced immunity. Vaccine. 2010;28:916–21.

9. Rautenschlein S, Aung YH, Haase C. Local and systemic immune responses following infection of broiler-type chickens with avian Metapneumovirus subtypes A and B. Vet Immunol Immunopathol. 2011;140:10–22.

10. Jones RC, Wolliams RA, Baxter-Jones C, Savage CE, Wilding GP. Experimental infection of laying turkeys with rhinotracheitis virus: distribution of virus in the tissues and serological response. Avian Pathol. 1988;17:841–50.

11. Cook JKA, Ellis MM, Dolby CA, Holmes HC, Finney PM, Huggins MB. A live attenuated turkey rhinotracheitis virus vaccine. 1.Stability of the attenuated strain. Avian Pathol. 1989;18:511–22.

12. Cook JKA, Holmes HC, Finney PM, Dolby CA, Ellis MM, Huggins MB. A live attenuated turkey rhinotracheitis virus vaccine. 2. The use of the attenuated strain as an experimental vaccine. Avian Pathol. 1989;18:523–34.

13. Jirjis FF, Noll SL, Halvorson DA, Nagaraja KV, Shaw DP. Pathogenesis of avian pneumovirus infection in turkeys. Vet Pathol. 2002;39:300–10.

14. Jones RC, Naylor CJ, al-Afaleq A, Worthington KJ, Jones R. Effect of cyclophosphamide immunosuppression on the immunity of turkeys to viral rhinotracheitis. Res Vet Sci. 1992;53:38–41.

15. Cha RM, Khatri M, Sharma JM. B-cell infiltration in the respiratory mucosa of turkeys exposed to subtype C avian metapneumovirus. Avian Dis. 2007;51:764–70.

16. Rubbenstroth D, Rautenschlein S. Compromised T-cell immunity in turkeys may lead to an unpredictable avian metapneumovirus vaccine responseand variable protection against challenge. Avian Pathol. 2010;39:349–57.

17. Cha RM. Immunopathogenesis of avian Metapnumovirus in the turkeys. PhD thesis. University of Minnesota, United States; 2009.

18. Cavanagh D, Mawditt K, Britton P, Naylor CJ. Longitudinal field studies of infectious bronchitis virus and avian pneumovirus in broilers using type-specific polymerase chain reactions. Avian Pathol. 1999;28:593–605.

19. Gough RE, Jones RC. Avian Metapneumovirus. In: Saif YM, editor. Diseases of poultry. 12th ed. Ames: Blackwell Publishing; 2008. p. 100–10.

20. Wiliams RA, Savage CE, Jones RC. Development of a live attenuated vaccine against turkey rhinotracheitis. Avian Pathol. 1991;20:45–55.

21. Patnayak DP, Sheikh AM, Gulati BR, Goyal SM. Experimental and field evaluation of a live vaccine against avian pneumovirus. Avian Pathol. 2002;31:377–82.

22. Kimpen JL, Rich GA, Mohar CK, Ogra PL. Mucosal T cell distribution during infection with respiratory syncytial virus. J Med Virol. 1992;36:172–9.

23. de Bree GJ, van Leeuwen EM, Out TA, Jansen HM, Jonkers E, van Lier RA.Selective accumulation of differentiated CD8+ T cells specific for respiratory viruses in the human lung. J Exp Med. 2005;202:1433–42.

24. Scott KG, Buret AG. Role of CD8+ and CD4+ T lymphocytes in jejunalmucosal injury during murine giardiasis. Infect Immun. 2004;72:3536–42.

25. Timms LM, Jahans KL, Marshall RN. Evidence of immunosuppression inturkey poults affected by rhinotracheitis. Vet Rec. 1986;119:91–2.

26. Chary P, Rautenschlein S, Njenga MK, Sharma JM. Pathogenic and immunosuppressive effects of avian pneumovirus in turkeys. Avian Dis.2002;46:153–61.

27. Chary P, Rautenschlein S, Sharma JM. Reduced efficacy of hemorrhagic enteritis virus vaccine in turkeys exposed to avian pneumovirus. Avian Dis. 2002;46:353–9.

28. Marien M, Decostere A, Martel A, Chiers K, Froyman R, Nauwynck H. Synergy between avian pneumovirus and Ornithobacterium rhinotracheale in turkeys. Avian Pathol. 2005;34:204–11.

29. Davidson F, Kaspers B, Schat KA. Avian immunology. 1st ed. Great Britain: Academic Press; 2008.

30. Parel Y, Chizzolini C. CD4+ CD8+ double positive (DP) T cells in health and disease. Autoimmune Rev. 2004;3:215–20.

31. Eterradossi N, Toquin D, Guittet M, Bennejean G. Evaluation of differentturkey rhinotracheitis viruses used as antigens for serological testing following live vaccination and challenge. Zentralbl Veterinarmed B.1995;42:175–86.

32. Mekkes DR, de Wit JJ. Comparison of three commercial ELISA kits for the detection of turkey rhinotracheitis virus antibodies. Avian Pathol. 1998;27:301–5.

33. Cook JKA. Avian rhinotracheitis. Rev Sci Tech. 2000;19:602–13.

CHAPTER 17

CARDIAC PATHOLOGY AND MOLECULAR EPIDEMIOLOGY BY AVIAN LEUKOSIS VIRUSES

Sayuri Nakamura, Kenji Ochiai, Akihiro Ochi, Hiroki Yabushita, Asumi Abe, Sayaka Kishi, Yuji Sunden, Takashi Umemura

INTRODUCTION

Retroviruses cause a variety of illnesses such as leukemia, acquired immunodeficiency syndrome and neurodegenerative diseases in humans and animals. Epidemiological studies [1–3] have indicated that the risk of dilated cardiomyopathy and nonsuppurative myocarditis rises in those infected with human immunodeficiency virus type 1 (HIV-1). As experimental evidence in non-human primates, non-suppurative myocarditis and dilated cardiomyopathy with myocardial hypertrophy have been described in rhesus monkeys infected with simian immunodeficiency viruses (SIVs) [4]. These may occur as a result of the retrovirus itself acting either directly or indirectly via immunological mechanisms, opportunistic infection of other cardiotropic viruses or a combination of these mechanisms [1–4]. HIV viral protein R (Vpr) has been recently suggested to be a unique polypeptide that causes atrial cardiomyocyte mitosis, mesenchymal tumor and dysrhythmia in the heart of transgenic mice with Vpr [5]. However, the information on the real cause and pathogenesis of retrovirus-induced cardiac disorders is limited. In animals other than humans and non-human primates, Maedi-Visna virus from the genus Lentivirus and subfamily Orthoretrovirinae induces primarily lymphocytic inflammation in the heart of sheep, but no alteration of cardiomyocytes [6]. In avian species, myocarditis associated with avian leukosis virus (ALV) belonging to Alpharetrovirus in the family Retroviridae has been reported in chickens [7–9]. ALVs in chickens are classified into six subgroups of A–E and J depending on envelope antigenicity, and subgroup A of ALV (ALV-A)-induced lymphoid leukosis is the most common neoplasm in infected birds. However, Rousassociated virus 1 (RAV-1) [8] and fowl glioma-inducing virus (FGV prototype) [9], both of which are ALV-A, could induce nonsuppurative myocarditis in chickens and the former resulted in chronic circulatory disturbance. The myocarditis induced by ALV-A has been considered to be associated with excessive viral replication in cardiomyocytes [8,9]. Additionally, dilated cardiomyopathy characterized by biventricular dilation and right ventricular hypertrophy has been described in chickens congenitally infected with ALV-J [10]. The authors speculated that the cardiomyopathy resulting from persistent high-level synthesis of viral products may usurp the cellular machinery and substrates in cardiomyocytes and Purkinje fibers. These reports imply that several strains of ALV are pathogenic to terminally differentiated cardiomyocytes. The causal relationship, however, is not completely established and the pathogenesis remains unclear. Here, we describe unusual hypertrophy and mitosis of cardiomyocytes, the morphology of which has not been previously reported in chickens infected with ALV-A. In addition, the causal relationship and pathogenesis are discussed on the basis of the molecular characteristics of isolates and the results of a reproducibility experiment with isolates.

Citation: Nakamura S, Ochiai K, Ochi A, Yabushita H, Abe A, et al. (2014) Cardiac Pathology and Molecular Epidemiology by Avian Leukosis Viruses. PLoS ONE 9(1): e86546. doi:10.1371/journal.pone.0086546.

MATERIALS AND METHODS

Ethics Statement

The Animal Care and Use Committee of Hokkaido University approved the protocols (Permit number 110071 and 120020), in accordance with the Act on Welfare and Management of Animals of Japanese government.

Animals, Histopathology and Immunohistochemistry

The chicken breeds used in this study are listed in Table 1. These fowls were collected from the chicken population in which ALVs are widespread. Feather pulps and cloaca swabs were collected from 18 Japanese native chickens and stored at 280uC until use. For histological examination, the chickens were euthanized humanely according to a procedure approved by Hokkaido University. The heart and other organs, including the liver, spleen, kidneys, lungs and brain, were fixed in 20% neutralbuffered formalin, routinely processed and embedded in paraffin wax. Sections (4 mm) were cut and stained with hematoxylin and eosin. Masson's trichrome staining was performed in order to score myocardial fibrosis. Immunostaining using the labeled streptavidin-biotin method (Nichirei Corp., Tokyo, Japan) was performed. The primary antibodies used were polyclonal antibody against ALV common antigens (1:5000, courtesy of Dr. K. Tsukamoto, Azabu University, Japan) and monoclonal antibodies against ARV p17 antigen (1:50, Genesis Biotech Inc., Taipei, Taiwan), PCNA (no dilution, DAKO, Copenhagen, Denmark), phosphorylated Akt (p-Akt) (1:100, Cell Signaling Technology, Danvers, MA) and phosphorylated tuberin (p-tuberin) (1:100, Cell Signaling Technology). The polyclonal antibody against ALVs was generated by immunization of rabbits with Rous-associated virus- 2, as previously described [11]. This antibody recognizes mainly gs proteins (p27) and glycoprotein 85 (gp85) of ALVs.

Rna Extraction and CDNA Synthesis

Total RNA was extracted from brains, kidneys and hearts using TRIzol reagent (Invitrogen Life Technologies, Carlsbad, California, USA). Reverse transcription (RT) was performed as previously described [12]. As an internal control for RNA extraction and cDNA synthesis, PCR amplification was performed for all cDNA samples using primers specific to chicken b-actin. The PCR for chicken b-actin was performed according to conditions described previously [12].

PCR Amplification

Two types of PCR amplification were carried out with cDNA using primer sets shown in Table 2. To detect the ALV genome, ALV-specific PCR was performed with a primer set of ALV #38 and ALV #39 according to conditions described previously [13]. The primers were designed for envSU (gp85), which is well conserved in ALV subgroups A to E. To detect the avian reovirus (ARV) genome, PCR for the S2 and S4 regions of ARV was performed with the primer sets of ARV sense primer (S2) and ARV antisense primer (S2), and ARV sense primer (S4) and ARV antisense primer (S4), respectively, using a procedure previously described with some modification [14]. The PCR mixture was a 10 ml reaction volume containing 1 ml of 10x Ex Taq Buffer(TaKaRa Bio, Hotsu, Japan), 0.8 ml of dNTP mixture, 2.5 mM each dNTP (TaKaRa Bio), 5 pmol each primer, 0.25 U TaKaRa Ex Taq HS (TaKaRa Bio) and 10 ng DNA. Touchdown PCR was performed. Denaturation was at 94uC for 1 min. The annealing temperature decreased in steps of 2uC, from 60uC down to 48uC. Two cycles were performed at each annealing temperature. Thirty-five cycles were then carried out at 95uC for 30 sec, 48uC for 30 sec and 72uC for 30 sec. Finally, a prolonged extension step was performed at 72uC for 10 min. The PCR products were analyzed by 2% agarose gel electrophoresis.

Virus Isolation and Identification

For virus isolation, the supernatants of cloaca swabs from 16 chickens were inoculated with 80% confluent DF-1 cells. DF-1 cells were obtained from the American Type Culture Collection (Manassas, Virginia, USA) and cultured as previously described. Cells were harvested 7 days after inoculation and were collected at that time. Concentrated culture supernatant was prepared according to the methods described previously [15].

Sequence Analysis and Phylogenic Analysis

The envSU regions of 15 strains isolated from field cases and the complete virus genome of Km_5892 were sequenced as previously described [12,16]. The complete genomic RNA (gRNA) of Km_5892 has been deposited in the DDBJ/GenBank (AB682778). The obtained sequences were analyzed with the Clustal W 1.8.3 program on the DDBJ website (http:// clustalw. ddbj.nig.ac.jp/top-j.html) and the NCBI BLAST program (http:// blast.ncbi.nlm.nih.gov/Blast.cgi). Phylogenic trees based on nucleotide sequences of envSU and pol were constructed on the DDBJ website by the neighbor-joining method implementing the Kimura method using isolated virus and other ASLVs. The topological accuracy of the tree was estimated by the bootstrap method with 1000 replicates.

Birds and Experimental Infection

Fertile eggs from specific-pathogen-free (SPF) White Leghorn strain WL-M/O (C/O) chickens were used for experimental infection. The fertile eggs were obtained from Nippon Institute for Biological Science (Yamanashi, Japan). This strain lacks both chicken helper factor and group-specific antigen and is susceptible to ALV subgroups A to E (International Registry of Poultry Genetic Stocks, Bulletin 476, March 1988, Ralph G. Somes, Jr., University of Connecticut, Storrs, Connecticut, USA). The presence of endogenous virus genes other than chicken helper factor and group-specific antigen is not known. Chicken embryos were inoculated via the yolk sac on the sixth day of incubation with 56104 Tissue Culture Infectious Units (TCIUs) of Km_5666 or Km_5892 strain and were euthanized at 35, 70 and 140 days of age. Seven control chicken embryos belonging to the uninfected group were similarly inoculated with 0.1 ml uninfected tissue culture medium and were euthanized at 35 days and 70 days of age. The hatched chickens were reared humanely according to the guidelines set by the Animal Care and Use Committee of our institute. No chickens received any vaccinations or medication.

RESULTS

Clinical signs and pathology in Japanese native fowls Eighteen Japanese native fowls were collected and examined. A low rate of egg laying was observed in four fowls and emaciation in two (Table 1). At necropsy, one bird each had a small renal cyst and egg yolk in the body cavity (yolk peritonitis). Other twelve fowls showed no apparent gross lesions. The main histopathological changes in hearts were lymphocytic myocarditis, matrix inclusion bodies in cardiomyocytes and Purkinje fibers, fibrosis and hypertrophy of cardiomyocytes with atypical nuclei (Figure 1A–D). These lesions were noted mainly in the left ventricular wall and the degree of severity varied (Table 1). Multifocal lymphocytic myocarditis with mild myocardial necrosis and mild to severe interstitial myocardial fibrosis were observed in all cases. These foci contained a small number of plasma cells and macrophages. Intracytoplasmic, round to oval, unstained to basophilic inclusions (matrix inclusion bodies) were detected in cardiomyocytes and Purkinje fibers of all cases (Figure 1E). Additionally, individual hypertrophied cardiomyocytes or small clusters of cardiomyocytes with atypical, large nuclei were observed in seven (39%) fowls and these cardiomyocytes occasionally contained matrix inclusion bodies. The atypical nuclei were frequently arranged in long chains (1–11/high-power field) in the cytoplasm, and mitotic figures were also observed in these cardiomyocytes. The degree of cardiomyocyte hypertrophy was not correlated to the degree of myocarditis and the frequency of matrix inclusion bodies in the cytoplasm (Table 1). Other eleven chickens had neither hypertrophied cardiomyocytes nor atypical nucleus. Ten (56%) of eighteen fowls developed gliomas in their brains. No significant lesion was observed in other organs of the examined cases, except for a renal cyst and yolk peritonitis grossly noted in one bird each. These cardiomyocytes, Purkinje fibers and matrix inclusion bodies were immunohistochemically positive for ALV common antigen (Figure 1F) and negative for avian reovirus (ARV) p17 antigen. Nuclei in hypertrophied cardiomyocytes were positive for proliferating cell nuclear antigen (PCNA) (Figure 1G).

Table 1: Clinical signs, gross findings, cardiac histopathology and results of ALV isolation in chickens

					Degree of cardiac lesions				Viral isolation [f]		
No.	Breed	Sex (M/F) [a]	Age (year)	Gross findings	Hypertrophied myocardium with atypical nucleus [b]	Non-suppurative myocarditis [c]	Fibrosis [d]	Matrix inclusion body [e]	Cloaca swab	Brain	Isolated strain
1	Japanese Bantam	M	2	Anemia	−	+	+	109	+	ND	Km_5880
2	Japanese Bantam	M	1	Atrophic comb	+++	+++	+	39	+	ND	Km_5960
3	Japanese Bantam	M	2	None	−	++	+	7	+	ND	Km_5622
4	Japanese Bantam	M	3	None	−	++	++	99	+	ND	Km_5967
5	Japanese Bantam	M	1	None	−	+++	+	5	+	ND	Km_6042
6	Japanese Bantam	F	1	Emaciation	−	+	+	6	+	ND	Km_5623
7	Japanese Bantam	F	3	Emaciation	−	+	+	1	+	ND	Km_5625
8	Japanese Bantam	F	4	Yolk peritonitis	+	++	++	57	+	ND	Km_5968
9	Japanese Bantam	F	1	Renal cyst	+	++	++	1	+	ND	Km_6045
10	Japanese Bantam	F	1	None	−	+	+	129	+	ND	Km_5621
11	Japanese Bantam	F	3	None	−	++	+	3	+	ND	Km_5624
12	Japanese Bantam	F	1	None	+	++	++	2	+	ND	Km_5843
13	Japanese Bantam	F	2	None	+	++	+	43	+	ND	Km_5852
14	Japanese Bantam	F	3	None	+++	+++	+++	27	+	ND	Km_5892
15	Japanese Bantam	F	1	None	−	+	+	49	ND	+	Km_5897
16	Japanese Bantam	F	3	None	−	+++	+	1	+	ND	Km_6209
17	Sebright Bantam	F	3	None	++	++	++	19	ND	+	Km_5900
18	Kumamoto Long Tail	F	4	None	−	+++	++	12	+	ND	Km_6039

[a]M: male, F: female

[b]The degree of hypertrophied cardiomyocytes with atypical nuclei was defined as follows: +, individual cells; ++, multiple and/or small clusters; +++, multiple and extensive growth.

[c]The degree of non-suppurative myocarditis was defined as follows: +, focal or multifocal (under 25 μm in diameter); ++, multifocal (from 25 μm to 200 μm); +++, multifocal (over 200 μm).

[d]The degree of fibrosis was defined as follows: +, interstitial; ++, multifocal interstitial (under 400 μm in diameter); +++, multifocal interstitial (over 400 μm in diameter).

[e]Total number in 10 high-power fields.

[f]+: positive, −: negative,ND: no data.

doi:10.1371/journal.pone.0086546.t001

Table 2: Sequence of the oligonucleotide primers and targets

Primer	Sequence (5′-3′)	Accession No.
Chicken β-actin forward	TAT CCG TTA GGA TCT GTA TG	L08165
Chicken β-actin reverse	ATC TCG TCT TGT TTT ATG CG	L08165
ALV #38	TTA GGT TCC CAG TCT CTC CC	AB112960
ALV #39	ATT GCG GGT GGT AGC GCT TT	AB112960
ARV sense primer (S2)	CCC ATG GCA ACG ATT TC	AF104311
ARV antisense primer (S2)	TTC GGC CAC GTC TCA AC	AF104311
ARV sense primer (S4)	GTG CGT GTT GGA GTT TC	U95952
ARV antisense primer (S4)	ACA AAG CCA GCC ATG AT	U95952

doi:10.1371/journal.pone.0086546.t002

Viral Isolation and Polymerase Chain Reaction for ALV and ARV

Feather pulps from all cases showed positivity for ALV-specific polymerase chain reaction (PCR) and ALVs were isolated from cloaca swabs or brains of all affected chickens (Table 1). The hearts from 5 chickens (No. 2, 5, 9, 14 and 16), 2 chickens (No. 2 and 14) and the kidneys from other 13 birds showed negativity for ARV-specific PCR.

Phylogenic Analysis of Isolated ALV Strains

The sequences of envSU genes of sixteen strains isolated from the affected chickens were determined. The phylogenic analysis based on this region revealed that these strains could be classified into three clusters: MAV-1, FGV prototype and Km_ clusters (Figure 2). Five strains, including Km_5843, Km_5852, Km_5892, Km_5960 and Km_6045, isolated from fowls with hypertrophied cardiomyocytes belonged to two different clusters: FGV prototype and Km_ clusters. Next, the full-length genome of Km_5892, an isolate from the fowl with the most extensive distribution of hypertrophied cardiomyocytes, was determined. The env gene of Km_5982 showed homology of 97% with that of FGV mutant strain Sp-40, and the 59UTR, gag-pol and 39LTR domains of Km_5982 showed 96%, 96% and 90% homology with those of TymS_90, respectively. Phylogenic analysis based on the pol gene revealed that Km_5892 could be classified into the same cluster as the other Km_ strains, including Km_5666, Km_5843 and Km_5845 (Figure 3). Km_5666 (accession number AB669896) is an ALV-A that was isolated from a Japanese Bantam kept in the same flock and the sequence of its full-length genome was previously analyzed [17].

Pathology of Chickens Experimentally Inoculated with KM_Strains

Km_5666 and Km_5892 were picked up from two different clusters in phylogenic tree based on envSU region of ALV and we performed infectious experiment using these two strains. C/O SPF chicken eggs, inoculated in ovo with 56104 TCIU of Km_5666 and Km_5892 on the sixth day of incubation, were euthanized at 35, 70 and 140 days of age and examined pathologically (Table 3). Paralysis of the left leg was observed in one of three (33%) Km_5666-inoculated chickens at 70 days of age. Other chickens showed no clinical signs. Multifocal lymphocytic myocarditis was observed in ten (83%) of twelve birds inoculated with Km_5666 and seven (100%) of seven with Km_5892. In these foci, lymphocytes and plasma cells admixed with a few macrophages and heterophils also infiltrated to various degrees. Hypertrophied cardiomyocytes with atypical nuclei were observed in two (17%) chickens with Km_5666 and one (14%) with Km_5892. One of the Km_5666-inoculated birds showed extensive multiple growths of atypical cardiomyocytes at 35 days of age (Figure 4A–C). Interstitial myocardial fibrosis associated with the inflammatory foci was observed in one (8%) Km_5666-inoculated chicken and two (29%) Km_5892-inoculated birds. Matrix inclusion body was detected in five (42%) and seven (100%) chickens inoculated with Km_5666 and Km_5982, respectively. In terms of immunohistochemistry, the cytoplasm of cardiomyocytes and Purkinje fibers and matrix inclusion bodies were positive for ALV common antigen and nuclei in atypical cardiomyocytes were positive for PCNA (Figure 4D). The cytoplasm of hypertrophied cardiomyocytes was also positive for phosphorylated Akt (p-Akt) and phosphorylated tuberin (p-tuberin) (Figure 4E–F). By using RTPCR and sequencing, the hypervariable regions of Km_5666 and Km_5982 were detected from the brain and heart of inoculated birds, respectively. No significant lesion was observed in the control birds inoculated with medium.

DISCUSSION

The cardiac lesions of native fowls were histologically characterized by atypical hypertrophied cardiomyocytes with mitosis and non-suppurative myocarditis. ALV strains were isolated from all affected cases and similar cytological changes were reproduced by experimental infection with Km_5666 and Km_5982. In addition, extensive multiple growths of atypical cardiomyocytes were also recognized in a Km_5666-inoculated bird. These results demonstrate that ALVs can cause atypical hypertrophied cardiomyocytes and mitosis and on rare occasions induce extensive growth. The distribution and cytological features of cardiomyocytes clearly differed from those of physiological cardiac hypertrophy and those of idiopathic primary myocardial diseases, including human and feline cardiomyopathies [18–20]. To our knowledge, such individual and/or massive hypertrophy of cardiomyocytes has not been previously reported in any avian disease, except for in one chicken experimentally infected with FGV [21]. ARV causes a cardiac abnormality grossly characterized by right atrial dilation and microscopically non-suppurative myocarditis with multifocal necrosis and multinuclear giant cell formation [22]. However, ARV infection occurs mainly in turkeys rather than in chickens, and the histological changes were easily differentiated from those of our cases,especially on the basis of atypical hypertrophy of cardiomyocytes. In addition, our cases were immunohistochemically negative for ARV p17 antigen and negative for ARV S2 and S4 genome by PCR. Cardiac rhabdomyomas are often found incidentally in human infants affected by tuberous sclerosis [23] and in swine juveniles [24], and are considered as a malformation or a hamartoma rather than a true neoplasm [18–20]. Rhabdomyomas are morphologically characterized by multiple nodules composed of enlarged cardiomyocytes with a distinct border and the appearance of "spider cells", which are defined as large cardiomyocytes with numerous vacuoles in their cytoplasm due to glycogen accumulation. In contrast, their malignant counterpart, rhabdomyosarcoma, is quite rare in avian species and there has been just one report of rhabdomyosarcoma induced by the MC29 strain [25,26], which is a replication-deficient avian myelocytomatosis virus with v-myc oncogene and belongs to the group of acutely transforming viruses. Morphologically, rhabdomyosarcoma is characterized by the proliferation of large vacuolated cardiomyocytes with severe cellular and nuclear atypia and shows invasive growth. Hypertrophied cardiomyocytes showing extensive growth in a Km_5666- inoculated bird had glassy cytoplasm with a few vacuoles. Although these morphological features were different from those of naturally occurring cardiac neoplasms, the cardiac lesions consisting of atypical cardiomyocytes in a Km_5666-inoculated chick (35 days of age) were considered to be in the category of hamartoma. ALVs show high tropism to myocardial fibers and matrix inclusion bodies are frequently formed in association with ALV replication in infected cardiomyocytes [27–29]. Non-suppurative myocarditis was found to occur in experimental cases with some ALV strains, including RAV-1 and 7 [7,30], and these inflammatory changes are interpreted as being induced by viral replication in cardiomyocytes. On the basis of these findings, four possible mechanisms are considered to explain the ALV-induced cardiomyocyte hypertrophy in the present study. First, these lesions may develop as reactive changes secondary to non-suppurative myocarditis. The frequency of hypertrophied cardiomyocytes, however, was not correlated with the degree of myocarditis and we have never experienced such changes in any myocarditis cases in chickens. It seems unlikely that the abnormal cardiomyocytes are merely a reactive change. Second, the lesions may develop due to physical stimuli of matrix inclusion bodies. Hypertrophy of myocardial fibers is known to be induced by the formation of matrix inclusion bodies [10]. However, there are no findings that atypical large nuclei and mitosis develop in association with the formation of matrix inclusion bodies in any previous reports on ALV-infected heart. Third, the formation of atypical cardiomyocytes may be associated with insertional mutagenesis of ALV [31]. However, transformation of host cells usually take more than 5 months in avian lymphoid leukosis by ALV-A because multiple rounds of infection must occur in an infected animal before a provirus inserts itself in the vicinity of a cellular proto-oncogene with appropriate activation. Insertional mutagenesis cannot explain the development of cardiomyocyte hypertrophy and hamartoma in 35-day-old birds in this study. Three oncogenic mechanisms of retroviruses other than insertional mutagenesis are known: 1) transformation by viral oncogenes, 2) oncogenesis by accessory proteins such as Tax of human T-cell leukemia virus type I and 3) envelope protein-induced transformation [31]. The viral genomes of Km_5666 [11] and Km_5892, however, contain neither a viral oncogene nor any genes coding accessory proteins. The last possible mechanism is that viral components such as envelope proteins may cause signaling pathway abnormality in host cells, resulting in a change of cellular function and morphology [31]. Recently, Jaagsiekte sheep retrovirus (JSRV) has been reported to cause transformation of host cells by its transmembrane (TM) envelope domain [32,33]. Envelope protein of JSRV involves the PI3K/Akt pathway in transformation [33]. Once the PI3K/Akt pathway is activated, cellular genes associated with protein synthesis, proliferation and survival are up-regulated. Among these genes, tuberin and mammalian target of rapamycin (mTOR) are known to have a crucial role in the proliferation and hypertrophy of cells [34]. Dephosphorylated tuberin (activated form) was shown to regulate cellular status by suppressing the mTOR activator, Rheb. In the presence of stimulation, phosphorylated Akt (p-Akt) suppresses the activation of tuberin by phosphorylation, resulting in cellular hypertrophy and proliferation. Immunohistochemical results on p-tuberin and p-Akt in this study suggest that hypertrophy and proliferation of cardiomyocytes may be caused by activation of the mTOR signaling pathway

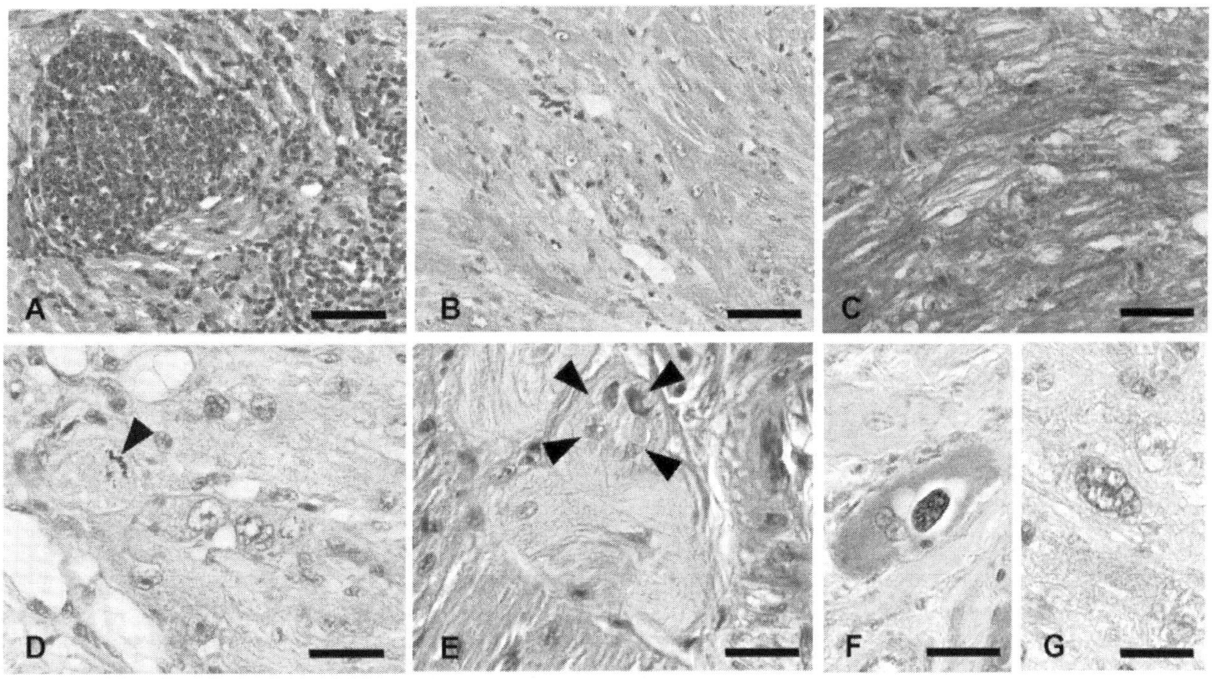

Figure 1: Histopathology and immunohistochemistry of hearts in Japanese native fowls. (A) Mononuclear cells infiltrate into myocardial fibers. Chicken No. 5. Bar = 40 mm. (B) Disarrangement of myocardial fibers with interstitial fibrosis. Chicken No. 9. Bar = 90 mm. (C) Interstitial myocardial fibrosis confirmed by Masson's trichrome stain. Chicken No. 14. Bar = 25 mm. (D) Hypertrophied cardiomyocytes with multiple atypical nuclei. Mitosis (arrowhead) is occasionally noted in these cardiomyocytes. Chicken No. 14. Bar = 20 mm. (E) Matrix inclusion bodies (arrowhead) in the cytoplasm of Purkinje fibers. Chicken No. 14. Bar = 20 mm. (F) Purkinje fibers and matrix inclusion bodies positive for ALV common antigen. Chicken No. 14. Bar = 20 mm. (G) A large atypical nucleus in a hypertrophied cardiomyocyte is positive for PCNA. Chicken No. 14. Bar = 20 mm. doi:10.1371/journal.pone.0086546.g001

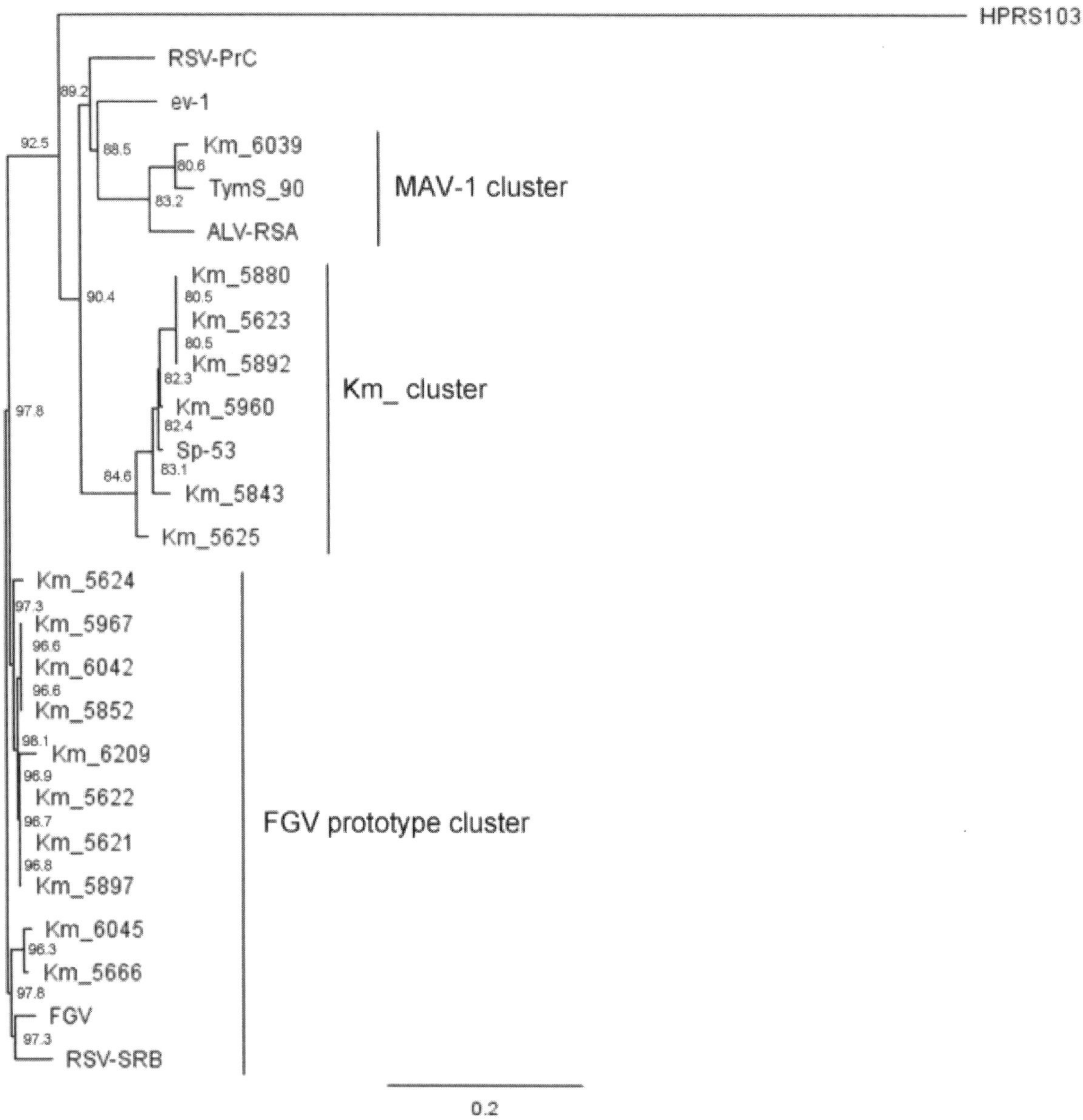

Figure 2: Phylogenic tree based on envSU region of the isolates. Phylogenic tree constructed by the neighbor-joining method showing the relationships among the isolates, FGV, Sp-53 (FGV variant) and standard ALSVs based on the SU region of the env gene. The amino acid sequences were aligned with the Clustal W 1.8.3 program. Bootstrap values of 1000 trials using the neighbor-joining method. Scale bar corresponds to a distance of 0.2 as the frequency of amino acid substitutions in the pairwise comparison of two sequences according to the Kimura two-parameter method.

The env gene and LTR are mainly involved in the tumorigenesis of ALVs. A previous study indicated that the env gene of Km_5666 is very similar to the FGV prototype, whereas other regions of this strain, including 59UTR, gag-pol, 39UTR and 39LTR, show high homology with TymS_90 [17]. On the other hand, the env gene of Km_5892 was found to be closely related to that of FGV variant (Sp-40) [35], whereas other regions, including 59UTR, gag-pol and 39LTR, were shown to be similar to those of TymS_90. Km_5666 and the other isolates from birds affected by cardiomyocyte hypertrophy in the current study belonged to two clusters in the phylogenic analysis based on the envSU region. These results suggest that there is genome diversity among ALVs showing the inducibility of cardiomyocyte hypertrophy, implying a high prevalence of cardiac abnormality among Japanese native fowls infected with ALV. Several reports have indicated the possibility that retroviruses have a direct role in the pathogenesis of myocarditis and dilated cardiomyopathy in humans, monkeys and other animals [3–9], although the relationship between retrovirus infection and cardiac abnormality remains obscure. Whether HIV-1 and SIV can infect cardiomyocytes is still controversial [36,37], whereas the fact that ALV has strong tropism for cardiomyocytes and Purkinje fibers is evidenced by the frequent formation of matrix inclusion bodies in these cells [27–29]. Despite these different degrees of tropism for cardiomyocytes, ALV-induced myocardial alteration is considered to be useful for clarifying the pathogenesis of retroviral cardiomyopathy.

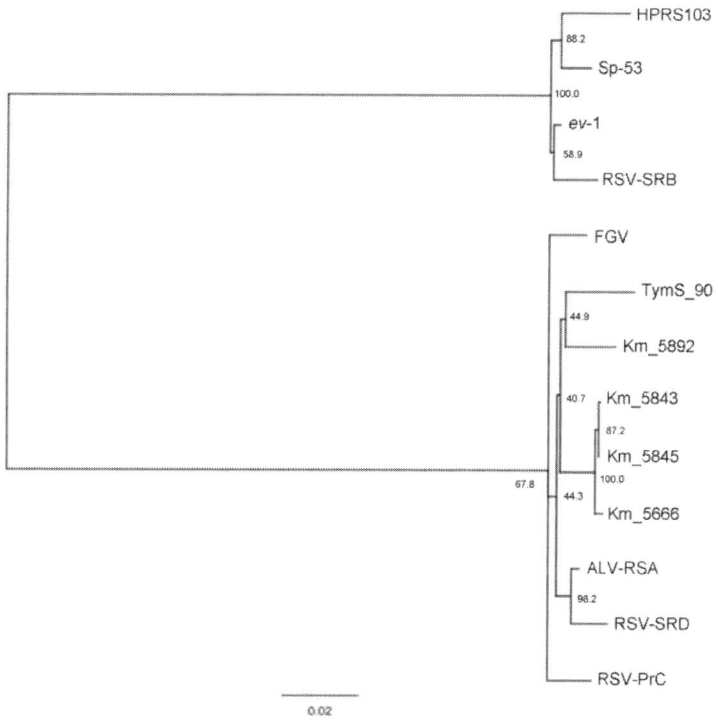

Figure 3: Phylogenic tree based on pol gene of the isolates. Phylogenic tree constructed by the neighbor-joining method showing the relationships among Km_5892, FGV prototype and standard ALSVs based on the pol gene. The nucleotide sequences were aligned with the Clustal W 1.8.3 program. Bootstrap values of 1000 trials using the neighbor-joining method. Scale bar corresponds to a distance of 0.03 as the frequency of amino acid substitutions in the pairwise comparison of two sequences according to the Kimura two-parameter method.

Table 3: The degree of cardiac lesions in chickens experimentally infected with Km_5666 and Km_5892

			Hypertrophied myocardium with atypical nucleus[a]				Non-suppurative myocarditis[b]				Fibrosis[c]		Matrix inclusion body[d]	
Inoculated strain	**Age (days)**	**N**	**−**	**+**	**++**	**+++**	**−**	**+**	**++**	**+++**	**−**	**+**	**−**	**+**
Km_5666	35	6	5	0	0	1	2	3	0	1	6	0	5	1
	70	3	2	1	0	0	0	1	2	0	3	0	1	2
	140	3	3	0	0	0	0	1	2	0	2	1	1	2
Km_5892	35	4	4	0	0	0	0	2	2	0	3	1	0	4
	70	3	2	0	1	0	0	1	2	0	2	1	0	3

[a]The degree of hypertrophied cardiomyocytes with atypical nuclei was defined as follows: +, individual cells; ++, multiple and/or small clusters; +++, multiple and extensive growth.
[b]The degree of non-suppurative myocarditis was defined as follows: +, focal or multifocal (under 25 μm in diameter); ++, multifocal (from 25 μm to 200 μm); +++, multifocal (over 200 μm).
[c]The degree of fibrosis was defined as follows: +, interstitial; ++, multifocal interstitial (under 400 μm in diameter); +++, multifocal interstitial (over 400 μm in diameter).
[d]Total number in 10 high-power field.
doi:10.1371/journal.pone.0086546.t003

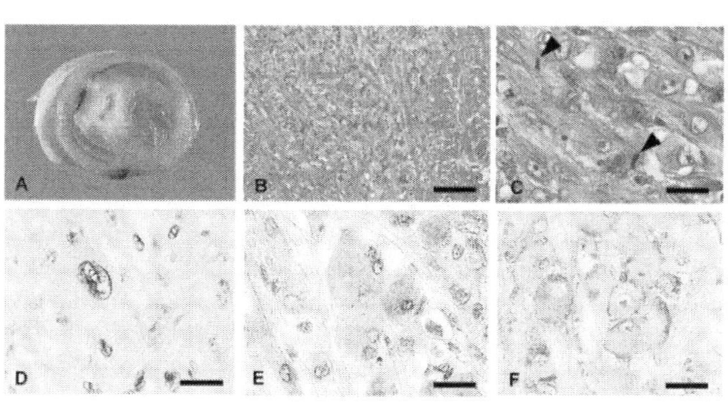

Figure 4: Gross pathology, histopathology and immunohistochemistry of chickens experimentally inoculated with Km_5666. (A) Grayish-white ill-defined tissue was located in the septum at 35 days old. (B) Locally extensive growth of hypertrophied cardiomyocytes. Bar = 75 mm. (C) Mitosis (arrowhead) observed in hypertrophied cardiomyocytes. Bar = 30 mm. (D) Numerous nuclei in atypical myocardium were positive for PCNA. Bar = 20 mm. (E–F) Hypertrophied cardiomyocytes positive for p-Akt (E) and p-tuberin (F). Bars = 20 mm.

In conclusion, our results indicate that several strains of ALV could promote growth activity of cardiomyocytes and induce cardiomyocyte hypertrophy by a mechanism other than insertional mutagenesis. This is the first report showing unique hypertrophy and growth of cardiomyocytes induced by ALV. This animal model may provide new insight into retrovirus-induced cardiac pathology

ACKNOWLEDGMENTS

The authors are very grateful for the participation and cooperation of the Fanciers and Breeders' Associations (Higo Chabo Club of Japan) participating in the study for their very good cooperation.

Author Contributions

Conceived and designed the experiments: SN KO HY AA. Performed the experiments: SN KO HY AA SK. Analyzed the data: SN KO AO HY AA. Contributed reagents/materials/ analysis tools: SN KO AO YS TU. Wrote the paper: SN KO.

REFERENCES

1. Jacob AJ, Sutherland GR, Bird AG, Brettle RP, Ludlam CA, et al. (1992) Myocardial dysfunction in patients infected with HIV: prevalence and risk factors. Br Heart J 68: 549–553.
2. De Castro S, d'Amati G, Gallo P, Cartoni D, Santopadre P, et al. (1994) Frequency of development of acute global left ventricular dysfunction in human immunodeficiency virus infection. J Am Coll Cardiol 24: 1018–1024.
3. Roldan EO, Moskowitz L, Hensley GT (1987) Pathology of the heart in acquired immunodeficiency syndrome. Arch Pathol Lab Med 111: 943–946.
4. Shannon RP, Simon MA, Mathier MA, Geng Y, Mankad S, et al. (2000) Dilated cardiomyopathy associated with simian AIDS in nonhuman primates. Circulation 101: 185–193.
5. Lewis W, Miller YK, Haase CP, Ludaway T, McNaught J, et al. (2005) HIV viral protein R causes atrial cardiomyocyte mitosis, mesenchymal tumor, dysrhythmia, and heart failure. Lab Invest 85: 182–192.
6. Brellou GD, Angelopoulou K, Poutahidis T, Vlemmas I (2007) Detection of maedi-visna virus in the liver and heart of naturally infected sheep. J Comp Pathol 136: 27–35.
7. Gilka F, Spencer JL (1990) Chronic myocarditis and circulatory syndrome in a White Leghorn strain induced by an avian leukosis virus: light and electron microscopic study. Avian Dis 34: 174–184.
8. Gilka F, Spencer JL, Chambers JR (1991) Response of meat-type chickens to infection with RAV-1 avian leukosis virus. Avian Pathol 20: 637–647.
9. Iwata N, Ochiai K, Hayashi K, Ohashi K, Umemura T (2002) Nonsuppurative myocarditis associated with so-called fowl glioma. J Vet Med Sci 64: 395–399.
10. Stedman NL, Brown TP (2002) Cardiomyopathy in broiler chickenscongenitally infected with avian leukosis virus subgroup J. Vet Pathol 39:161–164
11. Tsukamoto K, Hihara H, Kono Y (1991) Detection of avian leukosis virus antigens by the ELISA and its use for detecting infectious virus after cultivation of samples and partial characterization of specific pathogen-free chicken lines maintained in this laboratory. J Vet Med Sci 53: 399–408.
12. Hatai H, Ochiai K, Naganuma K, Imanishi S, Ochi A, et al. (2008) A recombinant avian leukosis virus associated with fowl glioma in layer chickens in Japan. Avian Pathol 37: 127–137.
13. Ono M, Tsukamoto K, Tanimura N, Haritani M, Kimura KM, et al. (2004) An epizootic of subcutaneous tumors associated with subgroup A avian leukosis/ sarcoma virus in young layer chickens. Avian Dis 48: 940–946.
14. Motitschke A, Ottiger HP, Jungba¨ck C (2010) Evaluation of the sensitivity of PCR methods for the detection of extraneous agents and comparison with in vivo testing. Biologicals 38: 389–392.
15. Bowles NE, Eisensmith RC, Mohuiddin R, Pyron M, Woo SLC (1996) A simple and efficient method for the concentration and purification of recombinant retrovirus for increased hepatocyte transduction in ovo. Hum Gene Ther 7: 1735–1742.
16. Tomioka Y, Ochiai K, Ohashi K, Ono E, Toyoda T, et al. (2004) Genome sequence analysis of the avian retrovirus causing so-called fowl glioma and the promoter activity of the long terminal repeat. J Gen Virol 85: 647–652.
17. Ochi A, Ochiai K, Kobara A, Nakamura S, Hatai H, et al. (2012) Epidemiological study of fowl glioma-inducing virus in chickens in Asia and Germany. Avian Pathol 41: 299–309.
18. Schoen FJ, Mitchell RN (2004) The heart. In: Kumar V, Abbas AK, Fausto N, Aster JC, editors. Robbins and Cotran Pathologic basis of disease. Eighth ed. Philadelphia: Saunders Elsevier. pp. 529–587.
19. Miller LM, Van Vleet JF, Gal A (2012) Cardiovascular system and lymphaticvessels. In: Zachary JF, McGavin MD, editors. Pathologic basis of veterinary diseases. Fifth ed. St. Louis: Elsevier Mosby. pp. 539–588.
20. GrantMaxie M, Robinson WF (2007) Cardiovascular system. In: Grant Maxie M, editor. Jubb, Kennedy, and Palmer's Pathology of domestic animals. Fifth ed. Philadelphia: Saunders Elsevier. pp. 1–103.
21. Tomioka Y, Ochiai K, Ohashi K, Kimura T, Umemura T (2003) In ovo infection with an avian leukosis virus causing fowl glioma: viral distribution and pathogenesis. Avian Pathol 32: 617–624.
22. Shivaprasad HL, Franca M, Woolcock PR, Nordhausen R, Day JM, et al. (2009) Myocarditis associated with reovirus in turkey poults. Avian Dis 53: 523–532.
23. Benyounes N, Fohlen M, Devys JM, Delalande O, Moures JM, et al. (2012) Cardiac rhabdomyomas in tuberous sclerosis patients: a case report and review of the literature. Arch Cardiovasc Dis 105: 442–445.
24. Tanimoto T, Ohtsuki Y (1995) The pathogenesis of so-called cardiac rhabdomyoma in swine: a histological, immunohistochemical and ultrastructural study. Virchows Arch 427: 213–221.

25. Beard JW (1980) Biology of avian oncornaviruses. In: Klein G, editor. Viral oncology. New York: Ravan Press. pp. 55–87.
26. Saule S, Me´rigaud JP, Al-Moustafa AE, Ferre´ F, Rong PM, et al. (1987) Heart tumors specifically induced in young avian embryos by the v-myc oncogene. Pros Natl Acad Sci USA 84: 7982–7986.
27. Gilka F, Spencer JL (1985) Viral matrix inclusion bodies in myocardium of lymphoid leukosis virus-infected chickens. Am J Vet Res 46: 1953–1960.
28. Nakamura K, Abe F, Hihara H, Taniguchi T (1988) Myocardial cytoplasmic inclusions in chickens with hemangioma and lymphoid leukosis. Avian Pathol 17: 3–10.
29. Nakamura K, Higashi T, Yamada M, Imai K, Yamamoto Y (2007) Basophilic intracytoplasmic viral matrix inclusions distributed widely in layer hens affected with avian-leukosis-virus-associated tumours. Avian Pathol 36: 53–58.
30. Heidrich JE, Adcock MA, Bolin C, Cheville NF, Smith RE (1987) Effect of Rous associated virus number 7 on lymphoid cells and tissues of the chicken. Vet Immunol Immunopathol 15: 267–283.
31. Maeda N, Fan H, Yoshikai Y (2008) Oncogenesis by retroviruses: old and new paradigms. Rev Med Virol 18: 387–405.
32. Hull S, Lim J, Hamil A, Nitta T, Fan H (2012) Analysis of jaagsiekte sheep retrovirus (JSRV) envelope protein domains in transformation. Virus Genes 45: 508–517.
33. Zavala G, Pretto C, Chow YH, Jones L, Alberti A, et al. (2003) Relevance of Akt phosphorylation in cell transformation induced by Jaagsiekte sheep retrovirus. Virology 312: 95–105.
34. Dorn GW II, Force T (2005) Protein kinase cascades in the regulation of cardiac hypertrophy. J Clin Invest 115: 527–537.
35. Nakamura S, Ochiai K, Hatai H, Ochi A, Sunden Y, et al. (2011) Pathogenicity of avian leucosis viruses related to fowl glioma-inducing virus. Avian Pahol 40: 499–505.
36. Grody WW, Cheng L, Lewis W (1990) Infection of the heart by the human immunodeficiency virus. Am J Cardiol 66: 203–206.
37. Rebolledo MA, Krogstad P, Chen F, Shannon KM, Klitzner TS (1998) Infection of human fetal cardiac myocytes by a human immunodeficiency virus-1-derived vector. Circ Res 83:738–742.

CHAPTER 18

THE PASSAGE OF CELLS CAN IMPROVE THE DETECTION RATE OF AVIAN LEUKOSIS VIRUS TO FACILITATE THE ELIMINATION OF AVIAN LEUKOSIS IN CHICKENS

Xiuzhen Wang, Bo Wang, Peipei Zhang, Hegang Cheng and Shuhong Sun

Keywords: Avian leukosis virus, Cells passage, Detection rate of p27 antigen

INTRODUCTION

Avian leukosis (AL) is caused by the avian leukosis virus (ALV), which is associated with a variety of malignant neoplasms, including lymphoid and myeloid leukosis, and other production problems in chickens (Witter, 1997). AL is found in most countries that have large-scale farms and is one of the most harmful diseases in the poultry industry, causing serious economic losses (Sun and Cui 2007; Zhang et al. 2011). The ALV p27 gene is a highly conserved gene among the different subtypes of ALV, and the p27 antigen is a serogroup-specific antigen. This antigen can be detected by ELISA for many kinds of samples for the diagnosis of avian leukosis, such as albumen, vaginal secretions, meconium, cloaca cotton swab, and also can be used to detect the propagation of the virus (Spencer et al. 1984; Spencer et al. 1976; De Boer et al. 1984).

At present, there are no effective vaccines or drugs to prevent or treat AL. Most countries control ALV primarily through the continual detection of the ALV p27 antigen and the elimination of antigen-positive chickens. Presently, there are three common schemes for eliminating ALV from breeding flocks: (1) Detect the p27 antigen using cloacal cotton swabs or albumen and eliminate the antigen-positive chickens. (2) Aseptically collect anticoagulated blood from the core group, use this blood to inoculate DF1 cells, and detect the p27 antigen after 9 d of culture. Then, detect the p27 antigen in the meconium of the next generation of chickens to continuously eliminate all antigen-positive chickens. This method is the international gold standard for AL elimination. (3) Combine the first and second methods to separate the p27 antigen-positive and negative chickens and aseptically collect anticoagulated blood or albumen from p27 antigen-positive chickens to inoculate DF1 cells for p27 antigen detection. Using this method, we can distinguish endogenous and exogenous ALV, just eliminating exogenous ALV-positive chickens and retaining the endogenous ALV-positive chickens (Spencer et al. 1984).

The collections of aseptic anticoagulated blood to inoculate DF1 cells and the detection of the p27 antigen in the supernatant after 9 d of culture play a vital role in the elimination of AL. This method complements the method of the direct detection of the p27 antigen in cloaca swabs or albumen. In this study, the influence of passaging cells on the detection rate and the titer of the ALV p27 antigen were investigated to explore a better method for timely and thorough AL elimination.

MATERIALS AND METHODS

Twenty-one chickens for which cloacal cotton swabs were positive for the p27 antigen were numbered from 01 # to 21 #. An aseptic anticoagulated blood sample was collected from each chicken, and then the samples were centrifuged at 1500 rpm for 5 min to isolate the leukocytes. The leukocytes

---

Citation: Xiuzhen Wang, Bo Wang, Peipei Zhang, Hegang Cheng and Shuhong Sun, The passage of cells can improve the detection rate of avian leukosis virus to facilitate the elimination of avian leukosis in chickens, DOI: 10.1186/2193-1801-2-138.

were then suspended in 200 ul and used to inoculate two sets of cell culture plates containing DF1 cells.

Each set of cell culture plates has 24 holes, with 21 holes inoculated and one hole inoculated with a blank to serve as the negative control, which was numbered 22#. Both the sets were dealt with uniformly. Thereafter, the inoculated DF1 cultures were incubated under CO2 at 37°C for 2 h, and 1% DMEM was substituted for the 5% DMEM during continuous culturing. After the cells had been cultured for 4 d, the cells of one set were passaged, and the cells in other set were not. The supernatants of both sets were collected for p27 antigen detection after 9 d of culturing.

All the experimental methods comply with current ethical consideration.

The Influence of Cell Passaging on the p27 Antigen-Positive Rate of the Supernatant

After 9 d of culturing, the DF1 supernatant from each sample was collected for ALV p27 antigen detection in one reaction plate using the Avian Leukosis Virus Antigen Test Kit® (IDEXX, USA) according to the manufacturer's instructions, and the p27 antigen-positive rates of the passaged and unpassaged cells were compared.

The Influence of cell Passaging on the titer of the p27 Antigen in the Supernatant

The S/P values determined using an ELISA for the p27 antigen in the 9-d supernatants of the passaged and unpassaged samples were compared, and the correlation was assessed using the matched-pair *T* test in SPSS 17.0.

RESULTS

The Influence of Cell Passaging on the p27 Antigen-Positive rate of the Supernatant

The results show that cell passaging can improve the p27 antigen detection rate when using cell supernatant, as the data was shown (Table 1). The ALV p27 antigen-positive rate for the supernatant of the passaged cells was 71.43% (15/21), which was far higher than that for the supernatant of the unpassaged cells, 42.86% (9/21). All of the positive unpassaged samples except 18# were also positive for the passaged samples, and the negative control 22# was negative for p27 antigen.

Table 1: The test results for the ALV p27 antigen in the 9-d cell supernatants

Number	1	2	3	4	5	6	7	8	9	10	11	12	13	14	15	16	17	18	19	20	21	positive rate
Passaged sample	-	+	+	+	+	+	-	-	-	+	+	+	+	+	+	+	+	-	+	+	-	15/21
Unpassaged sample	-	+	+	-	-	+	-	-	-	-	-	+	-	+	+	-	-	+	+	+	-	9/21

The Influence of Cell Passaging on the titer of p27 Antigen in the Supernatant

The S/P values of most of the passaged samples were higher than those of unpassaged samples. Only seven passaged samples had lower titers than the unpassaged samples, and these differences were small. When the S/P value of a sample was greater than 0.2, that sample was judged to be positive. The correlation coefficient of the S/P value between the passaged samples and the unpassaged samples was as high as 0.928. The results showed that there is a strong correlation between the passaged and unpassaged samples with respect to the S/P value, and the passaging of cells improved the p27 antigen titer in the 9 d culture supernatants (Figure 1).

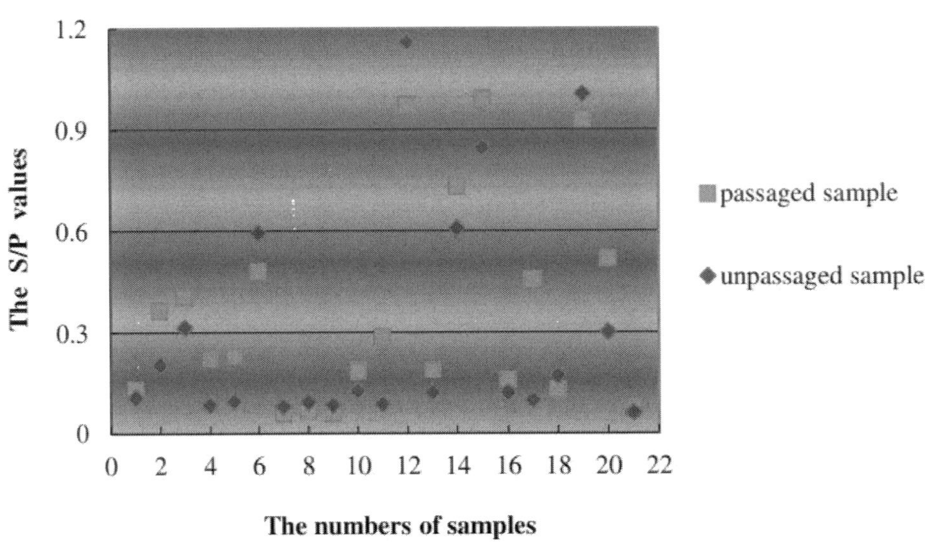

Figure 1: The S/P values of the supernatant of cells cultured for 9 d.

DISCUSSION

ALV has spread all over the world since the early 20th century and has resulted in serious harm to the bird farming industry (Ochi et al. 2012). At present, the primary means to control the disease in most countries combines direct ALV p27 antigen detection with detection of the ALV p27 antigen in the supernatant of cells directly cultured for 9 d, followed by the elimination of all positive chickens. This process is continually repeated to eliminate infected chickens. In China, ALV has spread to native varieties of poultry and is a serious threat to the maintenance of high-quality varieties of chickens (Wang et al. 2011; Zhang et al. 2010; Cheng et al. 2005; Qi et al. 2011; Zhao et al.2010; Qian et al. 2011). Therefore, it is important to remove all ALV p27 antigen-positive chickens from stock breeding groups as soon as possible.

Based on the ALV elimination scheme commonly used internationally for stock breeding flocks, this study explored the influence of passaging cells on the detection rate and titer of the ALV p27 antigen as part of a protocol for timely and thorough AL elimination. The results show that the passage of cells can improve the p27 antigen-positive detection rate in the 9-d cell culture supernatants, leading to a positivity rate of 71.43% (15/21), which was far higher than that of the unpassaged cells, 42.86% (9/21). In addition, all of the positive unpassaged samples except for 18# were positive for the passaged samples. There was a strong correlation between the passaged and unpassaged samples with respect to the S/P value, and the passage of cells improved the p27 antigen titer. These results suggest that, compared with the classic method in which the cells are cultured for 9 d without passaging, a protocol in which the cells are passaged in the middle of the culture period can identify more ALV p27 antigen-positive chickens in the same amount of time. Therefore, the passaging of cells can speed up the AL elimination process, which has positive practical results and objective economic significance.

ACKNOWLEDGMENTS

The authors would like to thank the biology post-doctoral stations at Shandong Agricultural University. The study was supported by Special Fund for Agro-scientific Research in the Public Interest, China (201203055).

COMPETING INTERESTS

The authors declare that they have no competing interests.

AUTHORS' CONTRIBUTIONS

WXZ and SSH conceived and designed the experiments. WXZ and WB performed the experiments. ZPP and CHG analyzed the data. WXZ and SSH wrote the paper. All authors read and approved the final manuscript.

REFERENCES

1. Cheng ZQ, Zhang L, Liu SD, Zhang LJ, Cui ZZ: Emerging of Avian leukosis virus subgroup J in a flock of Chinese local breed. *Acta Microbiol Sinica* 2005, 45(4):584-587.
2. De Boer GF, Gielkens ALJ, Boerrigter HM, Hartog L: Identification of avian leukosis virus shedding hens. Comparison of ALV GS-antigen levels in various test samples. *Avian Pathol* 1984, 13: 321-332. 10.1080/03079458408418534
3. Ochi A, Ochiai K, Nakamura S, Kobara A, Sunden Y, Umemura T: Molecular characteristics and pathogenicity of an avian leukosis virus isolated from avian neurofibrosarcoma. *Avian Disease* 2012, 56(1):35-43. 10.1637/9830-060711-Reg.1
4. Qi XY, Zhang WY, Ju HB, Deng B, Zhou JP, Liu J, Wang QZ: Isolation and identification of the subgroup J avian leukosis virus associated with clinical multiple tumours in Sanhuang layers and pathological observation. *Chinese J Prevent Veterinary Med* 2011, 33(2):101-104.
5. Qian K, Zhu YF, Shen HY, Jin WJ, Qin AJ: Full length cDNA sequence analysis of avian leukosis virus subgroup A isolated from local layer chickens flocks. *Chinese Veterinary Sci* 2011, 41(10):1005-1010.
6. Spencer JL, Crittenden LB, Burmester BR: Lymphoid leukosis viruses and gs antigen in unincubated chicken eggs. *Avian Pathol* 1976, 5: 221-226. 10.1080/03079457608418188
7. Spencer JL, Gilka F, Gavora JS: Distribution of group specific antigen of lymphoid leukosis virus in tissues from laying hens [J]. *Avian Diseases* 1984, 28: 358-373. 10.2307/1590343
8. Sun SS, Cui ZZ: Epidemiological and pathological studies of subgroup J avian leukosis virus infections in Chinese local yellow chickens [J]. *Avian Pathology* 2007, 36(3):221-226. 10.1080/03079450701332345
9. Wang B, Li QY, Liu SQ, Zhang YG, Cui ZZ, Sun SH: Evaluation on ALV infection in fertilized eggs from a Wan-nan yellow-feather parent broiler breeder flock. *Acta Veterinaria Et Zootechnica Sinica* 2011, 42(2):224-227.
10. Witter RL: Avian tumor viruses: persistent and evolving pathogens [J]. *Acta Veterinria Hungarica* 1997, 45(3):261-266.
11. Zhang SJ, Wu TW, Wei P, Wu D, Wu ZJ, Cui ZZ, Lu LF, Quan CY, Feng SW, Yu ZP, Wang GR: Investigation of avian leukosis virus infections and eradication practices in local lines of chickens. *China Poultry* 2010, 32(12):11-16.
12. Zhang ZJ, Liu SQ, Wang B, Cui ZZ, Zhang YG, Sun SH: Isolation and identification co-infection status of ALV-J and REV and molecular characterization analysis of Anhui native chickens yellow. *Scientia Agricultura Sinica* 2011, 44(11):2379-2386.
13. Zhao DM, Zhang QC, Cui ZZ: Isolation and Identification of a Subgroup B Avian Leukosis Virus from Chickens of Chinese Native Breed Luhua. *Chinese Journal of Virology* 2010, 26(1):53-57.

INDEX

M

N

O

P

R

S

T

U

V

W